ATMOSPHERIC MODEL APPLICATIONS

Edited by **Ismail Yucel**

Atmospheric Model Applications
Edited by Ismail Yucel

Published by InTech
Janeza Trdine 9, 51000 Rijeka, Croatia

Publishing Process Manager Martina Blecic

Technical Editor Teodora Smiljanic

Cover Designer InTech Design Team

Additional hard copies can be obtained from orders@intechopen.com

Atmospheric Model Applications, Edited by Ismail Yucel
p. cm.
ISBN 978-953-51-0488-9

For EU product safety concerns:
IN TECH d.o.o., Prolaz Marije Krucifikse Kozulić 3, 51000 Rijeka, Croatia,
info@intechopen.com or visit our website at intechopen.com.

Meet the editor

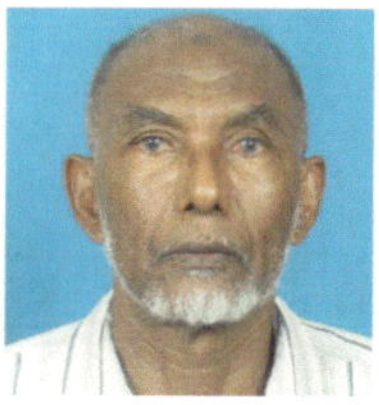

Professor Bashir has received his PhD in Electronics and Telecommunication Engineering, University of Bradford, West Yorkshire, England, 1984. His research interest areas include antennas em radio wave propagation (at the full radio spectrum), optical communication systems, microwave and satellite communication systems, em theory, em scattering computation methods. Professor Bashir has been invited as an expert by the ITU for presenting "Contribution to ITU-R (Antennas & Propagation), Study Group 3-WP3J contribution 108, accepted on 8 Sept. 2010. The following paper has been chosen as the Best-Of-Book: Electromagnetic Scattering Computation Methods for very small particles: Part II, Inern. Conference on Modeling & Simulation (France based) in Petra, Jordan, 18-20 November, and automatically published in MS Journal,Vol.9, No. 1-2-3, 2008. Two novel papers written by Professor Bashir have been accepted in Springer and will be in press in 2012.

Contents

Preface

This book covers comprehensive text and reference work on atmospheric models for methods of numerical modeling and important related areas of data assimilation and predictability. It incorporates various aspects of environmental computer modeling including an historical overview of the subject, approximations to land surface and atmospheric physics and dynamics, radiative transfer and applications in satellite remote sensing, and data assimilation. With individual chapters authored by eminent professionals in their respective topics, Advanced Topics in application of atmospheric models try to provide in-depth guidance on some of the key applied in atmospheric models for scientists and modelers.

Dr. Ismail Yucel
Middle East Technical University,
Civil Engineering Department,
Water Resources Lab, Ankara,
Turkey

1

Improving Atmospheric Model Performance on a Multi-Core Cluster System

Carla Osthoff[1], Roberto Pinto Souto[1], Fabrício Vilasbôas[1], Pablo Grunmann[1], Pedro L. Silva Dias[1], Francieli Boito[2], Rodrigo Kassick[2], Laércio Pilla[2], Philippe Navaux[2], Claudio Schepke[2], Nicolas Maillard[2], Jairo Panetta[3], Pedro Pais Lopes[3] and Robert Walko[4]

[1]*Laboratório Nacional de Computação Científica (LNCC)*
[2]*Universidade Federal do Rio Grande do Sul (UFRGS)*
[3]*Instituto Nacional de Pesquisas Espaciais (INPE)*
[4]*University of Miami*
[1,2,3]*Brazil*
[4]*USA*

1. Introduction

Numerical models have been used extensively in the last decades to understand and predict weather phenomena and the climate. In general, models are classified according to their operation domain: global (entire Earth) and regional (country, state, etc). Global models have spatial resolution of about 0.2 to 1.5 degrees of latitude and therefore cannot represent very well the scale of regional weather phenomena. Their main limitation is computing power. On the other hand, regional models have higher resolution but are restricted to limited area domains. Forecasting on limited domain demands the knowledge of future atmospheric conditions at domain's borders. Therefore, regional models require previous execution of global models.

OLAM (Ocean-Land-Atmosphere Model), initially developed at Duke University (Walko & Avissar, 2008), tries to combine these two approaches to provide a global grid that can be locally refined, forming a single grid. This feature allows simultaneous representation (and forecasting) of both the global and the local scale phenomena, as well as bi-directional interactions between scales.

Due to the large computational demands and execution time constraints, these models rely on parallel processing. They are executed on clusters or grids in order to benefit from the architecture's parallelism and divide the simulation load. On the other hand, over the next decade the degree of on-chip parallelism will significantly increase and processors will contain tens and even hundreds of cores, increasing the impact of levels of parallelism on clusters. In this scenario, it is imperative to investigate the scale of programs on multilevel parallelism environment.

This chapter is based on recent works from *Atmosfera Massiva Research Group*[1] on evaluating OLAM's performance and scalability in multi-core environments - single node and cluster.

Large-scale simulations, as OLAM, need a high-throughput shared storage system so that the distributed instances can access their input data and store the execution results for later analysis. One characteristic of weather and climate forecast models is that data generated during the execution is stored on a large amount of small files. This has a large impact on the scalability of the system, especially when executing using parallel file systems: the large amount of metadata operations for opening and closing files, allied with small read and write operations, can transform the I/O subroutines in a significant bottleneck.

General Purpose computation on Graphics Processing Units (GPGPU) is a trend that uses GPUs (Graphics Processing Units) for general-purpose computing. The modern GPUs' highly parallel structure makes them often more effective than general-purpose CPUs for a range of complex algorithms. GPUs are "many-core" processors, with hundreds of processing elements.

In this chapter, we also present recent studies that evaluates a implementation of OLAM that uses GPUs to accelerate its computations. Therefore, this chapter presents an overview on OLAM's performance and scalability. We aim at exploiting all levels of parallelism in the architectures, and also at paying attention to important performance factors like I/O.

The remainder of this chapter is structured as follows. Section 2 presents the Ocean-Land-Atmosphere Model, and Section 3 presents performance experiments and analysis. Related works are shown in Section 4. The last section closes the chapter with final remarks and future work.

2. The Ocean-Land-Atmosphere Model – OLAM

High performance implementation of atmospheric models is fundamental to operational activities on weather forecast and climate prediction, due to execution time constraints — there is a pre-defined, short time window to run the model. Model execution cannot begin before input data arrives, and cannot end after the due time established by user contracts. Experience in international weather forecast centers points to a two-hour window to predict the behavior of the atmosphere in coming days.

In general, atmospheric and environmental models comprise a set of Partial Differential Equations which include, among other features, the representation of transport phenomena as hyperbolic equations. Their numerical solution involves time and space discretization subject to the Courant Friedrichs Lewy (CFL) condition for stability. This imposes a certain proportionality between the time and space resolutions, where the resolution is the inverse of the distance between points in the domain mesh. For a 1-dimensional mesh, the number of computing points n is given by L/d, where L is the size of the domain to be solved and d is the distance between points over this domain. In our case, the mesh is 4-dimensional (3 for space and 1 for time). The computational cost is of $O(n^4)$ if the number of vertical points also increases with n, where n is the number of latitude or longitude points in the geographical domain of the model. The resolution strongly influences the accuracy of results.

[1] http://gppd.inf.ufrgs.br/atmosferamassiva

Operational models worldwide use the highest possible resolution that allow the model to run at the established time window in the available computer system. New computer systems are selected for their ability to run the model at even higher resolution during the available time window. Given these limitations, the impact of multiple levels of parallelism and multi-core architectures in the execution time of operational models is indispensable research.

This section presents the Ocean-Land-Atmosphere Model (OLAM). Its characteristics and performance issues are discussed. We also discuss the parameters used in the performance evaluation.

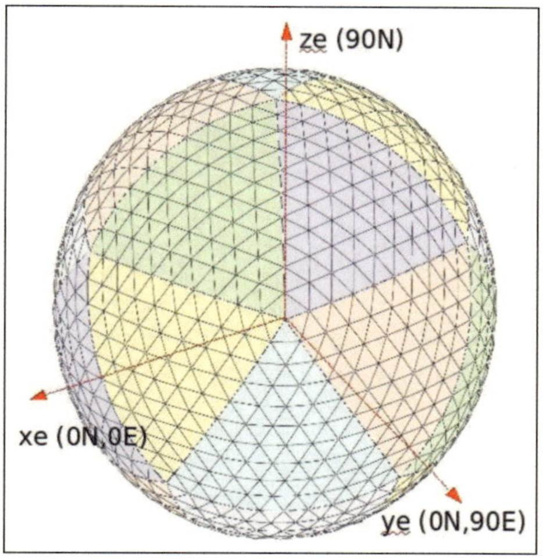

Fig. 1. OLAM's subdivided icosahedral mesh and cartesian coordinate system with origin at Earth center.

OLAM was developed to extend features of the Regional Atmospheric Modeling System (RAMS) to the global domain (Pielke et al., 1992). OLAM uses many functions of RAMS, including physical parameterizations, data assimilation, initialization methods, logic and coding structure, and I/O formats (Walko & Avissar, 2008). OLAM introduces a new dynamic core based on a global geodesic grid with triangular mesh cells. It also uses a finite volume discretization of the full compressible Navier Stokes equations. Local refinements can be defined to cover specific geographic areas with more resolution. Recursion may be applied to a local refinement. The global grid and its refinements define a single grid, as opposed to the usual nested grids of regional models. Grid refined cells do not overlap with the global grid cells - they substitute them.

The model consists essentially of a global triangular-cell grid mesh with local refinement capability, the full compressible nonhydrostatic Navier-Stokes equations, a finite volume formulation of conservation laws for mass, momentum, and potential temperature, and numerical operators that include time splitting for acoustic terms. The global domain greatly expands the range of atmospheric systems and scale interactions that can be represented in the model, which was the primary motivation for developing OLAM.

OLAM was developed in FORTRAN 90 and parallelized with Message Passing Interface (MPI) under the Single Program Multiple Data (SPMD) model.

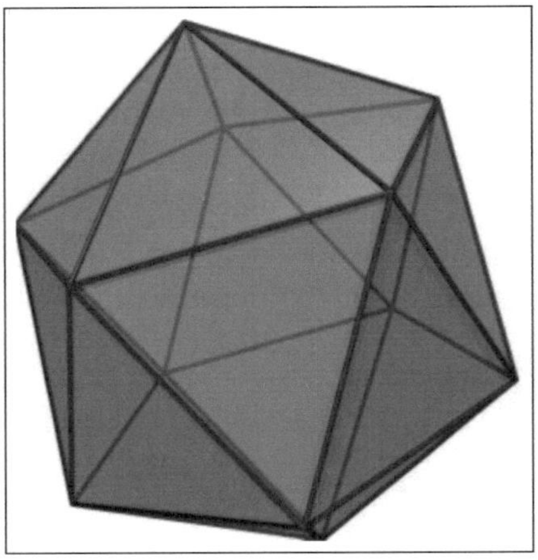

Fig. 2. Icosahedron object.

2.1 OLAM's global grid structure

OLAM's global computational mesh consists of spherical triangles, a type of geodesic grid that is a network of arcs that follow great circles on the sphere (Walko & Avissar, 2008). This can be seen in Figure 1.

The geodesic grid offers important advantages over the commonly used latitude-longitude grid. It allows mesh size to be approximately uniform over the globe, and avoids singularities and grid cells of very high aspect ratio near the poles. OLAM's grid construction begins from an icosahedron inscribed in the spherical earth, as is the case for most other atmospheric models that use geodesic grids. Icosahedron is a regular polyhedron that consists of 20 equilateral triangle faces, 30 triangle edges, and 12 vertices, with 5 edges meeting at each vertex, as represented in Figure 2. The icosahedron is oriented such that one vertex is located at each geographic pole, which places the remaining 10 vertices at latitudes of $\pm tan^{-1}(1/2)$, as shown in Figure 1. The numerical formulation allows for nonperpendicularity between the line connecting the barycenters of two adjacent triangles and the common edge between the triangles.

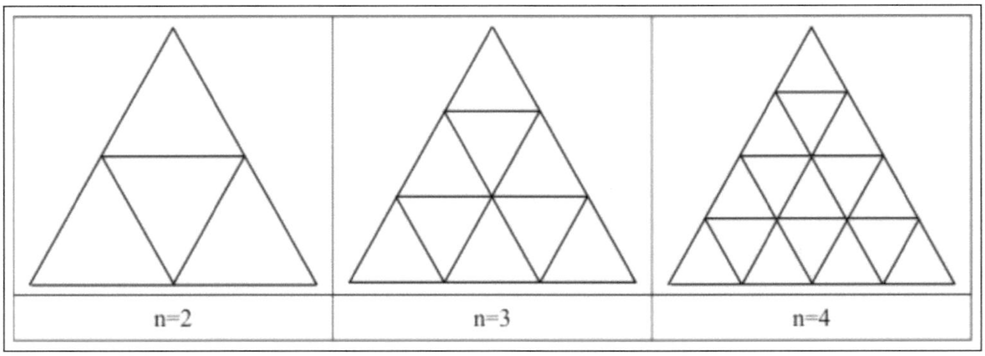

Fig. 3. Uniform subdivision of an icosahedron.

An uniform subdivision is performed in order to construct a mesh of higher resolution to any degree desired. This is done by dividing each icosahedron triangle into N^2 smaller triangles, where N is the number of divisions. The subdivision adds $30(N^2 - 1)$ new edges to the original 30 and $10(N^2 - 1)$ new vertices to the original 12, with 6 edges meeting at each new vertex. This situation is represented in Figure 3. All newly constructed vertices and all edges are then radially projected outward to the sphere to form geodesics.

Figure 1 shows an example of the OLAM subdivided icosahedral mesh and cartesian coordinate system with origin at Earth center, using $N = 10$.

OLAM uses an unstructured approach and represents each grid cell with single horizontal index (Walko & Avissar, 2008). Required information on local grid cell topology is stored and accessed by means of linked lists. If a local horizontal mesh refinement is required, it is performed at this step of mesh construction. The refinement follows a three-neighbor rule that each triangle must share finite edges length with exactly three others.

Fig. 4. Local mesh refinement applied to South America.

An example of local mesh refinement is shown in Figure 4, where resolution is exactly doubled in a selected geographic area by uniformly subdividing each of the previously existing triangles into $2x2$ smaller triangles. Auxiliary edges are inserted at the boundary between the original and refined regions for adherence to the three-neighbor rule. Each auxiliary line in this example connects a vertex that joins 7 edges with a vertex that joins 5 edges. More generally, a transition from coarse to fine resolution is achieved by use of vertices with more than 6 edges on the coarser side and vertices with fewer than 6 edges on the finer side of the transition. For more mesh refinement procedure details, refer to (Walko & Avissar, 2011), that presents OLAM's method for constructing a refined grid region for a global Delaunay triangulation, or its dual Voronoi diagram, that is highly efficient, is direct (does not require iteration to determine the topological connectivity although it typically does use iteration to optimize grid cell shape), and allows the interior refined grid cells to remain stationary as refined grid boundaries move dynamically. This latter property is important because any shift in grid cell location requires re-mapping of prognoses quantities, which results in significant dispersion.

The final step of the mesh construction is the definition of its vertical levels. To do this, the lattice of surface triangular cells is projected radially outward from the center of the earth to a series of concentric spheres of increasing radius, as in Figure 5. The vertices on consecutive spheres are connected with radial line segments. This creates prism-shaped grid cells having

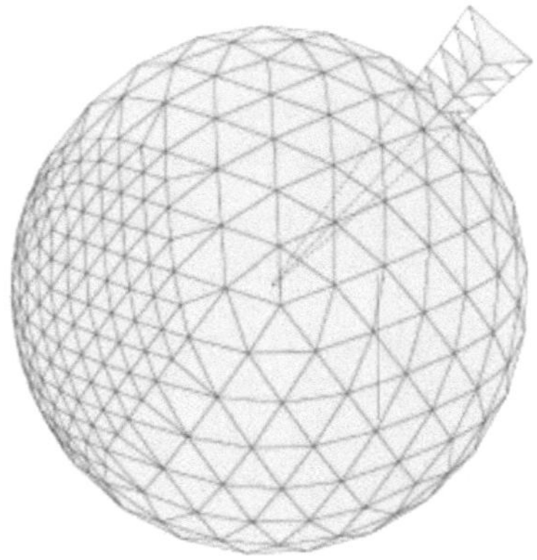

Fig. 5. Local mesh refinement (in the left portion of the image) and projection of a surface triangle cell to larger concentric spheres (in the right portion of the image).

two horizontal faces (perpendicular to gravity) and three vertical faces. The horizontal cross section of each grid cell and column expands gradually with height. The vertical grid spacing between spherical shells may itself vary and usually is made to expand with increasing height.

OLAM uses a C-staggered grid discretization for an unstructured mesh of triangular cells (Walko & Avissar, 2008). Scalar properties are defined and evaluated at triangle barycenters, and velocity component normal to each triangle edge is defined and evaluated at the center of each edge. The numerical formulation allows for nonperpendicularity between the line connecting the barycenters of two adjacent triangles and the common edge between the triangles.

Control volume surfaces for horizontal momentum are the same as for scalars in OLAM. This is accomplished by defining the control volume for momentum at any triangle edge to be the union of the two adjacent triangular mass control volumes. This means that no spatial averaging is required to obtain mass flux across momentum control volume surfaces.

OLAM uses a rotating Cartesian system with origin at the Earth's center, z-axis aligned with the north geographic pole, and x- and y-axes intersecting the equator at 0 deg and 90 deg E. longitude, respectively, as shown in the image of the Figure 1. The three-dimensional geometry of the mesh, particularly relating to terms in the momentum equation and involving relative angles between proximate grid cell surfaces, is worked out in this Cartesian system.

For more details regarding the OLAM's physics parameterizations (radiative transfer, bulk microphysics, cumulus parameterizations, turbulent transfer and surface exchange, and water and energy budgets for the vegetation canopy and multiple soil layers) see (Cotton et al., 2003) paper where most of RAMS development took place. The key point is that these parameterizations are all programmed as column-based processes, with no horizontal communication except possibly between adjacent grid columns. The importance of this is that no horizontally-implicit equation needs to be solved, and in fact the same is true of the dynamic core. This fact can have a huge impact on the communication between MPI processes and the overall efficiency of this communication. Basically, avoidance of horizontal elliptic solvers makes MPI much easier and more efficient.

2.2 OLAM's implementation

OLAM is an iterative model, where each timestep may result in the output of data as defined in its parameters. Its workflow is illustrated in Figure 6.

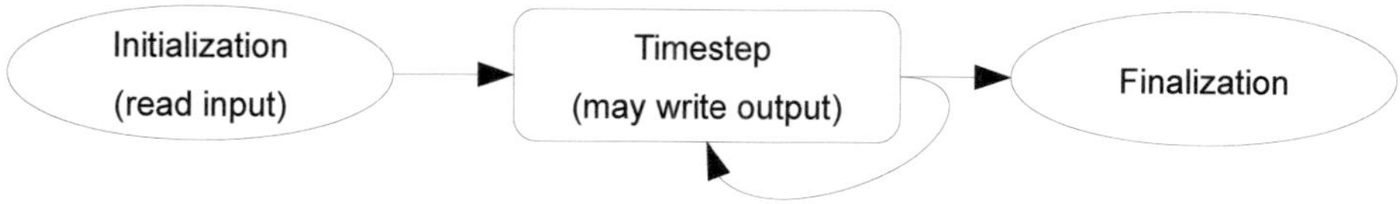

Fig. 6. OLAM's iterative organization.

OLAM input files are not partitioned for parallel processing, as each process reads the input files entirely. Typical input files are: global initial conditions at a certain date and time and global maps describing topography, soil type, ice covered areas, Olson Global Ecosystem (OGE) vegetation dataset, depth of the soil interacting with the root zone, sea surface temperature and Normalized Difference Vegetation Index (NDVI). Reading data over the entire globe is true only for files that are read at the initialization time. For those files that also need to be read during model integration, namely SST and NDVI, because their values are time dependent, the files are organized into separate geographic areas of 30 x 30 degrees each. An individual MPI process does not need to read in data for the entire globe.

After this phase, the processing and data output phases are executed alternately: during each processing phase, OLAM simulates a number of timesteps, evolving the atmospheric conditions on time-discrete units. After each timestep, processes exchange messages with their neighbors to keep the atmospheric state consistent.

After executing a number of timesteps, the variables representing the atmosphere are written to a history file. During this phase, each MPI process opens its own history file for that superstep, writes the atmospheric state and closes the history file. Each client executes these three operations independently, since there is no collective I/O implemented in OLAM.

These generated files are considered of small size for the standards of scientific applications: each file size ranges from 100KB to a few MB, depending on the grid definition and number of MPI processes employed. The amount of written data varies with the number of processes running the simulation (the number of independent output files increases with the processes count).

2.3 OLAM's configuration for the performance evaluation

In our experiments, the represented atmosphere was vertically-divided in 28 horizontal layers. Each execution simulates 24 hours of integration of the equations of atmospheric dynamics without any additional physical calculation (such as moisture and radiative processes) because we have interest only in the impact on the cost of fluid dynamics executions and communications. Each integration timestep simulates 60 seconds of the real time.

We executed tests with resolutions of 40km and 200km and with three implementations of OLAM:

1. **The MPI implementation**: The computation is divided among MPI processes.

2. **The Hybrid MPI/OpenMP implementation**[2]: Each MPI process creates OpenMP threads at the start of the timestep and destroy them after the results output. OpenMP threads execute the *do* loops from OLAM's highest cache miss/hotspot routine, named *progwrtu*.

   Therefore, this implementation uses a different level of parallelism, improving the application memory usage and generating less files. This happens because it maintains the same parallelism degree, but with a smaller number of MPI processes (each of them with a number of threads). As each output file correspond to one MPI Rank, the total number of generated files decreases compared to the MPI-only implementation.

3. **The Hybrid MPI/CUDA implementation**[3]: This implementation starts one MPI process on each core of the platform, and each MPI process starts threads on the GPU device. Due to development time reasons, we decided to implement for this work, only two CUDA kernels out of nine *do* loops from the hotspot routine. Therefore, each MPI process starts two kernel threads on the GPU device.

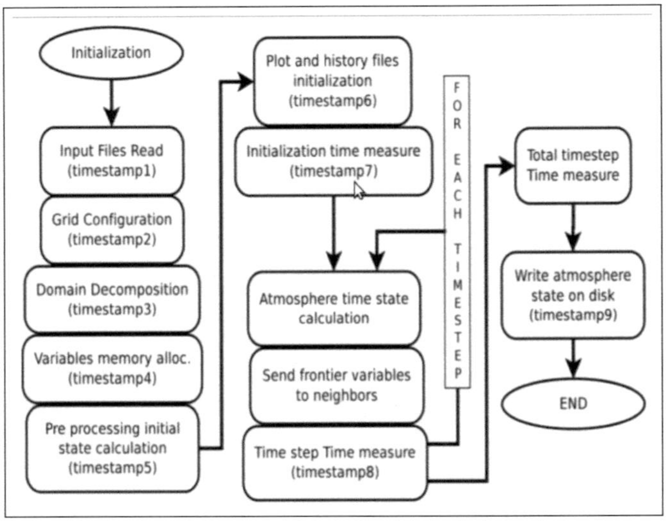

Fig. 7. OLAM's algorithm fluxogram.

We present some evaluations obtained with Vtune Performance Analyser. In order to obtain, them, we divided the OLAM's algorithm in three major parts: the initialization, the atmospheric time state calculation and the output. Figure 7 presents this algorithm in details. Finally, we inserted timestamps barriers on selected points of OLAM source (a few module boundaries) in order to correctly assign partial execution times to OLAM main modules.

3. OLAM's performance evaluation

The experiments evaluated the performance of the three implementations of OLAM version 3.3 (MPI, MPI/OpenMP, and MPI/CUDA). The tests were conducted in a multi-core cluster system and in a single multi-core node.

[2] Both the MPI and the Hybrid MPI/OpenMP implementations were developed by Robert Walko (Miami University).

[3] The Hybrid MPI/CUDA implementation was developed by Fabrício Vilasbôas and Roberto Pinto Souto (LNCC).

- The multi-core cluster environment used in the performance measurements considering the Network File System (NFS) in Section 3.1 is a multi-core SGI Altix-XE 340 cluster platform (denoted Altix), located at LNCC. Altix-XE is composed of 30 nodes, where each nodes has two quad-core Intel Xeon E5520 2.27GHz processors, with 128Kb L1 cache, 1024KB L2 cache, 8192KB L3 cache and 24GB of main memory. The nodes are interconnected by an *Infiniband* network. The used software includes MPICH version 2-1.4.1 and Vtune Performance Analyzer version 9.1. The results presented here are the arithmetic average of four executions. The Hyperthread system is active and Turbo-boost increases processor speed when only one core is active.
- The environments used for the file system (PVFS and NFS) tests in Sections 3.2 and 3.3 were part of the Grid'5000[4] infrastructure. The tests were executed on the clusters Griffon (Nancy) – equipped with 96 bi-processed nodes with Intel Xeon (quad-core), 16 GB of RAM and 320 GB of SATA II hard disks – and Genepi (Grenoble) – 34 bi-processed nodes with Intel Xeon (quad-core), 8GB of RAM and 160GB GB of local SATA storage. Both clusters have their nodes interconnected by Gigabit Ethernet.

  For the tests with PVFS file system, 30 clients were evenly distributed among 4 data servers servers, and accessed the file system through a Linux kernel module. We used PVFS version 2.8.2 obtained from http://www.pvfs.org/. We modified OLAM in order to obtain the time spent by each I/O. The tests with the I/O optimization had up to 26 nodes (208 cores) using the shared NFS infrastructure present on the Genepi cluster.
- The performance measurements for the Hybrid MPI/CUDA implementation in Section 3.4 were made on a multi-core/many-core node, denoted prjCuda, located at LNCC, composed of a dual Quad-Core Xeon E5550, 2.67GHz, with 8 MB of L3 cache, 1MB L2 cache, 256KB L1 cache and 24 GB of RAM memory, GTX285 and Tesla C2050. The software employed included MPICH version 2-1.2.p1, Vtune Performance Analyzer version 9.1, CUDA toolkit version 4.0. and PGI Fortran version 11.2 compiler. The experiments evaluate parallel performance of the three implementations of OLAM. As we are running in one single node system, OLAM input and output phase execution times are the same for all implementations. Therefore, they were not considered in our analysis. All single node tests implement a 40km OLAM configuration.

3.1 Experimental results and analysis with NFS on the multi-core cluster environment

We present two experimental analysis in two distinct OLAM workload configurations.

- In the first experiment we performed a 200km OLAM configuration analysis, a typical horizontal resolution most commonly used for climate simulation.
- The second experiment performed a 40km OLAM configuration. This is a typical resolution for global forecast. This experiment decreases 5 times the horizontal distance between points in the globe, hence more points are necessary to cover the same area. The other parameters remain the same as in the 200km configuration. Since it is necessary 25 times the number of points as before to cover the same area at 5 times shorter space intervals, the number of calculations per timestep is now increased 25 times with respect to the previous experiment. Furthermore, it implies a 20-fold increase in the memory workload, thus increasing the the proportion of computing time with respect to data transfers.

[4] http://www.grid5000.fr/

3.1.1 Experiments with resolution of 200km

Figure 8 presents the ideal and measured speedups from 1 to 80 cores for the experiments with a 200km resolution. The results show the performance of the MPI and the Hybrid MPI/OpenMP implementations of OLAM. We observe that the performance for both implementations are similar up to 16 cores. As we increase the number of cores, the Hybrid MPI/OpenMP implementation presents a better performance than the MPI-only implementation.

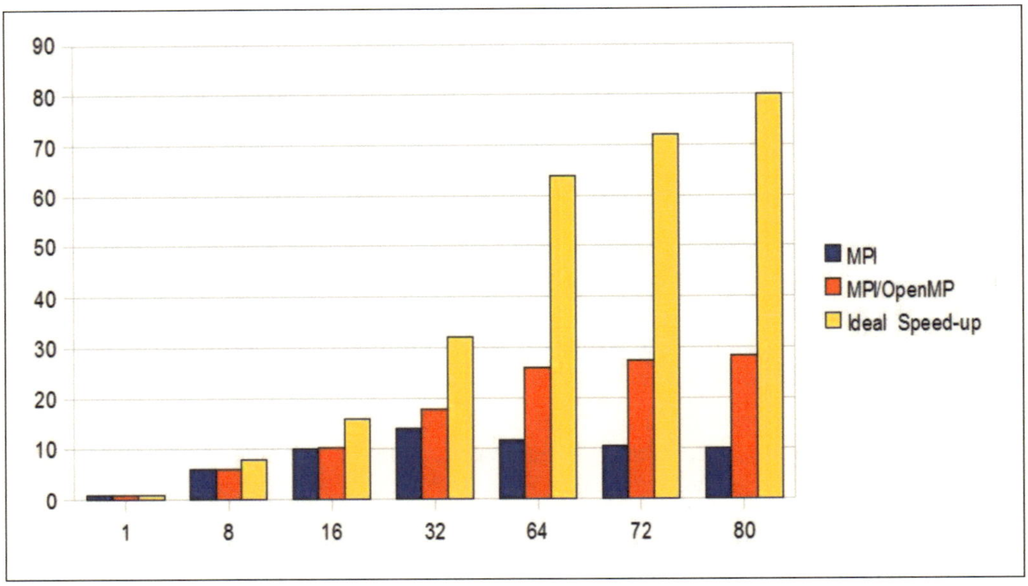

Fig. 8. 200km-resolution ideal and measured speedups for the MPI and the MPI/OpenMP implementations.

Previous works, (Osthoff et al., 2010) and (Schepke et al., 2010), evaluated the performance of OLAM on a multi-core cluster environment and demonstrated that the scalability of the system is limited by output operations performance. OLAM suffers significantly due to the creation of a large number of files and the small requests.

3.1.2 Experiments with resolution of 40km

Figure 9 presents the ideal and measured speedups from 1 to 80 cores for the experiments with a 40km resolution. The results show the performance of the MPI and the Hybrid MPI/OpenMP implementations of OLAM. We observe that the performance of the MPI implementation is better up to 32 cores. As we increase the number of cores, the Hybrid MPI/OpenMP implementation performs better than the MPI-only one. This test shows that as we increase the dependency relationship between computing time and data transfers, output operations overhead decreases overall system performance impact.

In order to explain why the MPI implementation performs better than the Hybrid MPI/OpenMP implementation at lower numbers of cores, we inserted VTUNE Analyser[5]

[5] http://www.intel.com

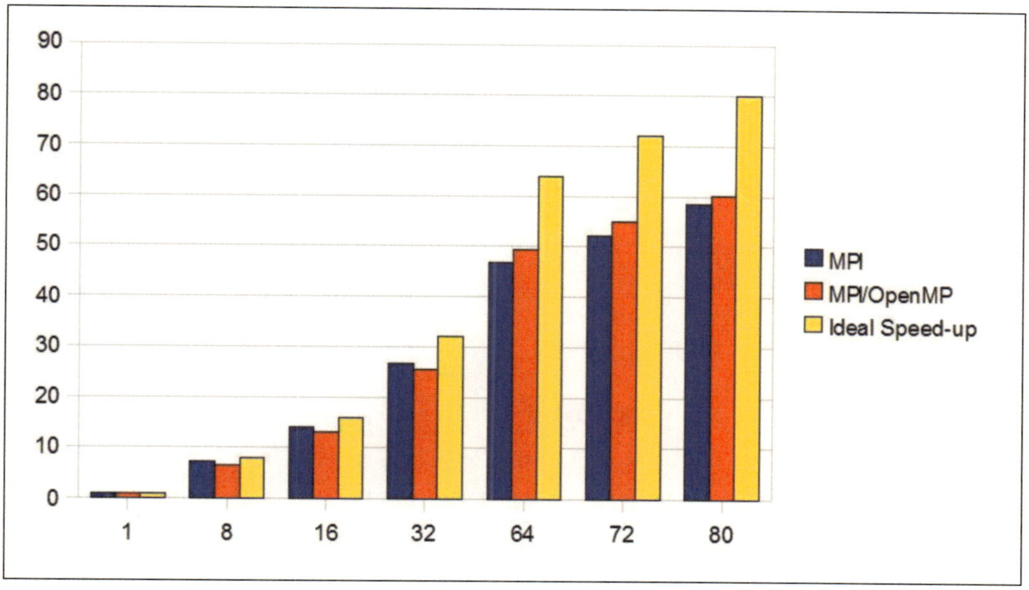

Fig. 9. 40km-resolution ideal and measured speedups for the MPI and the MPI/OpenMP implementations.

performance instrumentation in both codes and evaluated the memory usage in one node, varying the number of processes/threads. We observed that, for this configuration, the memory usage of the MPI implementation grows from 2GB for one process to 5GB to 8 processes. In the other hand, the memory usage of the Hybrid MPI/OpenMP implementation remains almost constant (2GB) as we increase the number of threads. These results confirm that the OpenMP global memory implementation improves OLAM's memory usage on a multi-core system.

The Hybrid MPI/OpenMP implementation parallelized the *do* loops from the subroutine *progwrtu*, This subroutine was responsible for up to 70% of the cpu time of each timestep (Osthoff et al., 2010; Schepke et al., 2010) and up to 75% of the cache misses. We observed that, after the parallelization with OpenMP, the subroutine became responsible for only 28% of the cpu time of each timestep. On the other hand, running with 8 cores, it means that only 28% of the code is running in parallel. This explains why we did not obtained the ideal speedup for the Hybrid MPI/OpenMP implementation.

3.2 Experimental results and analysis with PVFS on a multi-core cluster environment

This section presents the evaluation of the MPI and of the Hybrid MPI/OpenMP implementations in a multi-core cluster regarding I/O performance. We instrumented OLAM in order to obtain the time spent by each I/O operation. For each test instance, we considered on each step the greatest values between the processes. The results presented here are the arithmetic average of four executions. OLAM was configured to output history files at every simulated hour, resulting in 25 files per process (there is one mandatory file creation at the start of the execution).

3.2.1 The MPI implementation's analysis

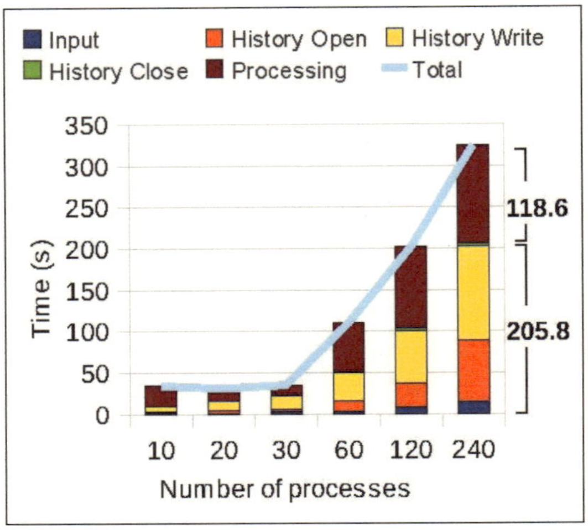

Fig. 10. Execution times for the MPI implementation with PVFS.

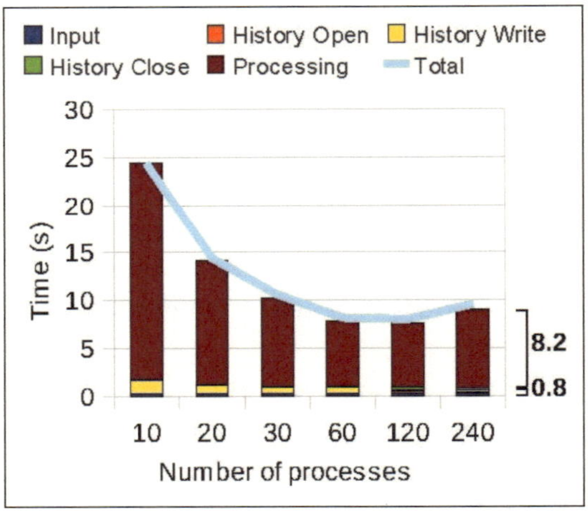

Fig. 11. Execution times for the MPI implementation using local files.

Figure 10 presents the test results for the MPI implementation using PVFS to store all the input and output files and Figure 11 presents the test results storing them in the local disks (Local files). For the second approach to work, the input files (that are previously created in only one node) have to be copied to all the nodes. Also, to work on a non-dedicated cluster, it may be needed to gather the output files on a master node.

Even when considering the times for scatter/gather operations on the input/output files, using the local disks had the best performance - around 36 times better (Figure 12). However, besides the disadvantage of the need to move all these files, the computing nodes do not have local disks in some clusters. This may happen for reasons like power consumption, price, etc. In the local files approach (Figure 11), we can see that, as the number of processes grows, both I/O and processing time decrease, benefited by the higher degree of parallelism. The lowest time is achieved by the 120 processes (4 processes per machine) configuration. PVFS results show a different behavior. In the results (Figure 10) from 10 to 30 processes (one process in each machine), we can see that the processing time decreases (as the degree of parallelism increases), but there is no speedup. This happens because I/O dominates the execution time as the concurrency in the PFS grows.

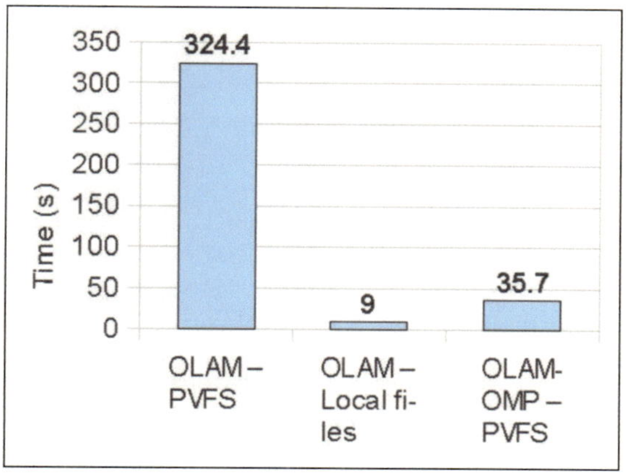

Fig. 12. Execution times for the MPI and the MPI/OpenMP implementations with PVFS and using local files.

When the number of processes in each machine becomes bigger than 1, both I/O and processing times increase. The first is affected by the high load on the file system and by the concurrent I/O requests from the processes on the same node. We believe that the rest of the simulation has its performance impacted by two factors. First, as small write operations are usually buffered, there may be concurrency in the access to the network when the processes communicate. Additionally, when more than one process is located in each machine, there may be competition for the available memory of the machine. We observed such phenomenon in previous works (Schepke et al., 2010).

3.2.2 The hybrid MPI/OpenMP implementation's analysis

Figure 13 shows the results for the Hybrid OLAM/OpenMP implementation with PVFS. The observed behavior is similar to the one shown in Figure 10 in the sense that there is no significant speedup. However, the processing time dominated the execution time in this test, not the I/O time. The time spent in I/O did not grow linearly with the number of processes, indicating that, with this number of clients, the total capacity of the parallel file system was not reached. It is important to highlight that the use of OpenMP incurs in a smaller number of processes to fully utilize the system, since 8 threads are being used per process. The system is fully utilized with only 30 processes (240 threads). Comparing this configuration with 240 processes with the MPI implementation, the MPI/OpenMP implementation is around 9 times faster, and around 20 times faster considering only the I/O time.

This increase in performance happened due to the generation of a smaller number of files (one per process). Besides, there is no intra-node concurrency in the access to the file system, because there is only one process in each machine. This gain comes also from a better memory usage by the application. Still, the time obtained for the the MPI/OpenMP implementation is almost 4 times greater than the MPI-only implementation without the use of the parallel file system.

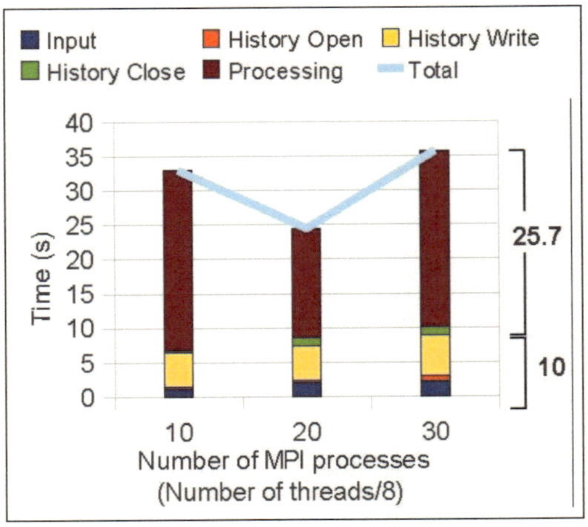

Fig. 13. Execution times for the MPI/OpenMP implementation with PVFS.

3.3 Trace visualization and close optimization for OLAM

To obtain the trace of OLAM we used the libRastro tracing library (Silva et al, 2003). LibRastro is used to generate execution traces of applications with a minimal impact on its behavior. To use the library, the source code of the target application must be modified at the points of interest in order to generate events. Beginning and end of these events are specified by two subroutine calls: `IN` and `OUT`, respectively. Each event has a name and optional parameters.

In the case of I/O operations of OLAM, the parameters are the name of the file, amount of data written/read, among others. Besides the I/O operations, we created events in all of the most important subroutines of OLAM, with the goal of identifying portions of the execution which are impaired by the I/O operations or other factors. Moreover, the detailed analysis of the application can identify the parts that do not scale.

During execution with libRastro, each process of the application generates a binary trace file. These trace files must be merged and converted to a higher level language by an application-specific tool, because the semantics of the events change from one application to the other.

One high-level event description language is Pajé (Kergommeaux et al, 2000). Pajé allows the developer to describe events, states and messages between distinct containers (a container being any element that may have states, events or be source or destination of a message).

The developer has a great flexibility to create containers and the associated events in a way which best describes his code.

In our OLAM's modeling, each MPI Rank was represented by one Pajé container. The states of this container are the events obtained from the trace. Therefore, there are states inside other states (when one subroutine calls another).

Each event must be of a predefined *type*. Pajé groups events of the same type in an execution flow and automatically stacks one state inside the other as in the case of function calls. We defined the `APP_STATE` type in which we map events related to the OLAM application. There are also the `P_STATE` type, which corresponds to I/O utility functions, and the `MPI_STATE` type to which MPI events are mapped.

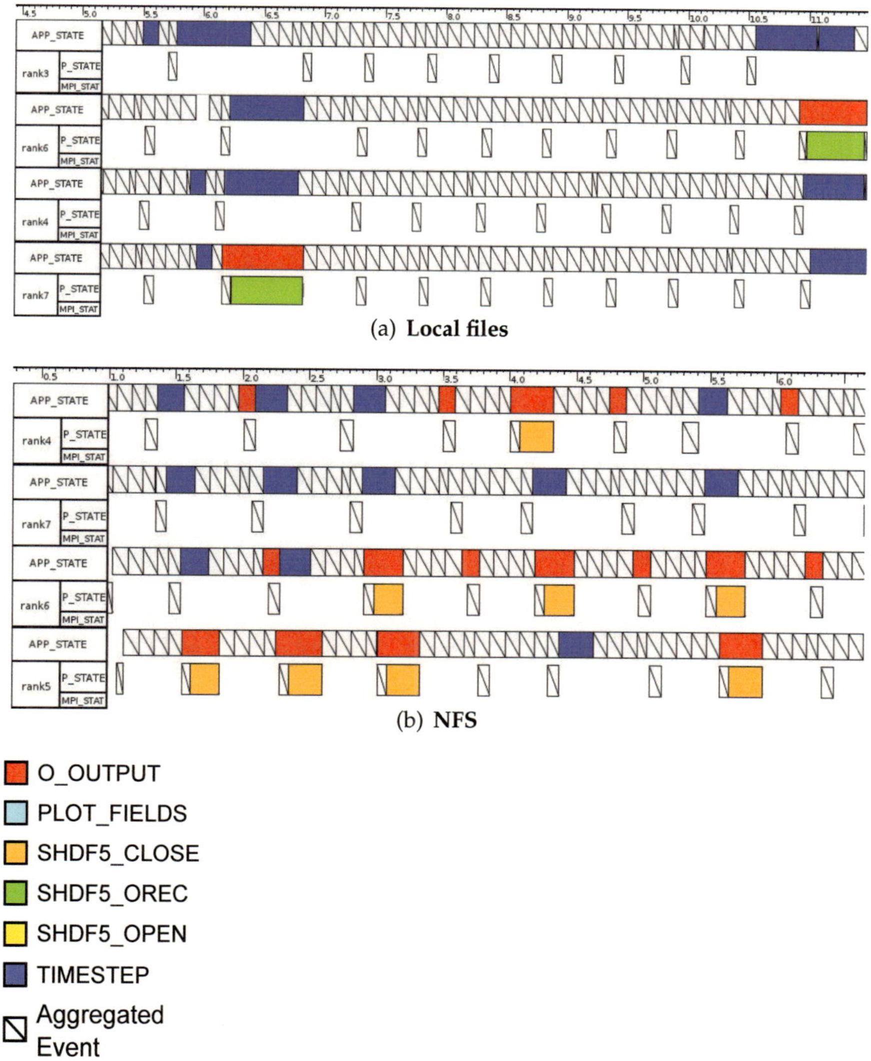

(a) **Local files**

(b) **NFS**

Fig. 14. OLAM execution with 8 Processes, 8 Threads each (64 cores)

The visualization was done via the Pajé Visualization Tool. It allows for a *gantt-chart* style, time based visualization of the events and states of the containers. The next section presents the results obtained.

In order to obtain traces from OLAM, we executed the instrumented version of the application on the clusters *Adonis* and *Edel* of Grid'5000. The tests were executed with 8 nodes using either the local file system or the shared NFS volume to store the execution output. We tested OLAM with and without OpenMP threads.

Figure 14 presents part of the Pajé visualization for the execution of OLAM with 8 processes, each with 8 threads, over local files and NFS. The rectangles on the left of the graph show the *APP_STATE* and *P_STATE*, as discussed in before. When the application enters the

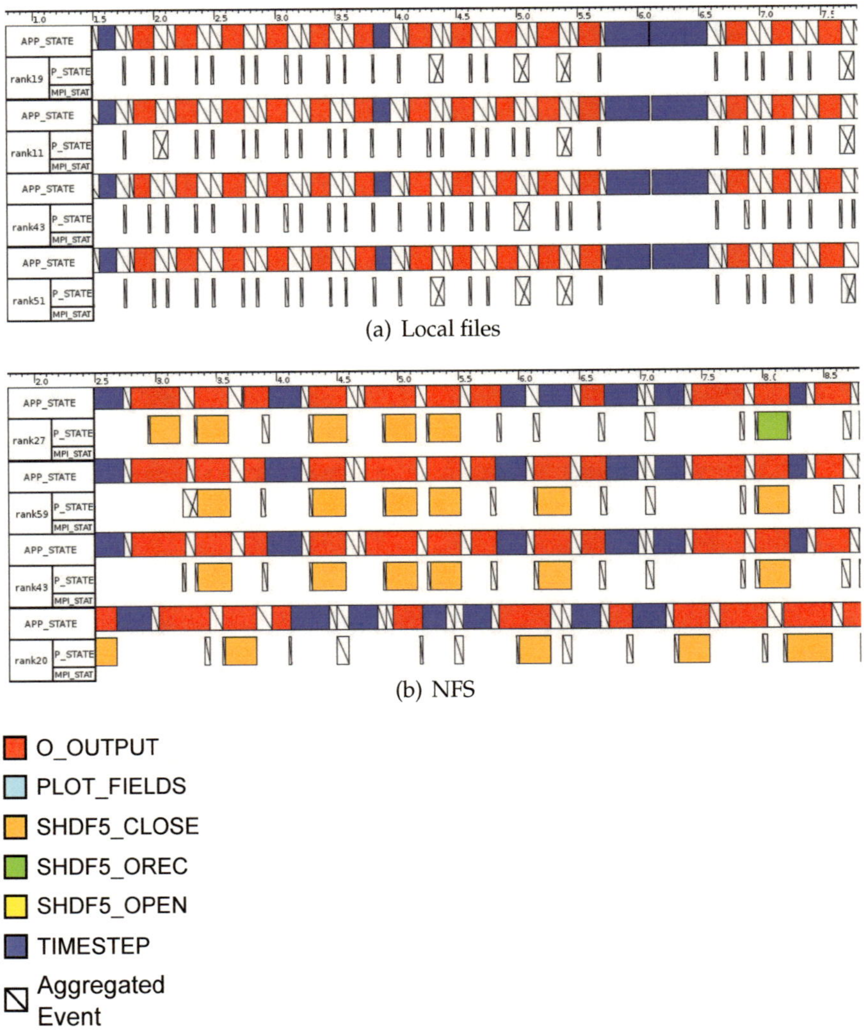

(a) Local files

(b) NFS

Fig. 15. OLAM execution with 64 Processes, 1 Thread each (64 cores)

OLAM_OUTPUT state, the underlying I/O functions are presented in the process container below. This can be more easily observed in Figure 14(a) at around 11 seconds: when the application enters the *O_OUPUT* state, the process calls a sequence of HDF5 helper functions, of which *SHDF5_OREC* takes the longest. This functions is responsible for writing the variables describing the atmospheric conditions to the output file of the process.

In Figure 14(a) we can see the first 9 seconds of execution for some of the processes of OLAM. In the *APP_STATE* flow, we can observe the execution of a sequence of *timesteps* (event TIMESTEP) after which the *olam_output* (event O_OUTPUT) function is called. At around 6.5 and 11 seconds of execution, we can observe that one O_OUTPUT event that takes longer to complete, something that can be observed in other parts of the execution and in other processes. The execution over NFS (Figure 14(b)) has a similar behavior, but the divergence between the normal I/O phases and the long ones is smaller. In these cases, we can observe that the function responsible for the divergence is *hdf5_close_write*, called from within *SHDF5_OREC*.

Figure 15 presents the visualization of traces generated when executing 64 processes of OLAM with no threads (classic MPI). In this case, since the data division becomes more fine-grained process-wise, the time of each timestep is smaller and there are more frequent I/O calls. Despite the overhead associated with creating more files for the same work, the performance was not penalized in the execution with local files (Figure 15(a)) due to the use of a fast local disk shared by each 8 processes. This overhead was made clear when output files are stored in the shared NFS volume (Figure 15(b)). In this case, most of the time spent in the P_STATE flow is in closing the file (SHDF5_CLOSE): small writes end up being delegated to the write-back mechanism in kernel, but the requests must be flushed by the time the file is closed.

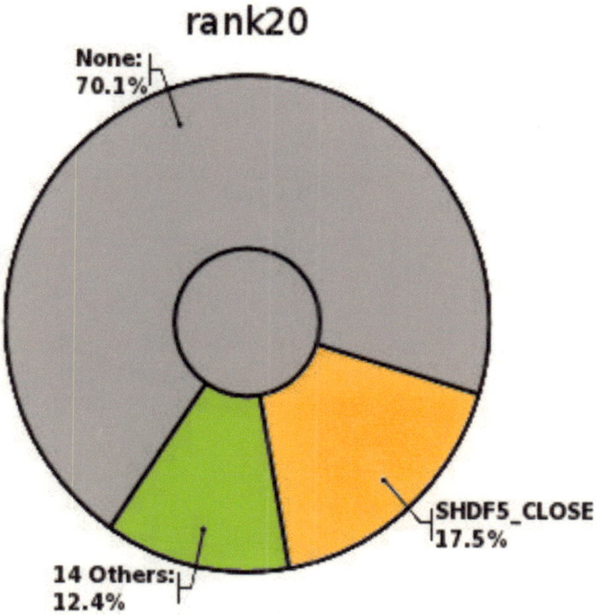

Fig. 16. Relative time spent on I/O states

Figure 16 presents the time spent on the I/O states of the application for a process from the trace of Figure 15(b). We have observed that I/O functions occupy around 25% of the execution time. Most of this time is spent on the *SHDF5_CLOSE_WRITE* function. This is due to the write-back mechanism, as said before, indicating that this is a good target to optimizations.

OLAM does not profit from the write-back mechanism due to the small size of it's I/O operations: by the time the process calls *SHDF5_CLOSE*, it's previous write requests have not been sent to the server. One way of forcing OLAM to profit from the background I/O operations offered by the file system is to change the order that the files are opened and closed. Instead of calling the close function after each process's I/O phase, the file can be closed before a new one is opened – i.e. during an *SHDF5_OPEN*.

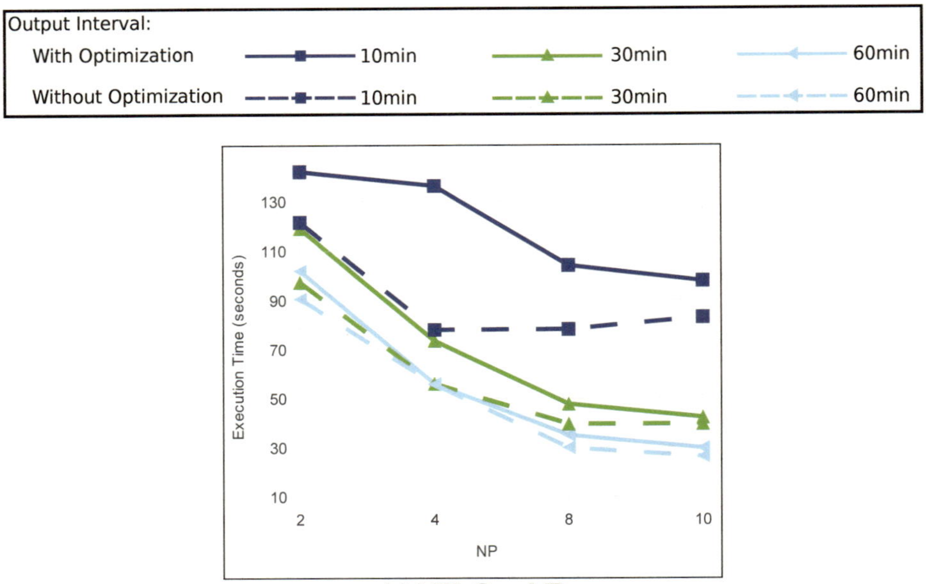

(a) MPI+OpenMP

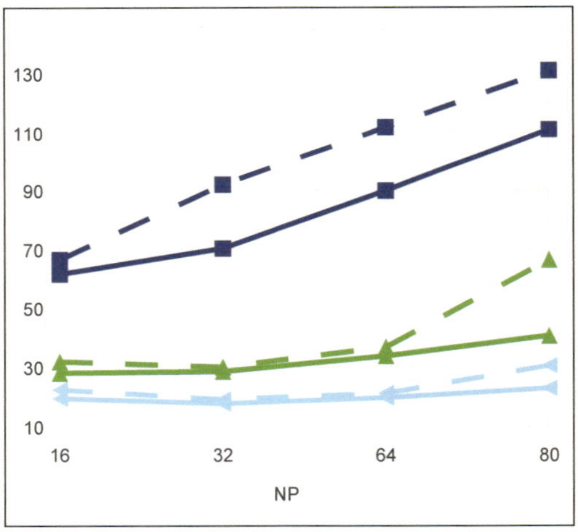

(b) Pure MPI

Fig. 17. Results for Close-on-Open optimization for MPI+OpenMP and pure MPI

This way, instead of having an order of events of *Computation – Open File – Write Data – Close File – Computation . . .*, the order is chained to *Computation – Close Previously Opened File – Open File – Write Data – Computation . . .*,

This change was evaluated in the *Adonis* cluster of Grid5000. We used the MPI+OpenMP version of OLAM and the pure MPI implementation, adjusting the number of OpenMP threads so that the number of cores used was the same as in the MPI version. OLAM was executed with three different output intervals: 10, 30 and 60 minutes of simulation. The smaller the intervals, the higher the number of generated files.

Figure 17 presents the execution time for this set of tests. The continuous lines are the execution times for OLAM with the proposed Close-on-open optimization, while the dashed lines represent the execution time with the standard implementation.

We have previously shown that the OpenMP implementation performs better than the pure MPI one due to the smaller amount of files to be created. Because of this, the optimization did not provide performance gains for this version. We can observe that, as the number of processes increases, the difference between the times with and without optimization decreases. We expect, therefore, an improvement in performance for larger number of files with the MPI+OpenMP version of OLAM.

With pure MPI the file system is stressed with the larger amount of files. With 80 cores the proposed optimization performed 15% better than the original one for the 10min and 37% better in the 30min interval. Gains were also seen for the 60min interval configuration. The smaller the interval, and therefore the larger the number of generated files, the sooner the gain was observed. This indicates that, as previously stated, the improvement obtained by the optimization increases with the number of generated files and, therefore, we can expect to observe these gains in the OpenMP version too (note that more processes incur in more files).

3.4 Experimental results and analysis on a single multi-core node

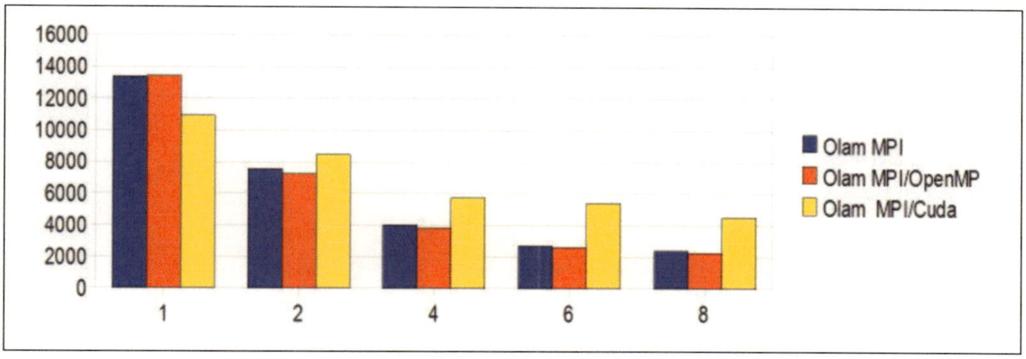

Fig. 18. Execution time for the three OLAM's implementations.

This section evaluates the three OLAM's implementations on a single multi-core node that has a GPU. Figure 18 presents OLAM's execution time in seconds as a function of cores for a resolution of 40km. The blue bar represents the MPI implementation's time, the orange bar represents the MPI/OpenMP implementation's execution time, and the yellow bar represents the MPI/CUDA implementation's execution time. This figure shows that the MPI and the MPI/OpenMP implementations perform quite similar for a single node and that the OLAM/CUDA implementation's performance decreases as the number of cores increases. The following sections discuss the results for each implementation.Figure 19 presents the three OLAMť's implementations speed-up. This figure shows that as we increase the number of cores OLAM MPI/CUDA speed-up gets worse than others implementations.

3.4.1 The MPI implementation's analysis

The MPI implementation starts one MPI process on each core. In order to explain OLAM MPI performance results we evaluate OLAM MPI performance with Vtune Performance Analyzer

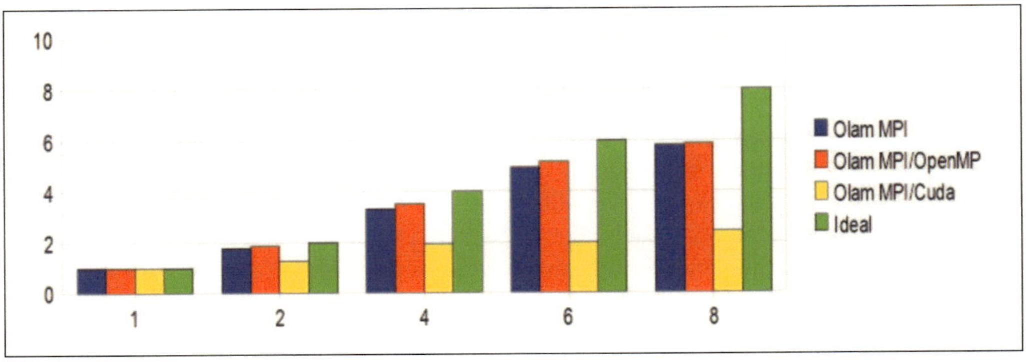

Fig. 19. Speed-up for the three OLAM's implementations.

[6]running with 8 processes on 8 cores. From previous work, (Osthoff et al., 2011a), we know that the OLAM MPI implementation scalability on 2 Quad-Core Xeon E5520 system is limited by memory contention, and that the hotspot routine presents the higher number of instruction cache miss. We observe that, for this system, the hotspot routine does not have the higher number of cache miss. Hotspot single system routine is a cpu-intensive calculation routine. Therefore, we conclude that the number of instruction cache miss has a lower performance impact in this processor architecture. Also, we observe that hotspot routine has no MPI communication overhead, therefore is a good candidate for future GPU code implementation. From Vtune analysis, we also observe that part of hotspot routines overhead are due to MPI communication routines execution time. We conclude that the speed-up is limited by, in decreasing order to: multi-core cpu processing power, multi-core memory contention and MPI communication routine overhead.

3.4.2 The hybrid MPI/OpenMP implementation's analysis

Hybrid OLAM MPI/OpenMP implementation starts one MPI process on the system, which then starts OpenMP threads on the cores of the platform. The OpenMP threads are generated on nine *do* loops from *higher number of cache miss* routine. From Vtune performance analyzer, we observed that the execution time and the number of cache misses of this routine decrease up to 50% in comparison to the MPI-only implementation. These results show that the use of OpenMP improved OLAM's memory usage for this architecture. On the other hand, we observe that CPI (Clock Per Instructions) increases from 1.2 on the MPI implementation to 2.7 on Hybrid MPI/OpenMP one, due to FORTRAN Compiler OpenMP routines overhead. We conclude OLAM MPI/OpenMP implementation still needs optimizations in order to obtain the desired speedup.

3.4.3 The hybrid MPI/CUDA implementation's analysis

The Hybrid MPI/CUDA implementation starts MPI process, and each process starts threads in the GPU device. As mention before, due to development time reasons, we implemented two CUDA kernels out of nine *higher number of cache miss* routine's *do* loops. Therefore, each MPI process starts two kernel threads on the GPU. In order to explain this implementation's results, we instrumented the two CUDA kernels to run with Computer Visual Profiler[7], which

[6] http://www.intel.com
[7] http://www.nvidia.com

Method	#Calls	GPU time (μs)	% GPU time
memcpyHtoD	1215	429677	0.83
memcpyDtoH	540	337662	0.65
kernel_loop1	135	185549	0.36
kernel_loop2	135	41151	0.08

Table 1. Execution time of the GPU kernels obtained from Compute Visual Profiler.

analyzes GPGPU systems' performance. We observed that these two loops account for up to 7% of each timestep's execution time.

First we run OLAM MPI/CUDA implementation with 1 MPI process starting 2 kernels for 5 timesteps with Compute Visual Profiler. Table 1 shows the kernels' execution times in details with 1 MPI process for 5 timesteps. These results show that GPU/memory transfer time is around three times bigger than the kernel execution time. We then conclude that the MPI/CUDA implementation needs optimizations in the transfers between GPU and main memory in order to improve performance.

A new experiment consisted in running the MPI/CUDA implementation varying the number of MPI processes from 1 to 8 (starting 2 CUDA kernels each). We observed that, as we increased the number of MPI processes, the data transfered from main memory to GPU's memory decreased from nearly 2MB for 1 MPI process to about 250KB for 8 processes. This reduction in the workload explains the MPI/CUDA implementation's poor performance for greater numbers of processes. Indeed, the gain of performance obtained using GPUs depends on the size of the problem[8]. These results explain the reason why as we increase the number of cores OLAM MPI/CUDA speed-up gets worse than others implementations.

Moreover, we observed that the GPU's utilization increased from 2% for 1 process to 6.5% for 8 processes, indicating that there is no access contention as we increase the number of processes. We plan to study this implementation for higher resolutions, obtaining heavier workloads.

4. Related work

Parallel applications scalability is the focus of several papers in the literature. In (Michalakes et al., 2008) the authors execute the high resolution *WRF* weather forecast code over 15k cores and conclude that one of the greatest bottlenecks is data storage. This same code was used in (Seelam et al., 2009) to evaluate file system caching and pre-fetching optimizations to many-core concurrent I/O, achieving improvements of 100%. I/O contention on multi-core systems is also a known issue in the literature and few strategies to mitigate the performance loss can be found. The work of (Wolfe, 2009) presents the lessons learned from porting a part of the Weather Research and Forecasting Model (WRF) to the PGI Accelerator. Like OLAM, the application used in our work, the WRF model is a large application written in FORTRAN. They ported the WSM5 ice microphysics model and measured the performance. This measurement compared the use of PGI Accelerator with a multi-core implementation and with a hand-written CUDA implementation. Finally, the work of (Govett et al., 2010) runs the weather model from the Earth System Research Laboratory (ESLR) on GPUs and relies on the CPUs for model initialization, I/O and inter-processor communication. They have shown

8 case studies examples: http://www.culatools.com/features/performance

that the part of the code that computed dynamics of the model runs 34 times faster on a single GPU than on the CPU.

A scalability study with the NFS file system had shown that the OLAM model's performance is limited by the I/O operations (Osthoff et al., 2010). The work of (Schepke et al., 2010) presents an evaluation of OLAM MPI with VtuneAnalyzer and identifies a large amount of cache misses when using 8 cores. Then, experiments comparing the use of no distributed file system and of PVFS cleared that the scalability of these operations is not a problem when using the local disks (Boito et al., 2011). The concurrency in the access to the shared file system, the size of the requests and the large number of small files created were pointed as responsible for the bad performance with PVFS and NFS. Recent work from (Kassick et al, 2011) presented a trace-based visualization of OLAM's I/O performance and has shown that, when using the NFS File System to store the results on a multi-core cluster, most of the time spent in the output routines was spent in the close operation. Because of that, they propose to delay the close operation until the next output phase in order to increase the I/O performance. A new implementation of OLAM using MPI and OpenMP was proposed in order to reduce the intra-node concurrency and the number of generated files. Results have shown that this implementation has better performance than the only-MPI one. The I/O performance was also increased, but the scalability of these operations remains a problem. We also observe that the use of OpenMP instead of MPI inside the nodes of the cluster improves the application memory usage (Osthoff et al., 2011a).

5. Conclusion

This work evaluated the Ocean-Land-Atmosphere Model (OLAM) in multi-core environments - single multi-core node and cluster of multi-core nodes. We discussed three implementations of the model, using different levels of parallelism - MPI, OpenMP and CUDA. The implementations are: (1) a MPI implementation; (2) a Hybrid MPI/OpenMP implementation and (4) a Hybrid MPI/CUDA implementation.

We have shown that a Hybrid MPI/OpenMP implementation can improve the performance of OLAM multi-core cluster using either the local disks, a single-server shared file system (NFS) and a parallel file system (PVFS). We observe that, as we increase the number of nodes, the Hybrid MPI/OpenMP implementation performs better than the MPI one. The main reason is that the Hybrid MPI/OpenMP implementation decreases the number of output files, resulting in a better performance for I/O operations. We also confirm that OpenMP global memory advantages improve the application's memory usage in a multi-core system.

In the experiments in a single multi-core node, we observed that, as we increase the number of cores, the MPI/OpenMP implementation performs better than others implementations. The MPI/OpenMP implementation's bottleneck was observed to be due to low routine parallelism. In order to improve this implementation's speed-up, we plan to further parallelize the OLAM's most cpu-consuming routine. Finally, we observed that the MPI/CUDA implementation's performance decreases for numbers of processes greater than 2 because of the workload reduction with the parallelism. Therefore, we plan to further evaluate the performance with resolutions higher than 40km.

We also applied a trace-based performance visualization in order to better understand OLAM's I/O performance. The libRastro library was used to instrument and obtain the traces of the application. The traces were analyzed and visualized with the Pajé. We have shown that, when using the shared file system to store the results, most of the time spent in the output routines was spent in the close operation. Then, we proposed a modification to delay this operation, obtaining an increase in performance of up to 37%.

For future works we plan to further parallelize OLAM timestep routines in order to improve both the MPI/OpenMP and the MPI/CUDA implementations. Also, we plan to study OLAM's performance for higher resolutions and for unbalanced workload on single nodes and on multi-core/many-core clusters.

In order to include further I/O optimizations, we also intend to evaluate the parallel I/O library. This will be done aiming to reduce the overhead on creation of files, improving the output phases' performance.

6. References

Adcroft,C.H.A. and Marshall,J. Representation of topography by shaved cells in a height coordinate ocean model, *Monthly Weather Review, 125:2293Ű2315, 1997.*

Boito, F.Z.; Kassick, R. V.; Pilla, L.L.; Barbieri,N.; Schepke,C.; Navaux,P.; Maillard,N.; Denneulin, Y.; Osthoff,C.; Grunmann, P.; Dias, P. & Panetta,J. (2011). I/O Performance of a Large Atmospheric using PVSF, *Proceedings of Renpar20 / SympA14/ CFSE8* INRIA, Saint-Malo, France.

Cotton, W.r.; Pielke, R.; Walko, R.; Liston, G.; Tremback, C.; Harrington, J. & Jiang, H. RAMS 2000: Current status and future directions, *Meteorol. and Atmos. Phys., 82, 5-29, 2003.*

Govett, M. et al. (2010). Running the NIM Next-Generation Weather Model on GPUs, *Proceedings of 10th IEEE/ACM International Conference on Cluster, Cloud and Grid Computing*, Melborne, Australia, pp. 729-796.

Michalakes, J., Hacker, J., Loft, R., McCracken, M. O., Snavely, A., Wright, N. J., Spelce, T., Gorda, B., & Walkup, R. WRF Nature Run., *Journal of Physics: Conference Series 125(1):012022*, URL http://stacks.iop.org/ 1742-6596/125/i=1/a=012022.

Ohta, K., Matsuba, H. & Ishikawa, Y. (2009). Improving ParallelWrite by Node-Level Request Scheduling, *Proceedings of the 2009 9th IEEE/ACM International Symposium on Cluster Computing and the Grid*, IEEE Computer Society, Washington, DC, USA, pp. 196£203

Osthoff,C.; Schepke, C.; Panetta, J.; Grunmann, P.;Maillard, N.; Navaux, P.; Silva Dias, P.L.& Lopes, P.P. I/O Performance Evaluation on Multicore Clusters with Atmospheric Model Environment, *Proceedings of 22nd International Symposium on Computer Architecture and High Performance Computing*, IEEE Computer Society, Petropolis,Brazil, pp. 49-55

Osthoff,C.; Grunmann, P.; Boito, F.; Kassick, R.; Pilla, L.; Navaux, P.; Schepke, C.; Panetta, J.; Maillard, N.& Silva Dias, P.L. Improving Performance on Atmospheric Models through a Hybrid OpenMP/MPI Implementation, *Proceedings of The 9th IEEE International Symposium on Parallel and Distributed Processing with Applications*, IEEE Computer Society, Busan, Korea, pp. 69-75

Pielke, R.A. et al. (1992). A Comprehensive Meteorological Modeling System - RAMS, in: *Meteorology and Atmospheric Physics.* 49(1), pp. 69-91

Schepke, C.; Maillard, N.; Osthoff, C.; Dias, P.& Pannetta, J. Performance Evaluation of an Atmospheric Simulation Model on Multi-Core Environments, *Proceedings of the Latin American Conference on High Performance Computing* , CLCAR, Gramado, Brazil, pp. 330-332.

Seelam, S., Chung, I.H., Bauer, J., Yu, H., & Wen, H.F. Application level I/O caching on Blue Gene/P systems, *Proceedings of IEEE International Symposium on Parallel Distributed Processing*, IEEE Computer Society, Rome, Italy, pp. 1-8.

Walko, R.L.& Avissar, R. The Ocean-Land-Atmosphere Model (OLAM). Part I: Shallow-Water Tests. *Monthly Weather Review 136:4033-4044, 2008.*

Walko, R.L.& Avissar, R. A direct method for constructing refined regions in unstructured conforming triangular-hexagonal computational grids: Application to OLAM. doi: 10.1175/MWR-D-11-00021.1 *Monthly Weather Review 139:3923-3937, 2011.*

Wolfe, M. The PGI Accelerator Programming Model on NVIDIA GPUs Part 3: Porting WRF. *In Thechnical News from Portland Group.* http://www.pgroup.com/lit/articles/insider/v1n3a1.htm

Da Silva, G.J; Schnorr, L. M. & Stein, B. O. Jrastro: A trace agent for debugging multithreaded and distributed java programs. *Computer Architecture and High Performance Computing, Symposium on*, 0:46, 2003.

Kassick, R. et al (2011). Trace-based Visualization as a Tool to Understand ApplicationsŠ I/O Performance in Multi-Core Machines, *Proceedings of of 23and International Symposium on Computer Architecture and High Performance Computing*, IEE Computer Society, Vitória, Brazil.

Kergommeaux,J.C.; Stein, B. & Bernard, P.E. Pajé, an interactive visualization tool for tuning multi-threaded parallel applications, *Parallel Computing*, 26(10):1253 – 1274, 2000.

2

Applications of Mesoscale Atmospheric Models in Short-Range Weather Predictions During Satellite Launch Campaigns in India

D. Bala Subrahamanyam[1] and Radhika Ramachandran[2]
[1]*Space Physics Laboratory, Vikram Sarabhai Space Centre, Indian Space Research Organization, Department of Space, Government of India, Thiruvananthapuram*
[2]*Indian Institute of Space Science and Technology (IIST), Department of Space, Government of India, Thiruvananthapuram*
India

1. Introduction

Knowledge of meteorology forms the basis of scientific weather forecasting, which evolves around predicting the state of the atmosphere for a given location. Weather forecasting as practiced by humans is an example of having to make judgments in the presence of uncertainty. Weather forecasts are often made by collecting quantitative data about the current state of the atmosphere and using scientific understanding of atmospheric processes to project how the atmosphere will evolve in future. Over the last few years the necessity of increasing our knowledge about the cognitive process in weather forecasting has been recognized. For its human practitioners, forecasting the weather becomes a task for which the details can be uniquely personal, although most human forecasters use approaches based on the science of meteorology in common to deal with the challenges of the task. The chaotic nature of the atmosphere, the massive computational power required to solve the equations that describe the atmosphere, error involved in measuring the initial conditions, and an incomplete understanding of atmospheric processes mean that forecasts become less accurate as the difference in current time and the time for which the forecast is being made (the range of the forecast) increases (Doswell, 2004; Ramachandran et al., 2006; Subrahamanyam et al., 2006; 2008).

In the Indian scenario where most of the agricultural industries depend on summer monsoon rainfall, several atmospheric models are run at different organizations to deliver regular forecasts to common man and media with a special emphasis on prediction of onset of monsoon and expected amount of rainfall. However, operational forecast models run by meteorological departments are not meant for prediction of local weather with the high spatial and temporal resolutions within a specific time window. Depending on the users requirements and event-based management, atmospheric models are tuned for delivering the forecast products. During satellite launch operations, accurate weather predictions and reliable information on the winds, wind-shears and thunderstorm activities over the launch site happens to be of paramount importance in the efficient management of launch time operations (Manobianco et al., 1996; Rakesh et al., 2007; Ramachandran et al., 2006;

Subrahamanyam et al., 2010; 2011). Therefore, additional procedures based on current observations (surface and upper air observations, Radar and satellite) are required to forecast imminent weather events and to warn if necessary. In this scenario, it is highly desired to develop a comprehensive, coordinated, and sustained Earth Observation System for collection and dissemination of improved data, information, and models to stakeholders and decision makers. The present chapter gives a detailed account on the utilization of two mesoscale atmospheric models in conjunction with other observational tools in providing short-range weather predictions for satellite launch campaigns in India.

2. Weather forecasting for decision making

In the present era, weather forecasting heavily relies on computer-based numerical weather prediction (NWP) models that take many atmospheric factors into account (Cox et al., 1998; Pielke, 1984). The evolution of operational NWP from larger to smaller scales partially reflects the increased computer power that has allowed global models to resolve more details of atmospheric flow fields. Using a consensus of forecast models, as well as ensemble members of the various models, can help reduce forecast error. There are a variety of end uses to NWP model-derived forecasts, many of which are directly linked to crucial decision making. Taking meaningful decisions has always been one of the demanding points of the forecasters job. Despite our dependence on NWP models, human input is anticipated to pick the best possible forecast model to base the forecast upon, which involves pattern recognition skills, teleconnections, knowledge of model performances and model biases. Because of the huge amounts of data that forecasters face, analytical decision-making processes are limited by time constrains. Therefore intuitive approaches are virtually mandatory (Doswell, 2004; Klein et al., 2006; Rasmussen et al., 2001; Stuart et al., 2007). Klein et al. 2006 describes the decision-making process of the experienced weather forecaster with a recognition primed decision model, that combines both analysis and intuition.

During satellite launch missions the human decision-making process is put under a lot of pressure and taking the right decision can become very difficult and require a lot of experience. In such situations we can maintain and reinforce the role of the human forecaster if we are able to give the forecaster the opportunity to find ways to continue to gain expertise in a rapidly changing technological environment. This is possible only if we devote more effort, time and research in order to better understand the demanding tasks of weather forecasting and - more important - if we are able to translate our knowledge into best practice. The launch weather guidelines involving the satellite launch vehicles and rockets are similar in many areas, but a distinction is made for the individual characteristics of each. The criteria are broadly conservative and assure avoidance of possibly adverse conditions. In this chapter, we attempt to give a detailed account on the forecasting requirements during satellite launch campaigns and the usage of mesoscale atmospheric models for catering to the needs of the mission team (Subrahamanyam et al., 2010; 2011).

Every satellite launch is related to several multi-disciplinary tasks which are performed in a serial as well as in parallel modes. All the activities in a way or other are related to each other and are linked with the success of the entire mission, thus none of the activities can be ignored at any point of the time. Having quite reliable and credible information on the local weather over the launch site is one of the mandatory requirements, which is gathered through all possible sources. Once the satellite payload is declared ready for its space journey, the mission team used to define a launch window extending over a few days with a specific time-slot for the launch. Definition of such a time-slot and launch window is one of the most crucial tasks and require adequate knowledge on local weather over the launch site in terms of the launch commit criteria (LCC) so as to avoid any hazardous events due to bad weather (Case et al., 2002; Manobianco et al., 1996). Each launch vehicle has a specific tolerance for wind shear, cloud cover, temperature, whereas lightening constraints are common for all types of vehicles. On occasions, the LCC requirements are confined to "go" and "no-go" kind of situations; thus the NWP model-derived forecast products play a very crucial role during each stage of the satellite launch campaigns. Thus, NWP model simulations during launch campaign are ultimately intended to provide short-range (< 48 hours) forecasts of winds, temperature, moisture, clouds and any hazardous weather event such as thunderstorm over the launch site with good accuracy, so as to help the mission team in decision making procedure.

3. Sriharikota: the Indian Satellite launch site

The Indian Satellite launch site is located at the Sriharikota High Altitude Range (also referred to as SHAR, 13.7167°N, 80.2167°E) is a barrier island off the coast of the southern state of Andhra Pradesh in India having a complex terrain with the Eastern Ghats located on its west and surrounded by the Pulicat Lake and Bay of Bengal on the east (Fig. 1). It houses India's only satellite launch centre in the Satish Dhawan Space Centre and is used by the Indian Space Research Organization to launch satellites using multi-stage rockets such as the Polar Satellite Launch Vehicle (PSLV) and the Geosynchronous Satellite Launch Vehicle (GSLV). This island was chosen in 1969 for setting up of a satellite launching station. Features like a good launch azimuth corridor for various missions, advantage of earth's rotation for eastward launchings, nearness to the equator, and large uninhabited area for the safety zone - all make SHAR an ideal spaceport. Presently, the SHAR has infrastructure for launching satellites into low earth orbit, polar orbit and geo-stationary transfer orbit. The launch complexes provide support for vehicle assembly, fuelling, checkout and launch operations. As per climatology of the Indian subcontinent, the ambient meteorological conditions are largely governed by monsoonal winds. The subcontinent experiences wet and rainy season during its summer monsoon (also referred to as south-west monsoon) in the month of June to September when the moist air from oceanic regions intrudes the continent resulting into large precipitation spells wide spread over the season. The month of March to May over the subcontinent is considered as pre-monsoon period when SHAR experiences quite hot and summer environmental conditions with very less rainfall resulting in a magnitude of about 5 to 7.5 cms (Indian Meteorological Department Website: http://www.imd.gov.in/). On the other hand, the period of October to December is considered as post-monsoon season, when SHAR experiences frequent rainfall amounting from 75 to 100 cms, almost an order larger than that observed during the pre-monsoon season. SHAR being one of the coastal stations on the eastern coastline of the Indian subcontinent very well falls in the vicinity of tropical cyclone landfall zone (Das et al., 2011; Rakesh et al., 2007; Singh et al., 2005), and therefore it is obvious that the underlying complex terrain could very well affect the local winds, resulting at times in the developments of hazardous thunderstorms. The depressions that develop over the Bay of Bengal can also affect the short-term weather of SHAR. Thus, the local weather over SHAR is highly dependent on atmospheric wind circulation prevailing over its adjoining Bay of Bengal Oceanic region and topography-induced convective activities.

Keeping the meteorological features over SHAR in background and to meet the forecasting needs for LCC, two numerical atmospheric models, namely High-resolution Regional Model

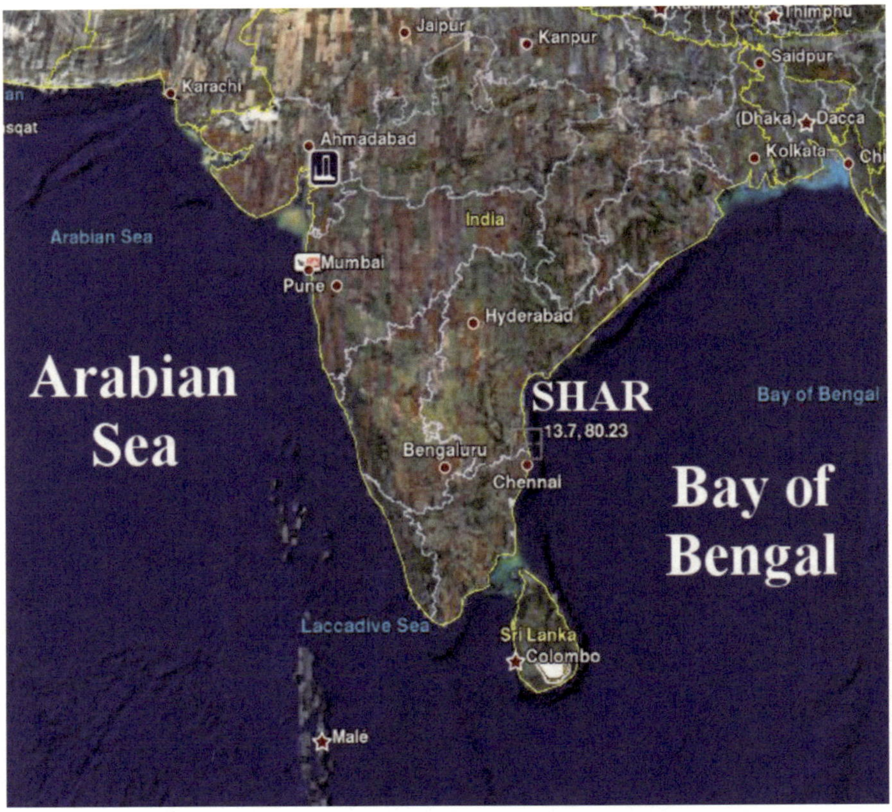

(a) Location Map of SHAR in the Indian sub-continent;

(b) Zoomed version of map showing the location of SHAR, Pulicat Lake and Bay of Bengal

Fig. 1. (a) Location Map of SHAR in the Indian sub-continent; (b) Zoomed version of map showing the location of SHAR, Pulicat Lake and Bay of Bengal (Courtesy: Google Earth Maps)

(HRM) and Advanced Regional Prediction System (ARPS) model are made operational for providing short-range weather predictions for the launch site (Ramachandran et al., 2006; Subrahamanyam et al., 2010; 2011). On one hand, the HRM simulations provide valuable information on meteorological conditions and mesoscale atmospheric circulation systems such as sea-breeze prevailing over the Indian sub-continent and its adjoining Oceanic counterpart with a special emphasis on formation of any hazardous weather event such as cyclonic storms and its movement (Ramachandran et al., 2006; Rani et al., 2010). On the other hand, ARPS is being run regularly at different time-intervals at very high horizontal and vertical resolution for alerting the mission team on probability of thunderstorm occurrence and rainfall time-slots over the launch site, if any (Subrahamanyam et al., 2010; 2011).

4. An overview of HRM and ARPS models

The basic idea of NWP models is to sample the state of the atmosphere at a given time and use the equations of fluid dynamics and thermodynamics to estimate the state of the atmosphere at some time in the future. In a broad sense, atmospheric models can be classified into two categories: (1) Hydrostatic; and (2) Non-hydrostatic. The first class of models assume hydrostatic equilibrium of atmosphere, while the second class are useful in dealing non-hydrostatic processes and their effects become more useful when the length of the feature is approximately equivalent to its height (Anthes & Warner, 1974; Bannoon, 1995; Pielke, 1984). Thus, non-hydrostatic models can be quite sensitive to small differences in atmospheric structure in vertical. Between the two models used in the present study, HRM assumes hydrostatic approximation, while ARPS is a non-hydrostatic model.

4.1 High-resolution Regional Model (HRM)

HRM is basically a hydrostatic regional model developed for shared memory computers at Deutscher Wetterdienst (DWD) of Germany that serve as one of the flexible tools for NWP for the usage in meso-α and meso-β scales and is widely used by several meteorological services, universities, and research institutes spanning a large domain of the globe (Majewski, 2010). Most of the technical features of HRM resemble with the German global model GME ((Majewski, 2010; Majewski et al., 2002)). The HRM consist of topographic data sets which can be obtained from DWD in regular or rotated latitude/longitude grid for any region of the world at mesh sizes from 30 km and 5 km. Several options exist to derive the initial state of the HRM. In the present study, however, we have used interpolation of the analysis of global model GME (grid spacing 30 km x 30 km, 60 vertical layers since 1 February 2010) to the HRM grid (Majewski et al., 2002). There is less adaptation of HRM to the fine scale topography, and local observations, not distributed on the global telecommunication system, may have a beneficial impact on the initial state of the HRM. As of now, two options are available for lateral boundary conditions of the HRM, namely (a) GME forecasts fields or (b) ECMWF forecast fields. We have made use of GME forecast fields for providing the lateral boundary conditions to the HRM. DWD provides the analyses and forecasts of GME on all 60 model layers and seven soil layers at a horizontal resolution of 30 km four times per day at 3-hourly intervals. These data are distributed by the DWD via the internet between 02:40 to 03:30 UTC for 00 UTC and between 14:40 to 15:30 UTC for 12 UTC.

Since the volume of the global data set (more than 30 GByte) is much too big for a timely transfer, DWD provides to each HRM user GME data sets tailored to the respective local HRM domain. Lateral boundary data may even be given for a frame, i.e. not covering

	High-resolution Regional Model (HRM)	Advanced Regional Prediction System (ARPS)
Dynamics	Hydrostatic	Non-hydrostatic
Usage	Different mesoscale processes	Storm model
Developed by	Deutscher Wetterdienst (DWD),Germany	Centre for Analysis and Prediction of Storms (CAPS) of Oklahoma State University, USA.
Model Domain	65°E to 95°E, 0 to 30°N	(~50 kms x 50 kms) centered around SHAR
Horizontal resolution	0.0625° (~5 kms)	~1 kms
Topography resolution	0.0625° (~5 kms) prepared by DWD	30 seconds (~1 kms), USGS
Vertical levels	60 levels with vertical stretching; Model ceiling is kept at 10 hPa (~31 kms)	35 levels with vertical stretching; Model ceiling is kept at 100 hPa (~16 kms)
Initialization and Lateral Boundary Conditions	Global Model Output for the HRM domain (horizontal resolution at 30 kms)	High-resolution GPS ascent data (maximum permissible levels = 600); symmetric initial perturbations for LBC
Prognostic variables	Surface Pressure,Cloud Ice, Temperature, Water Vapour, Cloud Water, Horizontal Wind (zonal and meridional)	Potential temperature,Horizontal wind (zonal and meridional), Water substance, subgrid scale turbulent kinetic energy
Time integration scheme	split semi implicit with $\Delta t = 30$ seconds	$\Delta t = 6$ seconds
Atmospheric boundary layer parameterization	Level-2 scheme vertical diffusion in the atmosphere, Similarity theory at the surface	Compute the time evolution of boundary layer depth in response to the surface heat fluxes or user specified boundary layer depth.
Turbulent Fluxes	Louis (1979) in the prandtl layer and Mellor and Yamada (1974) for the boundary layer and free atmosphere.	Modified Businger formulation (Businger et al., 1971)
Convective parameterization schemes	Tiedtke (1989)	Kuo (1965, 1974) Kain and Fritsch, (1993)
Soil Model	The Multilayer soil model (Heise and Schrodin, 2002) with seven layers.	Noilhan and Planton (1989) and Pleim and Xiu (1995)

Table 1. Comparison between HRM and ARPS Model (Subrahamanyam et al., 2011)

the full HRM domain but only a frame with a width of about 10 HRM rows and columns. Thus the amount of GME data which has to be transferred via the internet can be reduced drastically. For a typical HRM domain of 4000 x 4000 km^2 the GME data set (initial state plus 26 lateral boundary data sets at 3-hourly intervals) needed for a 78-h HRM forecast is in the order of (1 + 26)*12.6 MByte = 341 MByte for the full version and about 110 MByte for the frame version (Majewski, 2010). This particular option makes the HRM superior to other models, as we need not download the entire global data for generation of initial state and lateral boundary conditions of the regional model. Further technical details on the HRM can be found elsewhere (Majewski, 2010; Subrahamanyam et al., 2010; 2011)

4.2 Advanced Regional Prediction System (ARPS)

Since 1970s, we find mention of development of three-dimensional non-hydrostatic modeling for convection in the literature (Pielke, 1984; Xue et al., 2000). The development of non-hydrostatic models and rapid increase of computer powers have made the explicit prediction of thunderstorm into a reality. Advanced Regional Prediction System (ARPS) is a three dimensional, non-hydrostatic and fully compressible, primitive equation model designed for storm and mesoscale atmospheric simulation and real time prediction by the Center for Advanced Prediction of Storms, University of Oklahoma, USA (Xue et al., 1995). From the beginning of its development, it is serving as an effective tool for basic and applied research and as a system suitable for explicit prediction of convective storms as well as weather systems at other scales.

ARPS uses a generalized terrain following coordinate system with equal spacing in x- and y- directions and grid stretching in the vertical. The prognostic variables are Cartesian wind components, perturbation potential temperature and pressure, sub grid scale turbulent kinetic energy, mixing ratios for water vapour, cloud water, rain water, cloud ice, snow and hail. The ARPS includes its own data ingest, quality control and objective analysis packages, a data assimilation system which includes single-Doppler velocity and thermodynamic retrieval algorithms, the forward prediction component, and a self-contained post-processing, diagnostic and verification package (Xue et al., 2000; 2001; 2003). The model employs advanced numerical techniques, including monotonic advection schemes for scalar transport and variance conserving fourth-order advection for other variables. The model also includes state-of-the-art physics parameterization schemes that are important for explicit prediction of convective storms as well as the prediction of flows at larger scales (Case et al., 2002). The spatial discretization is achieved by second order quadratically conservative and fourth quadratically conservative finite differences for advection and second order differencing for other terms, while temporal discretization is treated with a second order leap-frog scheme for large time steps with Asselin time filter option. Depending on the choice of Users and availability of lateral boundary conditions, ARPS can be configured in one-, two- and three-dimensional modes (Subrahamanyam et al., 2008). This model has a salient feature that the base state of the model variables can be initialized through a single sounding profile and time dependent fields can be opted for self initialization using analytic functions.

One of the basic differences in the HRM and ARPS model simulations is that the initial and lateral boundary conditions of the HRM are derived from the analyses and forecast fields of a global model, whereas the base-state field for ARPS domain are generated through vertical column GPS ascents and potential temperature perturbations are generated through symmetric random option available within the model code (Subrahamanyam et al., 2006; 2010; 2011; 2008; Xue et al., 1995). Table 1 provides a glimpse of major differences between the two models.

5. Launch Commit Criteria (LCC) related requirements

The launch weather guidelines involving the satellite launch activities for different launch vehicles and rockets are similar in many areas, but a distinction is maintained for the individual characteristics of each. These guidelines include weather trends and their possible effects on launch day. More importantly, these guidelines, often referred to as *Launch Commit Criteria* are broadly conservative and assure avoidance of possible adverse conditions. The

LCC for a particular satellite launch activities essentially define the permissible limits of meteorological conditions prevailing over the site, and are reviewed for each individual launches.

5.1 Surface-layer meteorological conditions

During the assembly of the launch vehicle and its movement to the launch pad, surface-layer meteorological conditions are periodically reviewed. Magnitudes of horizontal wind speed measured through multi-level instruments mounted on a 100-m meteorological tower exceeding 20 m/s are not favourable. Similarly, a careful examination of wind gusts is also done during the movement of the vehicle.

5.2 Lightning and electric fields

For a perfect satellite launch, it is expected that there should neither be any precipitation or lightning (and electric fields with triggering potential) at the launch pad or within the flight path. If lightning is detected within 20 kms of the launch pad or planned flight path within 30 minutes prior to launch window, it is not advisable to go ahead for the launch. The absolute values of all electric field measurements over the launch site must be below 1000 V/m for 15 minutes during the liftoff period.

5.3 Clouds

As per the Kennedy Space Centre Press Release (2000) on LCC, it is advised that the satellite launch should be kept on hold, if any part of the planned flight path is through a layer of thick convective clouds. However, in case of the non-precipitating clouds or cirrus-like clouds which are not associated with convective events, launch can be shown green signal. In addition to this, the cloud cover over the launch site also helps in range safety operations.

A detailed description on LCC can be found online on NASA news release from Kennedy Space Centre (http://www.nasa.gov/centers/kennedy/news/facts/nasa_facts_toc.html). Keeping the criticalities involved during the launch activities, it is highly desired to have greater accuracy on local weather. For this purpose, various atmospheric and oceanic measurements are carried out on routine basis through different instruments such as Doppler Weather Radar, Field Mill, Balloon-borne GPS Sondes, Portable Automatic Weather Station and Oceanic Buoys to name a few. During the countdown, formal weather briefing occurs periodically at regular intervals with special emphasis on probability of adverse weather events on the launch site. In addition to routine measurements, different atmospheric models are run for providing short to medium-range weather predictions for the launch site. In the present chapter, we have confined our description to HRM and ARPS model simulations, which are being carried out at Space Physics Laboratory.

6. Results and discussion

With reference to the LCC conditions summarized in Section 5, the basic requirements for satellite launch missions from SHAR (India) can be summarized under the following classes:

- **Low Pressure Systems:** Since SHAR is a coastal station located on the eastern coastline of the Indian subcontinent, the local weather is severely affected by *low pressure systems* formed in the Bay of Bengal. Thus an adequate information on the probability of formation of any *low pressure system* over Bay of Bengal with sufficient lead time remain one of the basic requirements from the mission team.

- **Thunderstorm/Lightning:** Mesoscale convective systems formed over SHAR are hazardous to the launch activities, thus reliable information on the location and timings of thunderstorms and lightning are anticipated. In-situ measurements of electric field through a network of field mill systems provide quite valuable information to address these aspects.
- **Clouds and Precipitation:** During the period of liftoff, adequate information on the cloud fraction over SHAR as well as along the trajectory of launch vehicle in the lower altitudes is crucial. Two Doppler Weather Radars (DWR, one at SHAR and another at Chennai) provide very valuable information on the passing showers and precipitating clouds. DWR products are also being assimilated in ARPS model simulations.
- **Visibility:** Periodical observations on the visibility are required for range safety operations. Ideally, direct visual observation of the launch vehicle is required through the lower troposphere ($\approx$ 5000 m). In the case of overcast sky conditions, range safety measures depend on other remote sensing techniques.
- **Surface and Upper Winds:** During the movement of launch vehicle from assembly building to the launch pad, the surface wind speeds should not exceed a pre-designated threshold ($\approx 20m/s$). Thus, in-situ measurements of wind speed through 100-m tall meteorological tower in conjunction with balloon-borne GPS Sonde provide exact information on the surface and upper winds.

6.1 HRM simulations

With a view to catering to the needs of LCC related requirements of PSLV and GSLV missions, HRM forecast products are customized and are disseminated to mission team members during the launch campaigns. HRM-simulated total cloud cover information is very important for forecasting the circulation and dynamics of the atmosphere. In most cases, it has been seen that HRM simulations of total cloud cover for +6 to +30 hours forecast over the Indian subcontinent showed a good agreement with the Kalpana-1 satellite cloud observations in visible range with errors of about 20% to 25%. Fig. 2 shows a typical example of HRM-simulated +6 and +30 hours forecast of total cloud cover fraction valid for 17th October 2008, 0600 GMT with the Kalpana satellite imagery for the concurrent timings. It could be noticed that - while HRM is able to capture the broad features of cloud cover, there are some areas where the simulations suggest moderate to dense clouds not supported by the observations. This kind of ambiguities in forecast fields are attributed to the uncertainties in the variables like specific cloud liquid water and ice contents, relative humidity, convective activity and atmospheric stability which are the basis for defining cloudiness in HRM. A combination of these parameters provides threshold values for the probability of cloud formation. Nevertheless there was a good qualitative resemblance between the model-simulated cloud fractions and satellite observations. Accumulated rainfall is also one of the important parameters, which need to be simulated with high precision and less quantitative errors. Similar to total cloud fraction, HRM could simulate rainfall pattern also to a good extent. During PSLV-C12 launch campaign, the Bay of Bengal (BoB) underwent severe cyclonic storm named - BIJLI between 14 - 16 April 2009 and it was one of the prime concerns for launch related activities. In this regard, HRM simulated meteorological fields were found to be quite useful in inference of wind circulation prevailing over the launch site

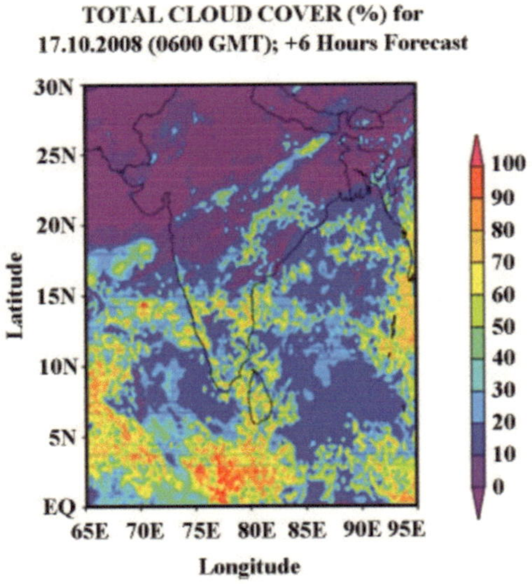

(a) HRM-simulated total cloud cover +6 Hrs in advance;

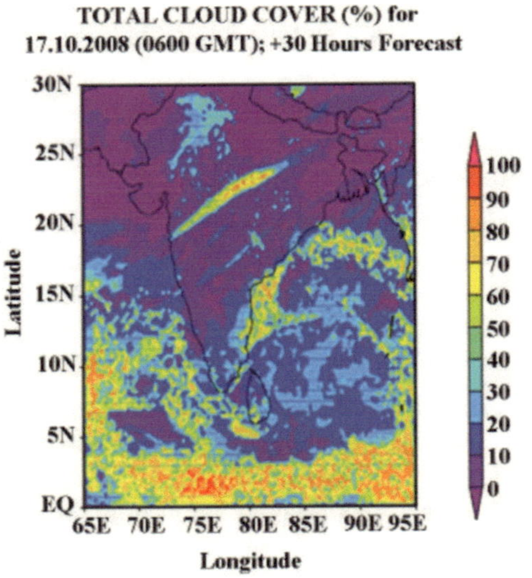

(b) HRM-simulated total cloud cover +30 Hrs in advance;

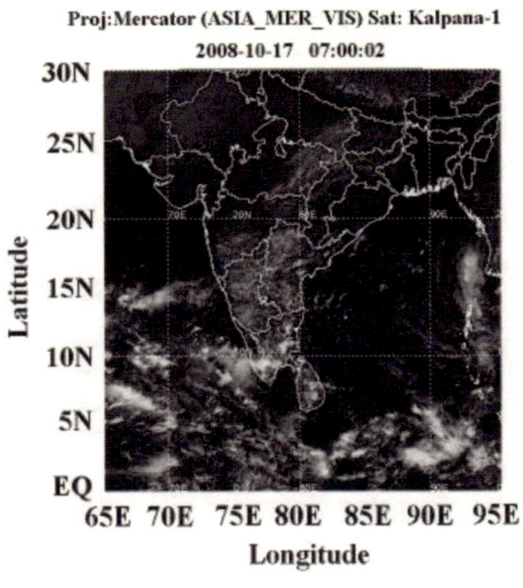

(c) Kalpana satellite imagery of cloud coverage in the visible range over the Indian sub-continent for 17th October 2008: 0700 GMT during Chandrayaan campaign;

Fig. 2. (a) and (b) HRM-simulated total cloud cover +6 Hrs and + 30 Hrs in advance respectively; (c) Kalpana satellite imagery of cloud coverage in the visible range over the Indian sub-continent for 17th October 2008: 0700 GMT during Chandrayaan campaign

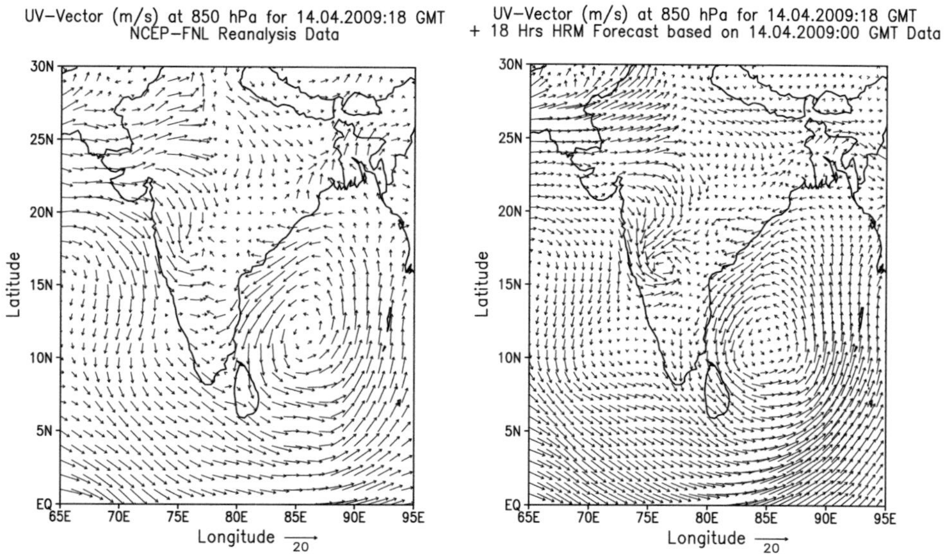

(a) NCEP-FNL Reanalysis (b) +18 Hrs HRM simulations

Fig. 3. (a) Horizontal (uv) wind vector field at 850 hPa observed on 14 April 2009: 1800 GMT through NCEP reanalysis; (b) HRM simulated uv wind vector field +18 hours in advance.

and adjoining BoB oceanic region. Fig. 3 depicts wind vector field at 850 hPa seen through NCEP reanalysis during formation of BIJLI cyclonic storm in the month of April 2009. HRM simulated wind vector field for the same time +18 hours in advance shown in Fig. 3 is well in tune with the NCEP reanalysis. A cyclonic circulation with low pressure over the BoB at 850 hPa levels are clearly captured in simulations almost 18 hours in advance; however there is a mismatch in the exact location of low pressure of the system and it can be attributed to lack of observational data in HRM model runs. The steering wind circulation at 200 hPa for the concurrent timings also showed a very good comparison between the simulations and Reanalysis (not shown here). From Fig. 3, it is very interesting to note that - almost 18 hours in advance HRM simulations clearly indicated formation of a cyclonic storm over the head BoB, and the wind circulation was regularly updated to the MET team at SHAR during the campaign. During the passage of cyclonic storm, the amount of moisture at about 3 kms (roughly 700 hPa) is often treated as one of the crucial parameters by the meteorologists to categorize the cyclone in terms of its intensity. Therefore, we show HRM simulation of relative humidity at 700 hPa on 16 April 2009 (1800 GMT) when BIJLI cyclonic storm had advanced towards north-east part of the Indian sub-continent in Fig. 4. Simulated field is also compared with the NCEP-FNL Reanalysis and results are quite encouraging, as almost all the gross features in relative humidity field with realistic magnitudes are nicely captured in +18 hours simulations, thereby providing a strong justification of usage of HRM simulated meteorological fields for LCC during launch activities. With special reference to the spatial heterogeneity in rainfall events around Sriharikota, a proper parameterization of convection is very crucial for improvements in the accuracy of short-range weather predictions. In order to test the impact of different convective parameterization schemes in HRM simulations, Tiedtke and Bechtold schemes were installed and their performance on rainfall forecasts and other

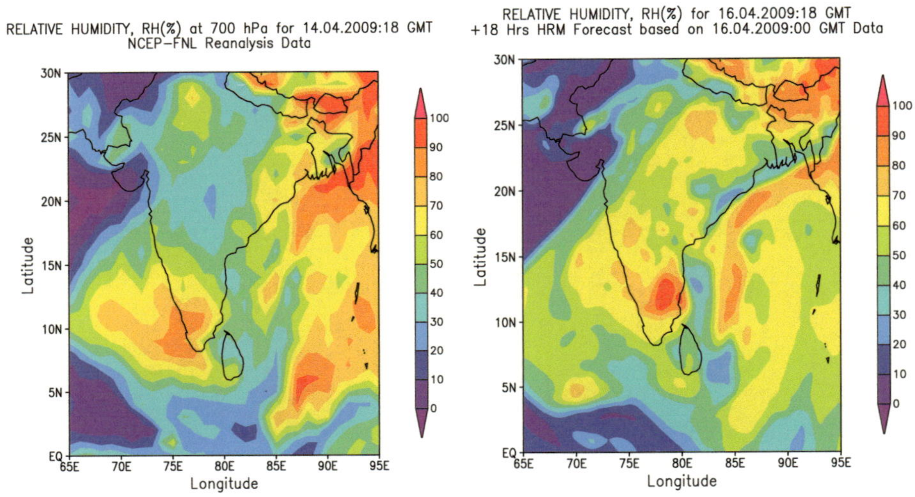

(a) NCEP-FNL Reanalysis; (b) +18 Hrs HRM simulations

Fig. 4. (a) Relative humidity at 700 hPa observed on 16 April 2009: 1800 GMT through NCEP reanalysis; (b) HRM simulated relative humidity at 700 hPa +18 hours in advance.

convective products were investigated. Fig. 5(a-b) shows daily accumulated rainfall forecast during GSLV-D3 campaign (April 2010) using Tiedtke and Bechtold convective schemes respectively. It can be seen that Bechtold convective scheme enhances the quantity of rainfall over cloudy regions. Both convection schemes are mass flux schemes and the main difference between them is that the closure assumption is based on moisture convergence in the Tiedtke scheme and CAPE used in the Bechtold scheme. Since Tiedtke convective scheme was originally developed for applications to the global scale for simulations with much smaller grid sizes, a convection closure based on convective available potential energy (CAPE) would be more suitable. In the Tiedtke scheme convection is triggered if the parcel's temperature exceeds the environment temperature by a fixed temperature threshold of 0.50 K, whereas in the Bechtold scheme, the onset of convection is decided by the large-scale vertical velocity. Both the Tiedtke and the Bechtold schemes distinguish penetrative and shallow convection. It is concluded that - Tiedtke scheme performs well with 0.25 deg. grid resolution, whereas Bechtold scheme is more suitable for smaller scale observations at 0.10 deg. grid resolution.

6.2 ARPS simulations

As part of PSLV and GSLV missions, various atmospheric models ranging from global scale, regional scale to mesoscale are made operational at SHAR for providing regional weather forecast during the period of launch campaign. With a view to in digesting the local in-situ information in an atmospheric model, ARPS is customized for SHAR domain and made operational for providing the probability of thunderstorm for +6 to +12 hours and forecast products were updated regularly in time during the launch window. The initial conditions to the ARPS model are provided by the vertical profiles of thermodynamic parameters obtained from balloon-borne GPS sondes, whereas the DWR products are assimilated for improvements in the initial conditions (Ramachandran, 2008). The DWR is perhaps one of the rare instruments that can sniff inside severe weather systems like tornado, tropical cyclone

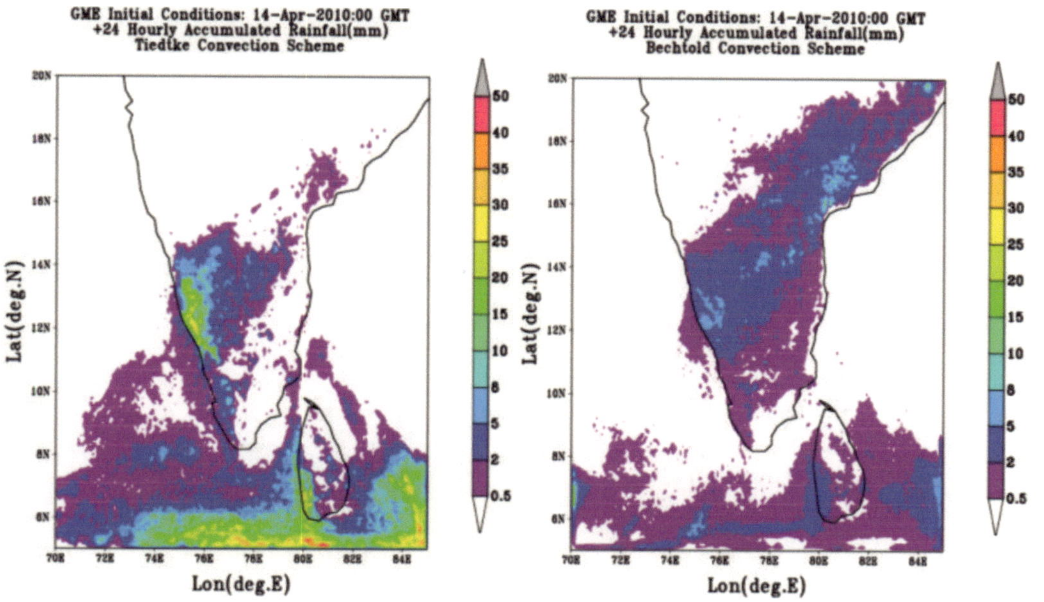

(a) Tiedtke convective scheme

(b) Bechtold convective scheme

Fig. 5. HRM simulated 24 hours accumulated rainfall using (a) Tiedtke and (b) Bechtold convective schemes during GSLV-D3 launch campaign.

and other convective systems such as severe thunderstorms. With the advent of operationally available Doppler radial wind data interest has increased in assimilating these data into NWP models. The limited area models require observations with high spatial-temporal resolutions for determining the initial conditions. Doppler radar wind measurements are one possible source of information, albeit over limited areas within about 100 km of each radar site. The resolution of raw data is however much higher than the resolution of the numerical models. Therefore these data must be preprocessed, to be representative of the characteristic scale of the model, before the analysis. When several observations are too close together then they will be more corrected and as a result the forecast error correlations at the observation points will be large. In contrast, the individual observations are less dependent and they will be given more weight in the analysis than observations that are close together. To reduce the representativeness error, as well as the computational cost, one may use (i) the vertical profiles of horizontally averaged wind in the form of Velocity Azimuth Display (VAD) technique, (ii) the observations of sparse resolution, or (iii) calculate spatial averages from the raw data to generate the so-called super-observations. The generated data correspond more closely to the horizontal model resolutions than do the raw observations. The 3D-Var has been used operationally to assimilate radar wind information in the form of VAD wind profiles (Ramachandran, 2008). The radar wind observations operator in this technique produces the model counterpart of observed quantity that is presented to the variational assimilation.

The radar reflectivity depends upon the size and density of the water droplet, whereas the radial velocities are one of the ways of measuring the horizontal and vertical wind pattern. While the DWR provides useful information on the reflectivity and radial winds over the region of interest within the radar-sight, it is equally important to make use of this information in the numerical weather prediction, so as to improve the prediction of severe weather events,

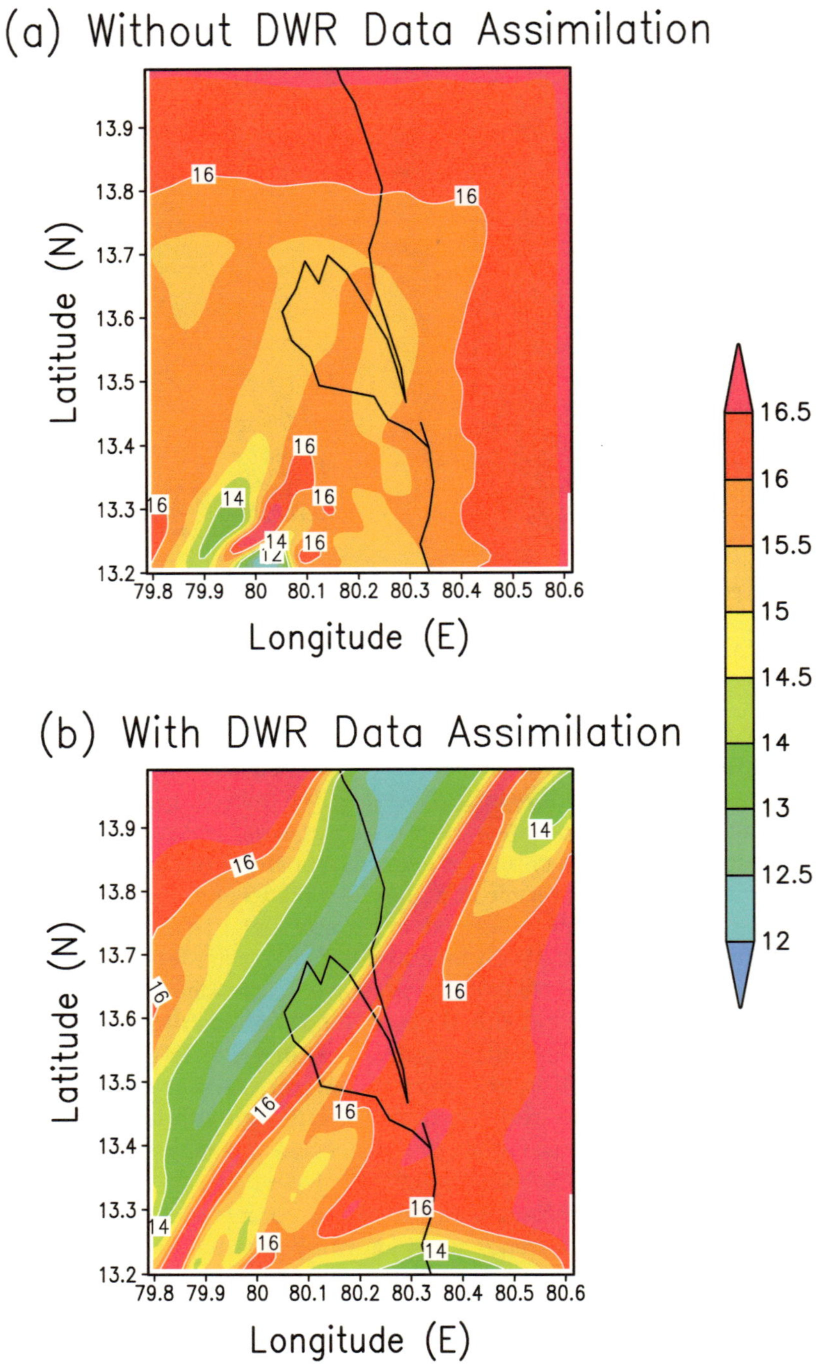

Fig. 6. Impact of DWR data assimilation on surface layer water vapour mixing ratio (g/kg) field over the SHAR, showing the refinement in magnitudes of mixing ratio in accordance with the DWR derived reflectivity.

such as the thunderstorms. The radial winds and reflectivity pattern obtained from the DWR of SHAR are assimilated in the ARPS for improvements in the initial conditions. The assimilation package consists of three codes: (1) in the very first step, the DWR products which are available on radar coordinates are converted to the model grids and a re-mapping package is activated so that the actual time of the observations and the concerned meteorological parameters (water vapour and wind patterns) are converted to the model grid coordinates; (2) after obtaining the radial velocity and reflectivity information from DWR in the model coordinates, the initial conditions are altered in accordance with the observational data and (3) finally the re-mapped data is used by activation of 3-D VAR assimilation package in the ARPS model. After incorporation of DWR derived radial velocities and reflectivity in the ARPS model, surface layer water vapour mixing ratio fields over the model domain are considerably changed in accordance with the DWR observations thereby improving the initial conditions of the model (Fig. 6).

During Chandrayaan mission, Subrahamanyam et al. (2011) made use of five different parameters obtained from ARPS simulations namely (1) hourly accumulated rainfall; (2) cloud and rain water mixing ratio; (3) vertical velocities; (4) low level wind shear and (5) magnitude of surface layer specific humidity as the probable indicators of thunderstorm occurrence, and ARPS model simulations were found to be of great importance, as a careful examination of model derived products could help in providing a logistic forecast on probability of thunderstorms. Through these products, ARPS could capture more than 70% of rainfall slots successfully; however, there were a few localized events, where model did not capture the rainfall properly. This model has also got very good potential in simulation of vertical profiles of winds. Fig. 7 shows one of the typical comparisons between ARPS model-simulated vertical profiles of zonal and meridional winds on 28 May 2008 with concurrent measurements through balloon-borne GPS sonde. Within the permissible constraints of the model, it was noticed that these parameters required hyper-tuning with greater accuracy and quantification. For nowcasting purpose, ARPS model- simulated forecast products are vastly used and are considered to be reasonably accurate up to +6 hours (Subrahamanyam et al., 2008).

7. Advantages of HRM and ARPS

There are various NWP models available for providing short-range weather predictions over a specific site, however in the present study we have confined our description to HRM and ARPS for the following reasons:

1. Very first, the initial and lateral boundary conditions for HRM are derived from the analysis and forecast fields of the German Global Model - GME which are made available to us very fast compared to other global models, thereby providing sufficient lead time for generation of forecast products.
2. Secondly, the required meteorological fields for running the HRM are tailored at DWD for our specific region. Thus, we need not download the entire global data which would obviously require relatively large number crunching.
3. ARPS model is probably one of the rare atmospheric models, which has got the potential of using highly localized observations such as upper-air meteorological data obtained through balloon-borne GPS Sondes. Also it assimilates the DWR-derived winds and reflectivity fields.

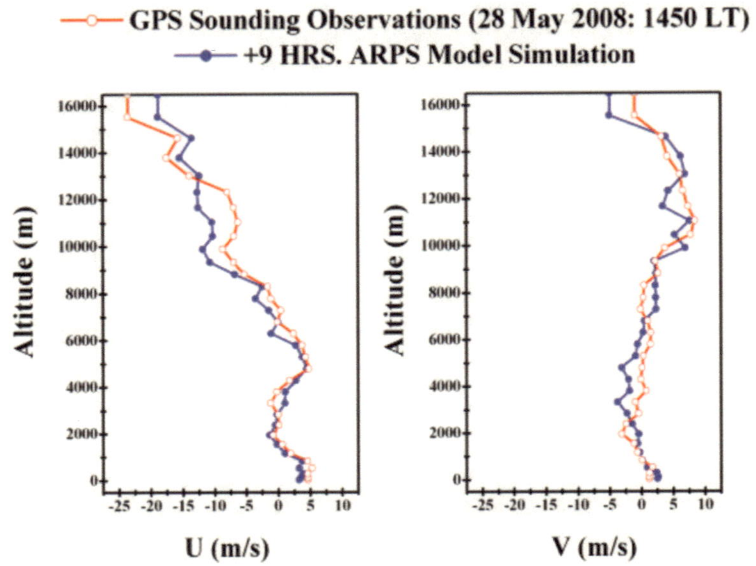

Fig. 7. A typical comparison of ARPS model-simulated zonal and meridional winds in vertical with balloon-borne GPS Sonde measurements.

4. By choosing HRM and ARPS, we take care of the possible differences in forecasting due to hydrostatic and non-hydrostatic approach. Thus, we make use of both the models simultaneously to provide short-range weather predictions over SHAR.

8. Scope for future work

In the present chapter, we described the potential of two atmospheric models, namely HRM and ARPS in providing valuable information with respect to *launch commit criteria* for satellite launch missions. Among various regional atmospheric models, HRM has got a definite advantage as the initial conditions and lateral boundary conditions are derived through a tailored dataset of the GME global model, or else, one need to depend on the whole global sets, thereby spending excessive time. Thus, the HRM simulations can be made available to PSLV and GSLV mission teams with a reasonable lead time. The HRM has also got a good potential to capture low pressure systems over the Bay of Bengal well in advance ($\approx$ 18 hours), thus severe weather threats can be provided at right time. Similarly, the potential of ARPS model is vastly exploited in simulations of mesoscale convective events, such as thunderstorms. It was also very useful in capturing the fine features in the vertical profiles of zonal and meridional winds to +9 hrs.

Having exploited the potential of these two models, one need to explore the possibility of assimilating the routine meteorological observations on a regular basis for continuous improvements in the initial and lateral boundary conditions of the model. Also there is need of generation of very good quality climatology of mesoscale convective events which are hazardous to the launch activities. In future, there may also be enough scope of validating a combination of different NWP models, which can lead to development of a statistical ensemble model. In a broad perspective, it may be indeed a difficult task to achieve 100% accurate model, but intelligent human intervention to the model-derived forecast products would lead to generation of error statistics of individual models, which can help the mission team in decision making tasks.

9. Acknowledgments

We greatly acknowledge the support and inspiring guidance rendered by Dr. K. Krishna Moorthy, Director, SPL. One of the authors DBS is very much thankful to all the members of "Weather Forecasting Expert Team" constituted by Satish Dhawan Space Centre, Sriharikota for useful discussion during several of the PSLV and GSLV launches. Special thanks are also due to Ms. T. J. Anurose and S. Indira Rani for their contribution in HRM and ARPS simulations during the launch campaigns. The NCEP-FNL Reanalysis data for this study are from the Research Data Archive (RDA) which is maintained by the Computational and Information Systems Laboratory at the National Center for Atmospheric Research (NCAR). NCAR is sponsored by the National Science Foundation. The original data are available from the RDA (http://dss.ucar.edu) in dataset number ds083.2. KALPANA Satellite images over the Indian sub-continent are downloaded from the Indian Meteorological Department (http://www.imd.gov.in/) and we duly acknowledge their services.

10. References

Anthes, R. A. & Warner, T. T. (1974). Development of hydrodynamical models suitable for air pollution and other mesometeorological studies, *Mon Weather Rev* 106: 1045 – 1078.

Bannoon, P. R. (1995). Hydrostatic adjustment: Lamb's problem, *J Atmos Sci* 52: 1743 – 1752.

Case, J. L., Manobianco, J., Oram, T. D., Garner, T., Blottman, P. F. & Spratt, S. M. (2002). Local data integration over East-Central Florida using the ARPS data analysis system, *Weather Forecast* 17: 3 – 26.

Cox, R., Bauer, B. L. & Smith, T. (1998). A mesoscale model intercomparisona mesoscale model intercomparison, *Bull. Amer. Meteor. Soc.* 79: 265 – 283.

Das, S. S., Sijikumar, S. & Uma, K. (2011). Further investigation on stratospheric air intrusion into the troposphere during the episode of tropical cyclone: Numerical simulation and MST radar observations, *Atmospheric Research* 101(4): 928 – 937.

Doswell, C. A. (2004). Forecasters's forum: Weather forecasting by humans-heuristics and decision making, *Weather and Forecasting* 19: 1115 – 1126.

Klein, G., Moon, B. & Hoffman, R. R. (2006). Making sense of sensemaking 2: A macrocognitive model, *Intelligent Systems* 21(5): 88 – 92.

Majewski, D. (2010). HRM-User's Guide: For Vrs. 2.8 or higher, Deutscher Wetterdienst, Germany, p. 121.

Majewski, D., Liermann, D., Prohl, P., Ritter, B., Buchhold, M., Hanisch, T., Wergen, G. P. D. W., & Baumgardner, J. (2002). The Operational Global Icosahedral-Hexagonal Gridpoint Model GME: Description and High-Resolution Tests, *Monthly Weather Review* 130.

Manobianco, J., Zack, W. & Taylor, G. E. (1996). Work station based real-time mesoscale modeling designed for weather support to operations at the Kennedy Space Center and Cape Canaveral Air Station, *Bullen of American Meteorological Society* 77: 653 – 672.

Pielke, R. A. (1984). *Mesoscale Meteorological Modeling*, Academic Press Inc,Orlando, Florida.

Rakesh, V., Singh, R., Pal, P. K. & Joshi, P. C. (2007). Sensitivity of Mesoscale Model Forecast During a Satellite Launch to Different Cumulus Parameterization Schemes in MM5., *Pure and Applied Geophysics* 164: 1617 – 1637.

Ramachandran, R. (2008). DWR Data Assimilation in Numerical Weather Prediction for Forecasts of Thunderstorms at Sriharikota, *Technology Development for Atmospheric Research and Applications* pp. 463 – 487.

Ramachandran, R., Mohanty, U. C., Pattanayak, S., Mandal, M. & Rani, S. I. (2006). Location specific forecast at Sriharikota during the launch of GSLV-01., *Current Science* 91(3): 285 – 295.

Rani, S. I., Ramachandran, R., Subrahamanyam, D. B., Alappattu, D. P. & Kunhikrishnan, P. (2010). Characterization of sea/land breeze circulation along the west coast of Indian sub-continent during pre-monsoon season, *Atmospheric Research* 95: 367 – 378.

Rasmussen, R., Dixon, M., Hage, F., Cole, J., Wade, C., Tuttle, J., McGettigan, S., Carty, T., Stevenson, L., Fellner, W., Knight, S. & andNancy Rehak, E. K. (2001). Weather support to deicing decision making (wsddm): A winter weather nowcasting system, *Bull. Amer. Meteor. Soc.* 82(4): 579 – 595.

Singh, R., Pal, P. K., Kishtawal, C. M. & Joshi, P. C. (2005). Impact of bogus vortex for track and intensity prediction of tropical cyclone, *J Earth Syst Sci* 114: 427 – 436.

Stuart, N. A., Schultz, D. M., G., G. K. & Doswell, C. A. (2007). Maintaining the role of humans in the forecast process, *Bull. Amer. Meteor. Soc.* 88: 1893 – 1898.

Subrahamanyam, D. B., Radhika, R. & Kunhikrishnan, P. K. (2006). Improvements in Simulation of Atmospheric Boundary Layer Parameters through Data Assimilation in ARPS Mesoscale Atmospheric Model, *Remote Sensing and Modeling of the Atmosphere, Oceans, and Interactions, Proc. of SPIE* 6404: 64040K.

Subrahamanyam, D. B., Ramachandran, R., Anurose, T. J. & Mohan, M. (2010). Short-to-Medium Range Weather Forecasting for Satellite Launches: Utilization of NWP Models in conjunction with the Earth Observing Systems, *Bulletin of National Natural Resources Management System* NNRMS(B) - 35: 37 – 49.

Subrahamanyam, D. B., Ramachandran, R., Rani, S. I., Sijikumar, S., Anurose, T. J. & Ghosh, A. K. (2011). Location-specific weather predictions for Sriharikota (13.72 N, 80.22 E) through numerical atmospheric models during satellite launch campaigns, *Natural Hazards* .

Subrahamanyam, D. B., Rani, S. I., Ramachandran, R. & Kunhikrishnan, P. K. (2008). Nudging of Vertical Profiles of Meteorological Parameters in One-Dimensional Atmospheric Model: A Step Towards Improvements in Numerical Simulations, *Ocean Science Journal* 43(4): 165 – 173.

Xue, M., Droegemeier, K. K. & Wong, V. (2000). The advanced regional prediction system (ARPS) - a multi-scale nonhydrostatic atmospheric simulation and prediction model. Part I: Model dynamics and verification, *Meteorol Atmos Phys* 75: 161 – 193.

Xue, M., Droegemeier, K. K., Wong, V., Shapiro, A. & Brewster, K. (1995). ARPS Version 4.0 UsersâĂŹs guide, p. 380.

Xue, M., Droegemeier, K. K., Wong, V., Shapiro, A., Brewster, K., Carrl, F., Weber, D., Liu, Y. & Wang, D. (2001). The advanced regional prediction system (ARPS)âĂŤa multi-scale nonhydrostatic atmospheric simulation and prediction tool. Part II: Model physics and applications, *Meteorol Atmos Phys* 76: 143 – 165.

Xue, M., Wang, D., Gao, J., Brewster, K. & Droegemeier, K. K. (2003). The advanced regional prediction system (ARPS), storm-scale numerical weather prediction and data assimilation, *Meteorol Atmos Phys* 82: 139 – 170.

3

Numerical Study on the Effect of the Ocean on Tropical-Cyclone Intensity and Structural Change

Akiyoshi Wada
Meteorological Research Institute
Japan

1. Introduction

The ocean is an indispensable source of energy for tropical cyclones (TCs). TCs enable extraction of heat and moisture from the sea surface through the transfer of turbulent heat energy in the atmospheric boundary layer. TCs are often generated where sea-surface temperature (SST) is higher than 26.5°C (Palmén, 1948), and they intensify in areas that have high SST and deep oceanic mixed layer, thus having high upper-ocean heat content. Previous studies reported that TC intensity is related to 'tropical-cyclone heat potential', which is oceanic heat content integrated from the surface to the depth of the 26°C-isotherm (Wada & Usui, 2007; Wada, 2010). However, most of people believe that TCs intensify when SST is higher than 26-28°C.

The oceans affect both the genesis and intensification of TCs, which in turn apply wind stresses to the ocean that induce sea-surface cooling (SSC) during their passages (Ginis, 1995). Numerous numerical modeling studies leave no doubt that SSC affects both the evolution of TCs and prediction of their intensities (Bender & Ginis, 2000; Wada et al., 2010). Nevertheless, uncertainties remain in the use of numerical models to predict TC intensity: these are related to initial atmospheric and oceanic conditions, the spatial and temporal resolution of the models, and the physical processes incorporated into the models (Wang & Wu, 2004; Wada, 2007). It may appear that a sophisticated model such as an atmosphere-wave-ocean coupled model can produce valid predictions of TC intensity when TC intensity predicted by that model matches best-track intensity derived from satellite observations. However, best-track intensity is not always valid, particularly in the western North Pacific where there is a lack of direct observation such as aircraft observations for measuring TC intensity directly. Furthermore, a coupled model that has successfully calculated TC intensities in one situation may provide erroneous results in another situation because of unexpected interactions among the specifications, physical processes, and initial and boundary conditions used in the model. One particular combination of model specifications and parameters may thus not always be valid for other TC predictions.

However, studies using a numerical model provide us scientific explanations on dynamical and physical processes associated with TC intensification. (Wada, 2009) explained the effects on TC intensification of atmospheric dynamics such as filamentation, the formation of mesovortices and vortex Rossby waves. Mesovortices and vortex merger events are directly affected by SSC, which slows the formation of an annular potential-vorticity (PV) ring, whereas SST changes have little effect on the radius of maximum wind speed (MWS) at the mature phase of a TC when the annular PV ring is completely formed, even though mature TCs continue to cool the underlying ocean (Wada, 2009).

TC-induced SSC is caused mostly by vertical turbulent mixing in the oceanic mixed layer and upwelling below a seasonal thermocline. In addition, strong wind stresses that accompany TCs cause variations in sea state or surface roughness length, leading to changes in frictional velocity and exchange coefficients for drag and enthalpy. Breaking surface waves are caused by variations of sea state under high winds and the resultant high waves. The breaking surface waves play an essential role in mechanical mixing near the surface (Wada et al., 2010). However, improvement of vertical turbulent mixing schemes and parameterizations in the ocean model is a challenging issue owing to a lack of *in situ* observations under high winds.

Changes in exchange coefficients lead to changes in surface wind stresses and turbulent heat fluxes from the ocean to the atmosphere. The change in turbulent heat fluxes, particularly latent heat flux, enhances the secondary circulation of a TC through a planetary-boundary-layer process (Emanuel, 1986; Smith, 2008). In particular, turbulent heat fluxes vertically transferred from the warm ocean affect cloud microphysics and atmospheric radiation in the middle to upper troposphere, causing latent heat release through condensation, thus resulting in the formation of a warm core within the inner core of a TC.

These atmospheric dynamics are confined to within the inner core of a TC. We need also to consider the role of atmospheric and oceanic environmental influences, such as vertical shear, synoptic thermal stratification, and warm-core oceanic eddies with a few hundred kilometers, in the evolution of a TC. (Wada & Usui, 2010) reported that changes in pre-existing oceanic conditions are synoptically related to the formation of spiral rainbands that accompanied Typhoon Hai-Tang in 2005 and are thus important for predictions of the intensity of Hai-Tang.

This chapter explores the effect of the ocean on TC intensity, intensification and structural change from the viewpoints of internal dynamical and pre-existing external atmospheric and oceanic environmental conditions. Section 2 describes the numerical model used in this study, and section 3 describes the experimental design. Section 4 provides the results of numerical experiments of an idealized TC and simulations of Typhoon Choi-Wan in 2009. The effect of the ocean on TC evolution at the mature phase and associated numerical issues are discussed in section 5. Section 6 presents the conclusion of this study.

2. Numerical model

A nonhydrostatic model (NHM) has been developed jointly by the Numerical Prediction Division (NPD) and Meteorological Research Institute (MRI) of the Japan Meteorological Agency (JMA) (hereafter JMANHM). The NHM, which has been equipped with two-way triply nested movable functions, was developed at MRI and was coupled with a multilayer ocean model (MRINHM) developed by (Wada, 2009). The physical processes in JMANHM and MRINHM include cloud physics expressed in an explicit three-ice bulk microphysics scheme based on the work of (Lin et al., 1983), a resistance law assumed for momentum and enthalpy fluxes in the atmospheric surface-boundary layer, exchange coefficients for momentum and enthalpy transfers over the sea based on the bulk formulas of (Kondo, 1975), a turbulent closure model in the atmospheric boundary layer formulated from the work of (Klemp & Wilhelmson, 1978) and (Deardorff, 1980), and an atmospheric radiation scheme based on (Sugi et al., 1990).

The JMANHM used here for numerical simulations of Typhoon Choi-wan in 2009 is an older version of the current nonhydrostatic mesoscale model of (Wada et al., 2010), but it is coupled with not only a multilayer ocean model but also a third-generation ocean-wave model developed for operational use at JMA (Wada et al., 2010). No cumulus parameterization was used in either MRINHM or JMANHM in this study.

A reduced-gravity approximation, a hydrostatic approximation, and Boussinesq fluid are assumed in the multilayer ocean model. The model has three layers and four levels. The uppermost layer represents a mixed layer where density is vertically uniform, the middle layer represents a seasonal thermocline where the vertical temperature gradient is greatest, and the bottom layer is assumed to be undisturbed by entrainment. Entrainment is calculated by using a multi-limit entrainment formulation proposed by (Deardorff, 1983) and modified by (Wada et al., 2009). The model calculates water temperature and salinity at the surface and at the base of the mixed layer, and calculates layer thickness and 2-dimentional flows for all three layers.

The JMA third-generation ocean-wave model is coupled with JMANHM to estimate changes in surface roughness lengths, drag coefficients and enthalpy coefficients. The method of (Taylor & Yelland, 2001) in which surface roughness length over the ocean depends on wave steepness is adopted here. The detailed wave-ocean coupling procedure is as described by (Wada et al., 2010).

The atmosphere-ocean coupling procedure in MRINHM and JMANHM is as follows. Short-wave and long-wave radiation, sensible and latent heat fluxes, wind stresses, and precipitation are provided to the ocean model for every time step of the ocean model. Land and sea distributions extracted from GTOPO30 digital elevation data from the US Geological Survey are provided from the atmosphere model to the ocean model only at the initial time in order to adjust the land and sea distributions between the atmosphere and ocean models. Oceanic topography is provided by ETOPO-5 data from the National Oceanic and Atmospheric Administration (NOAA) National Geophysical Data Center; these elevations are spaced at 5-minute intervals of latitude and longitude. Conversely, SST calculated by the ocean model is provided to MRINHM and JMANHM for every time step of the ocean model.

3. Experimental design

This study presents the results of two numerical experiments. One is a series of idealized numerical experiments applying MRINHM. MRINHM covers a 600 km x 600 km square computational domain with a horizontal grid spacing of 2 km. The other is a series of numerical simulations of Typhoon Choi-Wan in 2009 performed using JMANHM. JMANHM covers a 2700 km x 3600 km rectangular domain with a horizontal grid spacing of either 6 km or 12 km. Both MRINHM and JMANHM have 40 vertical levels with variable intervals from 40 m for the lowermost (near-surface) layer to 1180 m for the uppermost layer. Both MRINHM and JMANHM have maximum height approaching nearly 23 km. The time step of MRINHM is 6 s and that of JMANHM is 15 s (6 km grid spacing) or 30 s (12 km grid spacing). The length of the time step of the ocean model is six times those of both MRINHM and JMANHM. The Coriolis parameter is uniformly set to 5.0 x 10^{-5} (nearly 20°N) in MRINHM and varies in JMANHM depending on the grid latitude.

Water depth in the multilayer ocean model coupled with MRINHM is uniformly set to 1000 m. Initial SST is set to 30°C, the initial temperature at the base of the mixed layer to 29°C, the initial temperature at the base of the thermocline to 18°C and the initial temperature at the bottom to 5°C. Initial salinity is set to 35 at all levels. The initial mixed-layer depth is set to be 30 m, the initial thermocline thickness to 170 m and the initial third-layer thickness to 800 m. The third layer thickness is assumed to be unaffected by entrainment. In JMANHM, the initial depth of the mixed layer is determined from oceanic reanalysis data by assuming a difference in the value of density from the surface of no more than 0.25 kg m^{-3} and the depth of the mixed layer is limited to 200 m. The base of the thermocline is limited to 600 m and water depth is limited to 2000 m. The oceanic reanalysis data are calculated using the MRI ocean variational estimation (MOVE) system (Usui et al., 2006).

3.1 Idealized experiment

Table 1 summarizes the idealized numerical experiments performed using MRINHM with and without coupling with the ocean model. The initial TC-like vortex and thermal conditions were as given by (Wada, 2009). The integration time was 81 h with results output every 30 min. The sensitivity of vertical turbulent mixing in the ocean model was evaluated using two tuning parameters: m_d = 17.5 and m_d = 175. Parameter m_d is associated with turbulent kinetic energy flux produced by breaking surface waves.

Experiment	Ocean coupling	Beginning hour of coupled model	Drag coefficient	Length scale in Deardorff (1980)
AT	No	-	Kondo(1975)	$\Delta s=(dx*dy*dz)^{1/3}$
OC	Yes	27h	Kondo(1975)	$\Delta s=(dx*dy*dz)^{1/3}$
CTL	Yes	0h	Kondo(1975)	$\Delta s=(dx*dy*dz)^{1/3}$
D2DIM	Yes	0h	Kondo(1975)	$\Delta s=(dx*dz)^{1/2}$
K35	Yes	0h	Kondo(1975) except for a constant value at v = 35 m s^{-1} when v > 35m s^{-1}	$\Delta s=(dx*dy*dz)^{1/3}$
K35D2	Yes	0h	Kondo(1975) except for a constant value at v = 35 m s^{-1} when v > 35m s^{-1}	$\Delta s=(dx*dz)^{1/2}$

Table 1. Summary of key parameters of idealized MRINHM numerical experiments with and without coupling with the ocean model.

Experiment CTL was a control run. 'D2' in the experiment index indicates that the representative mixing-length scale was determined as a two-dimensional geometric mean, and 'K35' indicates that drag coefficients leveled off when 10-m wind speed exceeded 35 m s^{-1}, This saturation of drag coefficients has been reported by (Powell et al., 2003) and (Donelan et al., 2004). Drag coefficients and mixing-length scales were as given by (Deardorff, 1980).

3.2 Typhoon Choi-Wan in 2009

Typhoon Choi-Wan in 2009 was simulated during its mature and decaying phases. Choi-Wan was initiated when a tropical depression evolved into a TC around 15.4°N, 150.9°E at 18:00 UTC on 12 September 2009. Choi-Wan moved initially west-northwestward but changed to a northwestward track as it rapidly intensified. From 12:00 UTC on 15 September to 18:00 UTC on 16 September, the minimum central pressure (MCP) was 915 hPa and MWS was 105 knots (~54 m s^{-1}). At a location around 23.2°N, 138.9°E, Choi-Wan gradually slowed and changed to a north-northeastward track as it began to decay.

Experiment	Model	Horizontal resolution	Atmospheric environmental dataset	horizontal resolution of ocean data	cloud physics
A6G5I	NHM	6 km	JMA global analysis (20 km)	0.5°	ice phase
A6G1I	NHM	6 km	JMA global analysis (20 km)	0.1°	ice phase
C6G5I	Coupled NHM-wave-ocean	6 km	JMA global analysis (20 km)	0.5°	ice phase
C6G1I	Coupled NHM-wave-ocean	6 km	JMA global analysis (20 km)	0.1°	ice phase
A6G5W	NHM	6 km	JMA global analysis (20 km)	0.5°	warm rain
A6G1W	NHM	6 km	JMA global analysis (20 km)	0.1°	warm rain
C6G5W	Coupled NHM-wave-ocean	6 km	JMA global analysis (20 km)	0.5°	warm rain
C6G1W	Coupled NHM-wave-ocean	6 km	JMA global analysis (20 km)	0.1°	warm rain
A6J5I	NHM	6 km	JCDAS (1.25°)	0.5°	ice phase
A6J1I	NHM	6 km	JCDAS (1.25°)	0.1°	ice phase
C6J5I	Coupled NHM-wave-ocean	6 km	JCDAS (1.25°)	0.5°	ice phase
C6J1I	Coupled NHM-wave-ocean	6 km	JCDAS (1.25°)	0.1°	ice phase
A6J5W	NHM	6 km	JCDAS (1.25°)	0.5°	warm rain
A6J1W	NHM	6 km	JCDAS (1.25°)	0.1°	warm rain
C6J5W	Coupled NHM-wave-ocean	6 km	JCDAS (1.25°)	0.5°	warm rain
C6J1W	Coupled NHM-wave-ocean	6 km	JCDAS (1.25°)	0.1°	warm rain
A12G5I	NHM	12 km	JMA global analysis (20 km)	0.5°	ice phase
A12G1I	NHM	12 km	JMA global analysis (20 km)	0.1°	ice phase
C12G5I	Coupled NHM-wave-ocean	12 km	JMA global analysis (20 km)	0.5°	ice phase
C12G1I	Coupled NHM-wave-ocean	12 km	JMA global analysis (20 km)	0.1°	ice phase
A12G5W	NHM	12 km	JMA global analysis (20 km)	0.5°	warm rain
A12G1W	NHM	12 km	JMA global analysis (20 km)	0.1°	warm rain
C12G5W	Coupled NHM-wave-ocean	12 km	JMA global analysis (20 km)	0.5°	warm rain
C12G1W	Coupled NHM-wave-ocean	12 km	JMA global analysis (20 km)	0.1°	warm rain
A12J5I	NHM	12 km	JCDAS (1.25°)	0.5°	ice phase
A12J1I	NHM	12 km	JCDAS (1.25°)	0.1°	ice phase
C12J5I	Coupled NHM-wave-ocean	12 km	JCDAS (1.25°)	0.5°	ice phase
C12J1I	Coupled NHM-wave-ocean	12 km	JCDAS (1.25°)	0.1°	ice phase
A12J5W	NHM	12 km	JCDAS (1.25°)	0.5°	warm rain
A12J1W	NHM	12 km	JCDAS (1.25°)	0.1°	warm rain
C12J5W	Coupled NHM-wave-ocean	12 km	JCDAS (1.25°)	0.5°	warm rain
C12J1W	Coupled NHM-wave-ocean	12 km	JCDAS (1.25°)	0.1°	warm rain

Table 2. Key data used for numerical simulations of Typhoon Choi-Wan in 2009

The initial time for all simulations was 00:00 UTC on 17 September 2009, when Choi-Wan entered its mature phase. Two sets of atmospheric initial and boundary conditions were used. One was derived from six-hourly global objective analysis data (GA data hereafter) from JMA with a grid spacing of 20 km. The other was derived from six-hourly data from the JMA Climate Data Assimilation System (JCDAS hereafter) with latitude and longitude grid spacings of 1.25°. Daily oceanic reanalysis data with two grid spacings (0.1° and 0.5°) calculated using the MOVE system were used as initial oceanic conditions.

Table 2 provides the key specifications of the JMANHM numerical simulations of Choi-Wan. Both NHM and coupled NHM-wave-ocean models were used to investigate the effect of the ocean on TC simulations from viewpoints of existence or non-existence of Choi-Wan-induced SSC. Warm-rain experiments did not take snow and ice-cloud phases into consideration in cloud physics.

4. Results

4.1 Intensity and structural change

4.1.1 Vortex intensity

Figure 1a indicates that central pressure (CP) is high during the integration when the MRINHM is coupled with the ocean model. Figure 1b indicates that a higher value of m_d (175) greatly reduces intensification, resulting in high CP. A difference in CP between the OC and CTL experiments is evident after 24 h and is unrelated to the values of m_d. The vortex intensifies slowly in experiment CTL, whereas intensification is rapid in experiment OC. CP maintains their values within ranges of 950-960 hPa (m_d = 17.5) and 980-990 hPa (m_d = 175) during the mature phase (except in experiment AT). These results indicate minimal intensification of the mature vortex despite continued vortex-induced SSC.

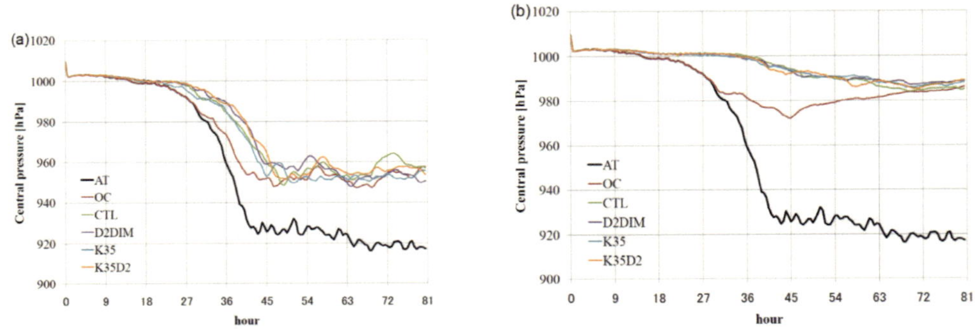

Fig. 1. Time series of CP of the idealized TC-like vortex at 30 min intervals for (a) m_d = 17.5 and (b) m_d = 175 for each experiment shown in Table 1.

SST is defined in this study as the temperature directly below the circulation center of the vortex. The circulation center is defined as the position where surface wind speed is at a minimum and the difference between CP and sea-level pressure is less than 4 hPa. The evolution of SST at the circulation center shows that the variation of SST is small during slow intensification. SST decreases rapidly during rapid intensification with m_d = 17.5 (Fig. 2a). The high value of m_d (175) produces a gradual decrease in SST in experiments CTL,

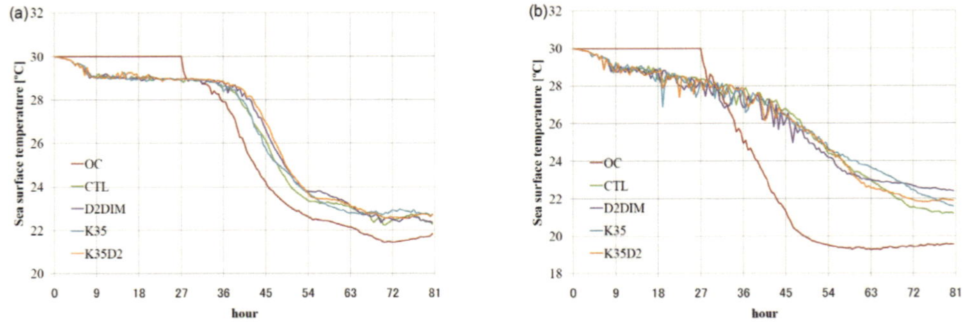

Fig. 2. Time series (as in Fig. 1) of SST immediately below the idealized TC-like vortex.

D2DIM, K35, and K35D2 but a rapid decrease in SST in experiment OC (Fig. 2b). The decrease in SST becomes small when the vortex reaches its mature phase. It should be noted that SST in experiment OC at 81 h (i.e., at the mature phase) is the smallest among all experiments (Table 1), but CP at that time is almost the same in all experiments except AT. This suggests that neither the evolution of the idealized vortex nor the final value of CP is dependent on SST directly below the circulation center.

Intensification of the vortex is more suppressed with m_d = 175 (Fig. 1b) than with m_d = 17.5. A difference in CPs for each experiment begins to appear after around 15 h, when the vortex undergoes slow intensification in experiment CTL, whereas rapid intensification of the vortex is apparent at this time in experiment OC. The vortices in experiments CTL, D2DIM, K35 and K35D2 begin to intensify rapidly at 36 h, later than in the experiments AT and OC. It is interesting that CP in experiment OC eventually reaches a value similar to those reached in experiments CTL, D2DIM, K35 and K35D2 at 81 h, suggesting that the evolution of SSC and the final value of SST may not be related to final TC intensity.

When the location of CP coincides with the circulation center, it is presumed that deep convection occurs easily there owing to enhancement of updraft. Deep convection leads to rapid intensification of the vortex. The distance between CP and the circulation center varies in an oscillatory manner from the initial time to the end of the period of rapid intensification (Fig. 3). On the other hand, variation of these distances is rarely evident from the end of the intensification phase through the mature phase (Fig. 3), indicating that rapid intensification occurs when the distance is small. Differences in the evolution of the distance between m_d = 17.5 (Fig. 3a) and m_d = 175 (Fig. 3b) imply that SSC has a negative effect on reducing the distance.

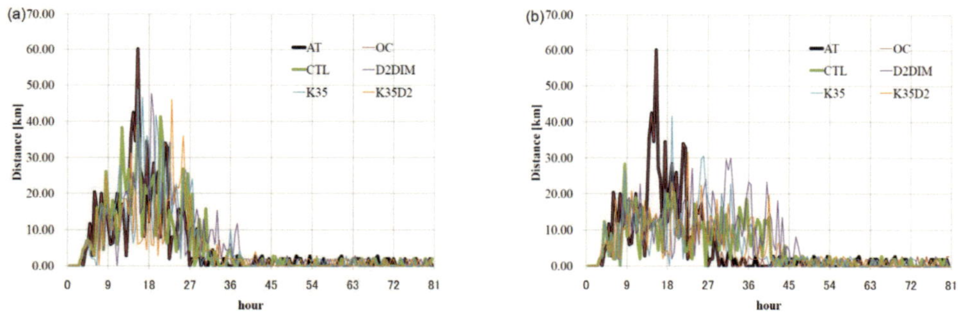

Fig. 3. Time series (as in Fig. 1) of distance between the CP and the circulation center of the vortex.

The results presented in Figs 1-3 indicate that SSC plays a crucial role in TC intensification. In contrast, the effects of the atmospheric boundary layer (experiment D2DIM) and drag coefficient (experiment K35) on TC intensification are smaller than the effect of vortex-induced SSC. To investigate the effect of the atmospheric boundary layer and drag coefficient on the vortex, the relationships between MWS and MCP were examined (Fig.4); the results show that the MWS-MCP relationship is remarkably different when the vortex is strong. For the same value of MCP, MWS tends to be high in experiments K35 and K35D2, implying that low surface friction causes not only a super-gradient inflow due to small drag coefficients for winds exceeding 35 m s^{-1} but also weak wind stresses, resulting in low

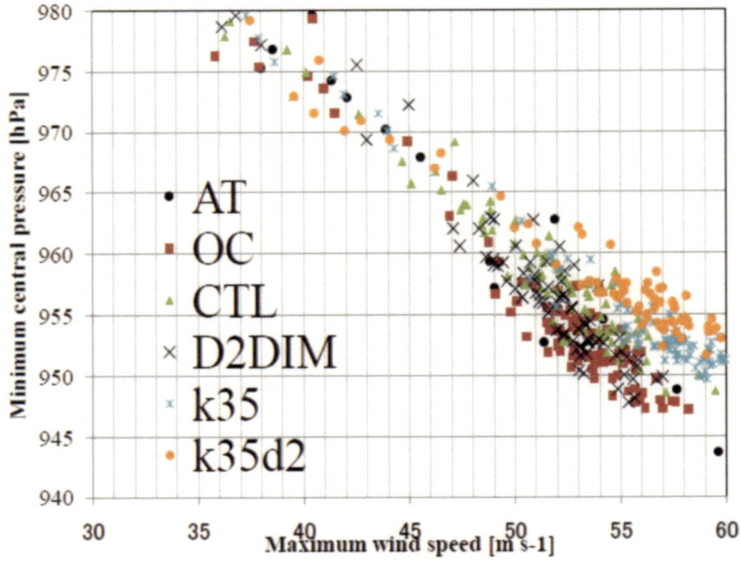

Fig. 4. Relationship between MWS and MCP in each of the numerical experiments shown in Table 1.

vortex-induced SSC. High MWS and small SSC, however, have little effect on the lowering of MCP in experiments K35 and K35D2.

In other words, differences in the MWS-MCP relationships have little effect on the evolution of CP, SST and the phases of the vortex (Figs. 1-2). In particular, the sensitivity of the tuning parameter m_d on the evolutions is noteworthy. The question then arises: How does vortex-induced SSC affect the evolutions of CP, the phase of the vortex and subsequent SSC?

4.1.2 Structural changes of the vortex

PV was used to investigate the structural changes of the vortex and the role of SSC in its evolution. PV is formulated as

$$PV = -g\left(\varsigma_\theta + f\right)\frac{\partial \theta}{\partial p}, \tag{1}$$

where g is the acceleration of gravity, ζ_θ the relative vorticity (Vor hereafter) on the isentropic surface (θ surface), and p the vertical pressure coordinate.

The time series of the vertical section of horizontally averaged PV for experiment AT (Fig. 5a) shows that there are four phases of the vortex: a spin-up phase from the initial time to 15 h, a slow intensification phase from 15 to 24 h, a rapid intensification phase from 24 to 43 h and a mature phase after 43 h (Fig. 1).

PV increases between altitudes of around 4 to 8 km during the slow intensification phase. Then at the start of the rapid intensification phase, PV above 1 km altitude increases and after 37 h, there are further increases above 10 km. After 37 h, the PV profile changes little, except above 10 km, where marked temporal oscillations are evident.

The time series of the vertical section of horizontally averaged Vor (Fig. 5b) shows that Vor below 1 km altitude increases after 15 h. Vor increases markedly at around 24 h when the vortex starts to rapidly intensify. The rapid increase of Vor continues until around 43 h, after which the profile changes little through the mature phase. These results indicate that variations of PV are controlled mainly by variations of Vor. In addition, the vertical gradient of horizontally averaged potential temperature becomes low from around 2 to 9 km altitude, corresponding temporally with the change of Vor (not shown). Thus, the evolution of the Vor profile depends on the phase of the vortex and affects both the thermal and PV profiles.

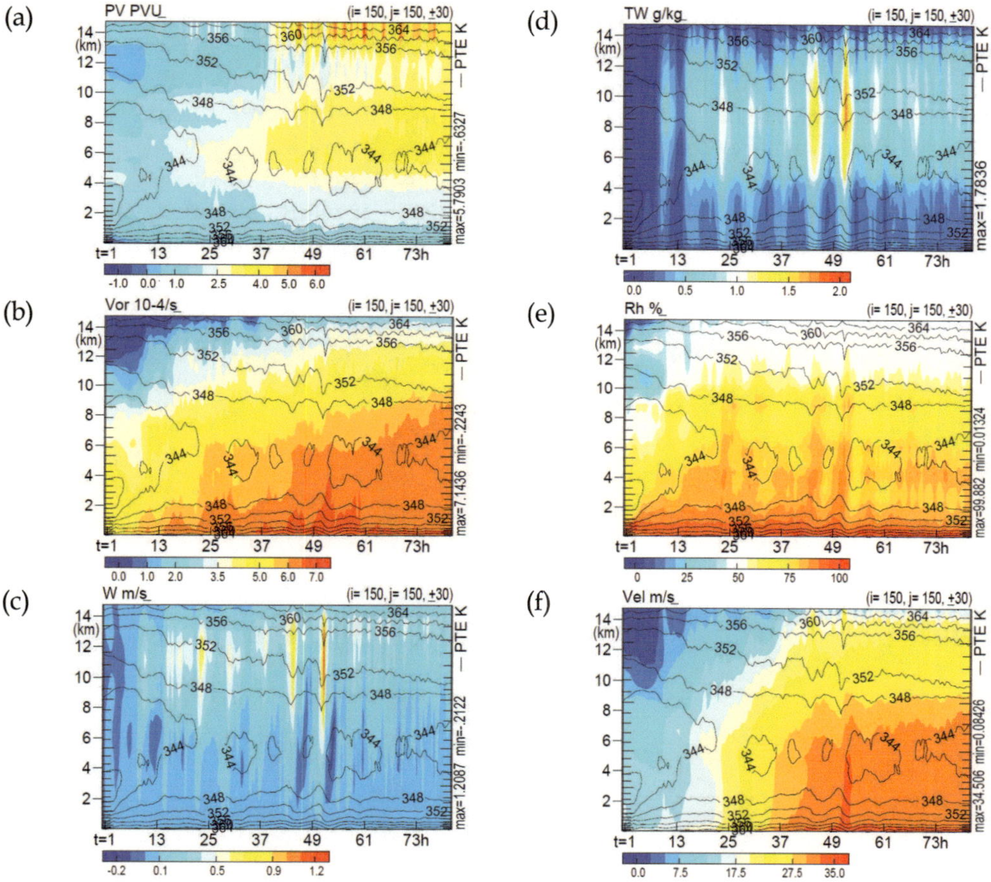

Fig. 5. Time series for experiment AT of the vertically averaged profiles of (a) PV, (b) Vor, (c) vertical velocity, (d) total water, (e) relative humidity and (f) horizontal wind speed. Averages are calculated over a 120 km x 120 km domain (corresponding to the scale of the initial vortex) centered in the computational domain. Black solid contours indicate equivalent potential temperature.

Upward vertical velocity increases suddenly between 10 to 12 km altitudes at 24, 43 and 51 h (Fig. 5c), corresponding temporally to sudden increases in total water (Fig. 5d) due to the production of cloud ice and snow. Total water in this study is defined as the sum of cloud water, rain, cloud ice, snow and graupel. The sudden increase in total water causes a sudden increase in relative humidity (Fig. 5e) at around 9 km altitude. The sudden increase in upward vertical velocity and relative humidity corresponds temporally to strengthening of horizontal wind speed (Fig. 5f). Because the increase in wind speeds is linked to increases in Vor, the intensification of the vortex can be explained by vertical transfers of heat and moisture due to the sudden increase of Vor and the production of cloud ice and snow around 10 to 12 km altitude, which in turn causes a sudden increase in Vor.

The horizontal distributions of Vor at altitudes of 1 and 4 km show clear differences for the slow intensification, rapid intensification and mature phases of experiment AT (Fig. 6). At the start of the slow intensification phase (Figs. 6a and 6d), Vor at the center of the vortex is low with mesovortices with horizontal diameters of up to 10 km scattered around the circulation center. At the start of the rapid intensification phase, the surviving mesovortices are concentrated closer to the circulation center at both 1 and 4 km altitudes (Figs. 6b and 6e) and those at 4 km altitude begin to show a spiral distribution as the vortex merger events far from the circulation center become enhanced by the eyewall-shrinking process (Fig. 3). After the rapid intensification phase, the vortex merger events cause the formation of an annular ring in the Vor distribution (Figs. 6c and 6f), with high Vor values also along the spirally bands at 4 km altitude. After formation of the annular ring, vortex intensification ceases during the mature phase (Fig. 1).

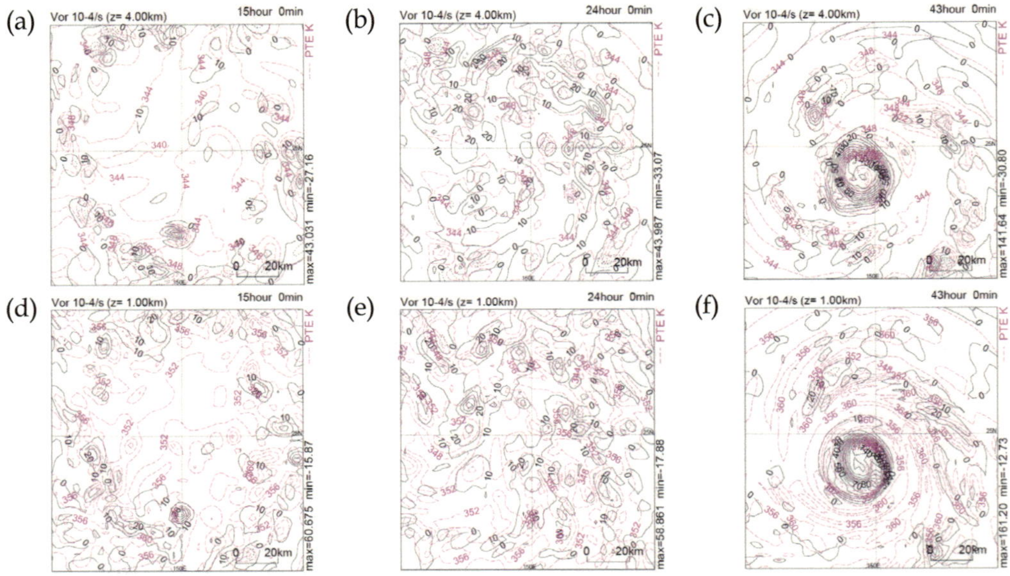

Fig. 6. Horizontal distributions of Vor at 4 km altitude for experiment AT at (a) 15 h, (b) 24 h, and (c) 43 h, and at 1 km altitude at (d) 15 h, (e) 24 h, and (f) 43 h. Black solid contours indicate positive Vor and black dashed lines indicate negative Vor. Pink contours indicate equivalent potential temperature at the same altitudes and times.

In other words, when the vortex merger events are extensive before formation of the annular ring of Vor, the mesovortices (high relative vorticity at 10 km scale accompanied by updraft) play a crucial role in the vertical transfer of heat and moisture. The locally-scattered horizontal distributions of equivalent potential temperature at 15 and 24 h are also strongly affected by the formation and enhancement of mesovortices (Fig. 6). Thus, the results of the numerical experiments suggest that the activity of the mesovortices that accumulated around the circulation center during the intensification phase contributes to formation of the complete annular ring, which is closely related to the warm-core structure of the vortex. This relationship is similar to that between TC intensity and accumulated tropical cyclone heat potential (Wada & Usui, 2007) in that the vortex merge effect coincides with accumulation of upper-ocean heat content directly below the vortex. The annular ring of Vor is accompanied by a robust warm-core structure even at 4 km altitude, which is lower than the general warm-core altitude (nearly 12 km).

In experiment CTL, the differences of horizontally averaged profiles of PV, Vor, vertical velocity, total water, relative humidity and horizontal wind speed are smaller during the spin-up phase than those in experiment AT (Fig. 7), except that the total water and wind speed in experiment CTL are even lower at around 10 h. A dry-air feature at around 6 to 8 km altitude at the initial time is maintained during the spin-up phase in both experiments AT and CTL (Figs. 5e and 7e). In contrast, distant differences between the AT and CTL experiments begin to appear after the slow intensification phase. Slow intensification continues until 36 h in experiment CTL (Figs 1 and 5), which is longer than that in experiment AT.

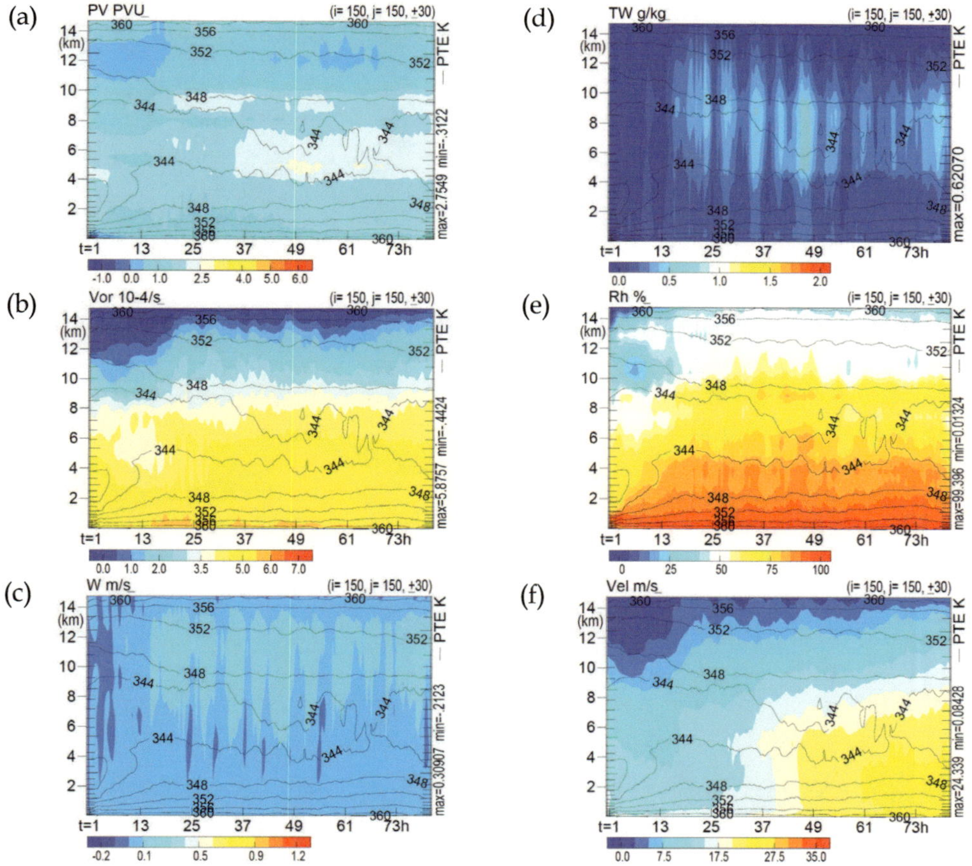

Fig. 7. As in Fig. 5, but for experiment CTL with m_d = 175.

After 37 h, the vortex intensifies gradually with an increase in PV at around 4 to 8 km altitude (Fig. 7a). The intensification produces a strong SSC effect (Fig. 2) and the subsequent decrease of SST suppresses the intensification of the vortex (Fig. 1). When the relative humidity becomes low at 5 to 10 km altitude (Fig. 7e), SST directly below the vortex continues to decrease (Fig. 2) and wind speed is maintained after 72 h (Fig. 7f). Therefore, SSC plays an essential role in determining the dynamic and thermal frameworks of the vortex and its intensification. However, it remains unclear which phases of vortex-induced SSC affect vortex intensification and control the maximum intensity reached. This may be because the vortex-induced SSC interacts with the vortex by different mechanisms according to the vortex phases.

Considering the effect of vortex-induced SSC on vortex intensification after 27 h in experiment OC with m_d = 175, an increase in PV above 10 km altitude is suppressed, which is associated with an increase in Vor is suppressed in the lower troposphere at altitudes lower than 2 km (Figs. 8a and 8b). The sudden increase in upward vertical velocity at altitudes of around 10 to 12 km in experiment AT (Fig. 5c) is not as marked as in experiment OC with m_d = 175 (Fig. 8c). This difference is related to the reduction of total water, particularly cloud ice and snow at the same altitude (Fig. 8d) and also to the reduction of relative humidity at altitudes of around 4 to 10 km (Fig. 8e). These reductions affect wind speed and the maximum intensity reached (Fig. 8f).

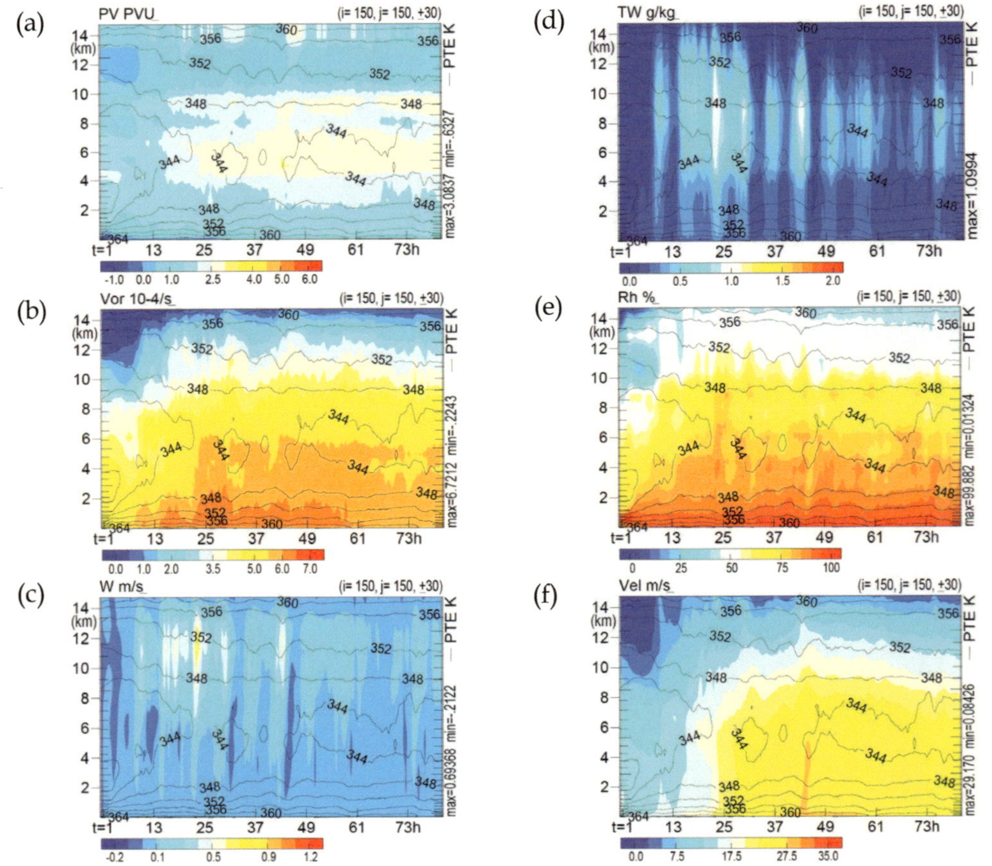

Fig. 8. As in Fig. 5, but for experiment OC with m_d = 175.

The value of m_d represents the amount of turbulent kinetic energy flux produced by breaking surface waves. When m_d is low, vortex-induced SSC becomes low. Therefore, the calculated SST may be relatively high directly beneath the vortex. Strengthening of the vortex due to low SSC is greater than that caused by the high SSC induced by vertical turbulent mixing enhanced by high m_d (Figs. 1 and 2). However, the evolution of the calculated SST is not determined simply by the value of m_d because the intensity of the vortex also affects the turbulent kinetic energy flux produced by breaking surface waves (Wada et al., 2010), which then affects subsequent vortex-induced SSC. The high value of m_d causes a rapid decrease in SST soon after the MRINHM-ocean coupled model starts to run.

In contrast, the low value of m_d causes a moderate decrease in SST, resulting in a higher SST around the vortex (Fig. 2). The high SST then leads to rapid vortex intensification followed by a rapid decrease in SST directly below the vortex. It should be noted that SSTs calculated by the coupled models are close to 22°C at 81 h and differs little among all numerical experiments except for experiments AT and OC, both with m_d = 175, although CPs differ between the numerical experiments with m_d = 17.5 and m_d = 175. The intensification processes shown in Fig. 5 correspond to the processes shown in Fig. 9. The numerical result in experiment OC with m_d = 17.5 suggests that CP is not easily determined by the SST

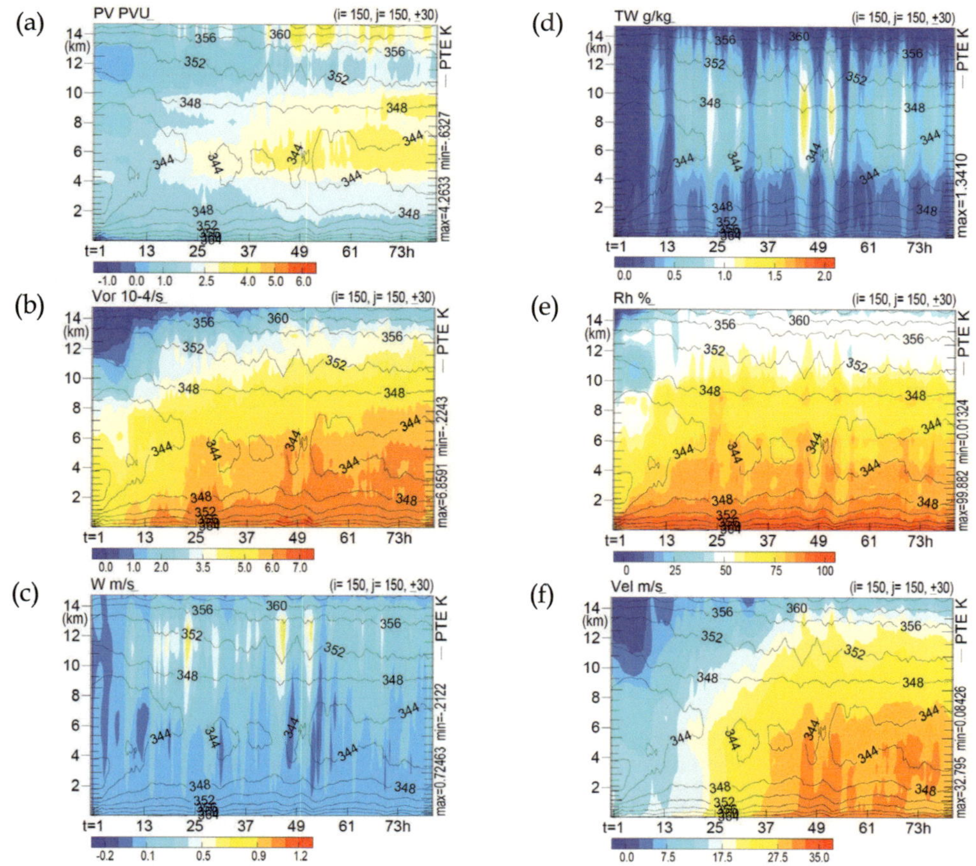

Fig. 9. As in Fig. 5 but for experiment OC with m_d = 17.5.

directly below the vortex, but is influenced by the background effects of CP and SST evolution.

The horizontal distributions of Vor at altitudes of 1 and 4 km show clearly that there are differences in TC phases between the CTL and OC experiments with m_d = 175 and the OC experiment with m_d = 17.5 (Fig. 10). An annular Vor-ring does not form in the CTL experiment with m_d = 175 (Figs. 10a and 10d), but it does form completely in the OC experiment with m_d = 17.5 (Figs. 10c and 10f). Comparison of the CTL experiment with m_d = 175 (Figs. 10a and 10d) with the OC experiment with m_d = 175 (Figs. 10c and 10f) clearly shows a difference of TC phases between the experiments of CTL and OC, implying that the slow formation of the annular ring is affected by SSC at both spin-up and slow-intensification phases (Figs. 7b and 8b). The effect of vortex-induced SSC on the annular Vor-ring at the mature phase differs from that during the other phases. The annular Vor-ring in experiment OC with m_d = 175 becomes small at 81 h (Fig. 11) as CP gradually increases during the mature phase (Fig. 1). The amplitude of Vor at 81 h at 4 km altitude in experiment CTL with m_d = 175 (Fig. 11a) is almost the same as that of experiment OC with m_d = 175 (Fig. 11b), even though the amplitude at 1 km altitude in experiment CTL with m_d = 175 (Fig. 11c) is higher than that in experiment OC with m_d = 175 (Fig. 11d). In addition, the SST in experiment OC with m_d = 175 is lower than that in experiment CTL with m_d = 175 (Fig. 2). These results are consistent with the small amplitude of Vor at 81 h in experiment OC with m_d = 175. Therefore, low vortex-induced SST has little effect on the structure of the vortex at 4 km altitude and that influence is limited to 1 km altitude during the mature phase.

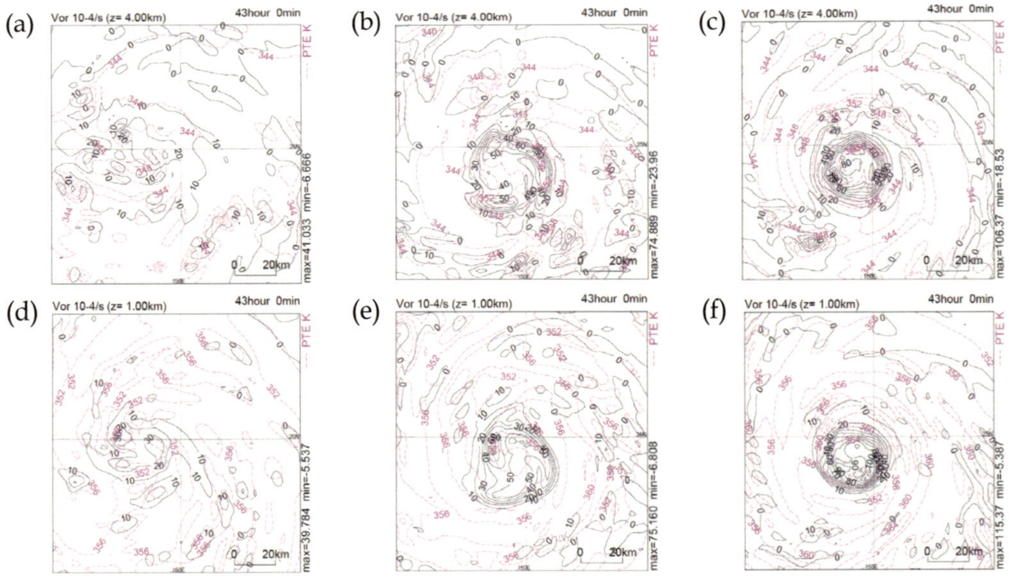

Fig. 10. Horizontal distribution of Vor at 4 km altitude at 43 h in experiments (a) CTL with m_d = 175, (b) OC with m_d = 175, and (c) OC with m_d = 17.5, and (d)-(f) at 1 km altitude at 43 h in the same experiments. Black solid contours indicate positive Vor and black dashed contours indicate negative Vor. Pink contours indicate equivalent potential temperature at the same altitudes and times.

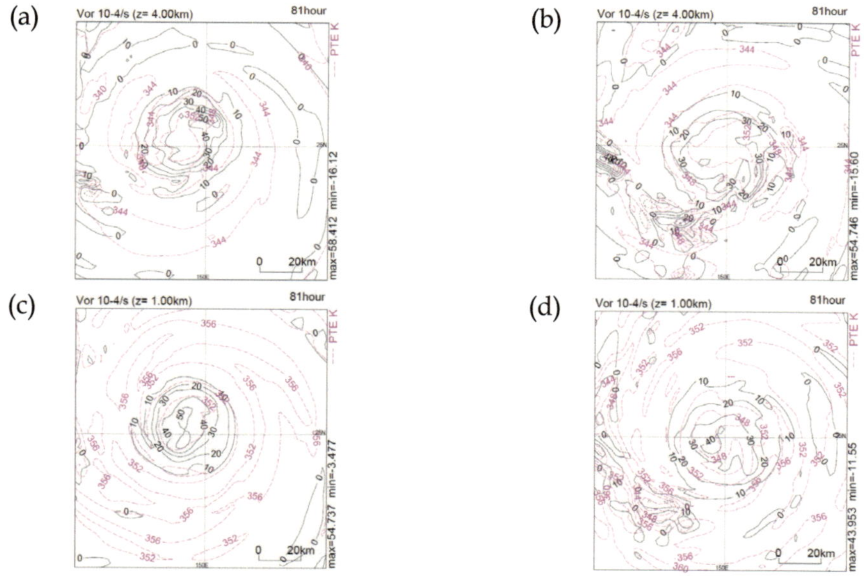

Fig. 11. Horizontal distribution of Vor at 4 km altitude at 81 h in experiments (a) CTL with m_d = 175 and (b) OC with m_d = 175 and (c)-(d) at 1 km altitude at 81 h in the same experiments. Black solid contours indicate positive Vor and black dashed contours indicate negative Vor. Pink contours indicate equivalent potential temperature at the same altitudes and times.

4.1.3 Absolute angular momentum analysis

The location of the circulation center differs clearly from that of CP during the slow intensification phase, but these locations are almost the same after the rapid intensification phase (Fig. 3). These spatial differences may be related to a change of the budget of absolute angular momentum (AAM) for the vortex with a diameter of almost 100 km, which is given as the initial TC-like vortex (Wada, 2009). This subsection addresses the relation of the deepening of CP and intensification of vortex circulation to the budget of AAM.

AAM (ς_a) is the sum of Vor and the Coriolis parameter (f):

$$\varsigma_a = \frac{\partial v}{\partial x} - \frac{\partial u}{\partial y} + f, \tag{2}$$

where x and y are zonal and meridional directions, and u and v are wind velocities in the x and y directions, respectively. The variation of AAM with time is expressed as

$$\frac{\partial \varsigma_a}{\partial t} = -\mathbf{v}_h \nabla_h \varsigma_a - w\frac{\partial \varsigma_a}{\partial z} - \varsigma_a\left(\frac{\partial u}{\partial x} + \frac{\partial v}{\partial y}\right) - \left(\frac{\partial w}{\partial x}\frac{\partial v}{\partial z} - \frac{\partial w}{\partial y}\frac{\partial u}{\partial z}\right), \tag{3}$$

where z is the vertical direction and w the vertical wind velocity. The upward direction for z and w are indicated by positive values. The four terms on the right-hand side of Eq. (3) indicate (from left to right) horizontal advection, vertical advection, divergence and titling.

The divergence term expresses vertical intensification (suppression) of a vertical vortex in response to stretching (compression); the tilting term expresses transformation from a horizontal vortex to a vertical vortex. The averages of these terms were calculated over a 120 km square at the center of the computational domain.

The contributions of each of the term on the right-hand side of Eq. (3) to the budget of AAM to an altitude approaching 15 km in experiment AT is shown in Fig. 12. Horizontally averaged horizontal advection is positive near the surface throughout the integration (Fig. 12a) and repetitive filament-like positive and negative features are evident during the spin-up and slow-intensification phases. However, horizontal advection is low over the entire integration, particularly after the rapid intensification phase.

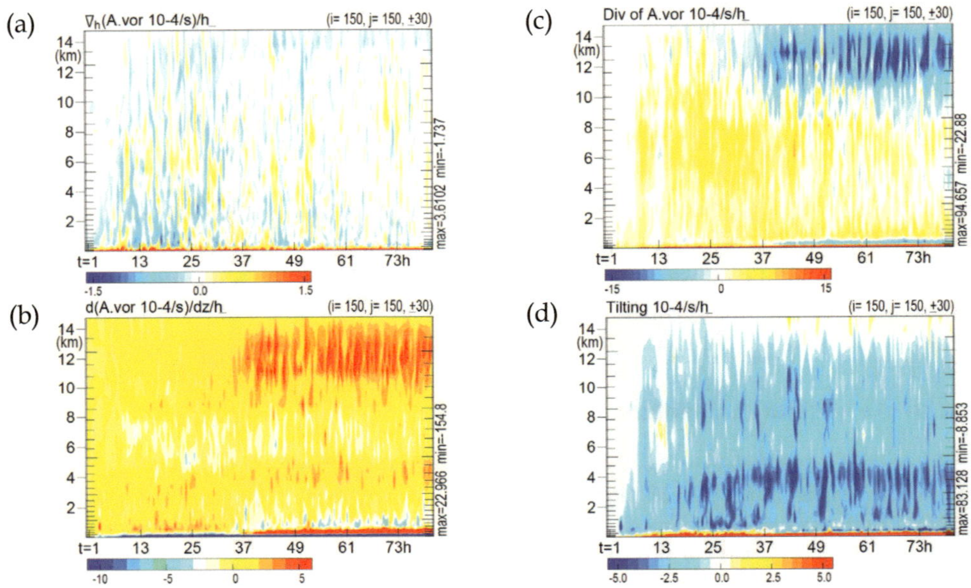

Fig. 12. Time series of the vertical profile of experiment AT for AAM terms averaged over a 120 km x 120 km domain (corresponding to the scale of the initial vortex) at the center of the computational domain: (a) horizontal advection, (b) vertical advection, (c) divergence, and (d) tilting.

Horizontally averaged vertical advection is positive at all levels during the spin-up phase, particularly below 2 km altitude (Fig. 12b). The time series of both vertical advection and stretching show that these processes enhance vortex intensification until the rapid intensification phase. After the rapid intensification phase, the increase in AAM due to vertical advection is almost balanced by the decrease of AAM above 10 km altitude due to compression (Fig. 12c).

The tilting effect is highly negative around 3 km altitude after the slow intensification phase (Fig. 12d). In addition, from the slow intensification phase to the mature phase, it is locally highly negative at around 8 km altitude during the period when total water is also locally high (Fig. 5d). The AAM decreases in response to tilting at around 4 km altitude, but this is partly offset by vertical advection. The AAM budget analysis suggests that both compression in the upper troposphere and tilting in the lower troposphere play significant roles in suppressing the vortex at the mature phase.

In experiment CTL with m_d = 17.5, SSC had a marked effect on the evolution of horizontally averaged AAM from the slow intensification phase onward, particularly the vertical advection and divergence terms of Eq. (3) (Fig. 13). These results indicate that vortex-induced SSC contributes to a decrease of AAM because of a reduction of vertical advection near the surface and a reduction of stretching at around 6 km altitude. These decreases decay the start of compression in the upper troposphere, which is compensated for by the reduction of vertical advection. In addition, vortex-induced SSC results in a decrease of AAM at around 3 km altitude in response to the effect of tilting during the spin-up phase. Therefore, vortex-induced SSC leads to suppression of the intensification of the vortex in the lower troposphere (Fig. 13).

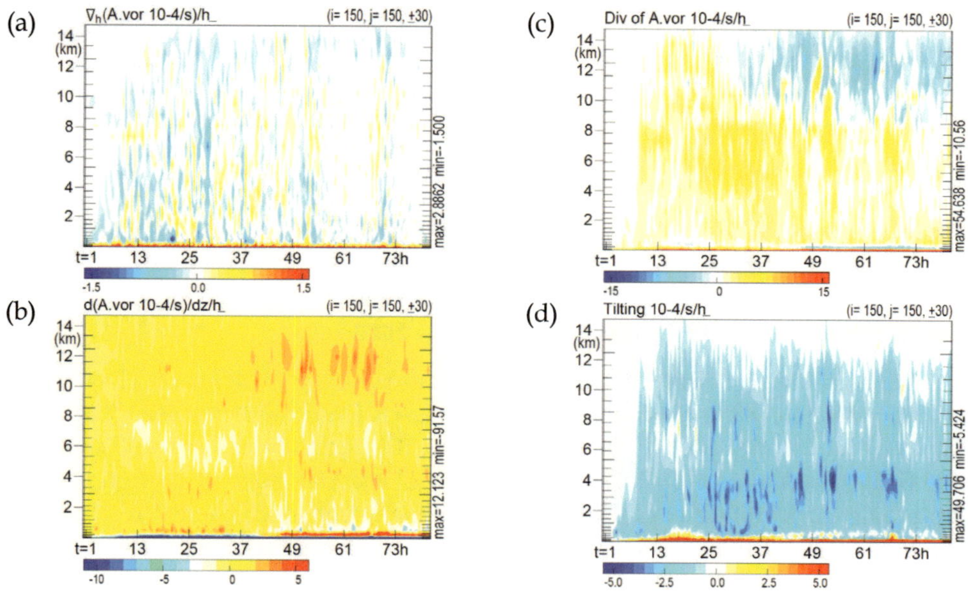

Fig. 13. As in Fig. 12, but for experiment CTL with m_d = 17.5.

In experiment CTL with m_d = 17.5, SSC has a marked effect on the evolution of horizontally averaged AAM from the slow intensification phase onward, particularly the vertical In experiment OC with m_d = 17.5, the effect of SSC on horizontally averaged AAM commences during the slow intensification phase with the evolution of horizontally averaged vertical advection in the lower troposphere and stretching at around 6 km altitude (Fig. 14). Decreases in the vertical advection and divergence terms of AAM at the mature phase lead to decay of the acceleration of vortex intensification, weakening of vertical advection, and suppression of AAM. Therefore, budget analysis of AAM suggests that the budget depends on the phase of the vortex. The effects of vertical advection and stretching are essential for intensification of the vortex, whereas tilting at around 3 km altitude and compression in the upper troposphere suppress vortex intensification at the mature phase. SSC plays a crucial role in the decrease of AAM in response to vertical advection and stretching. The decrease of AAM subsequently leads to decay of the acceleration of vortex intensification and

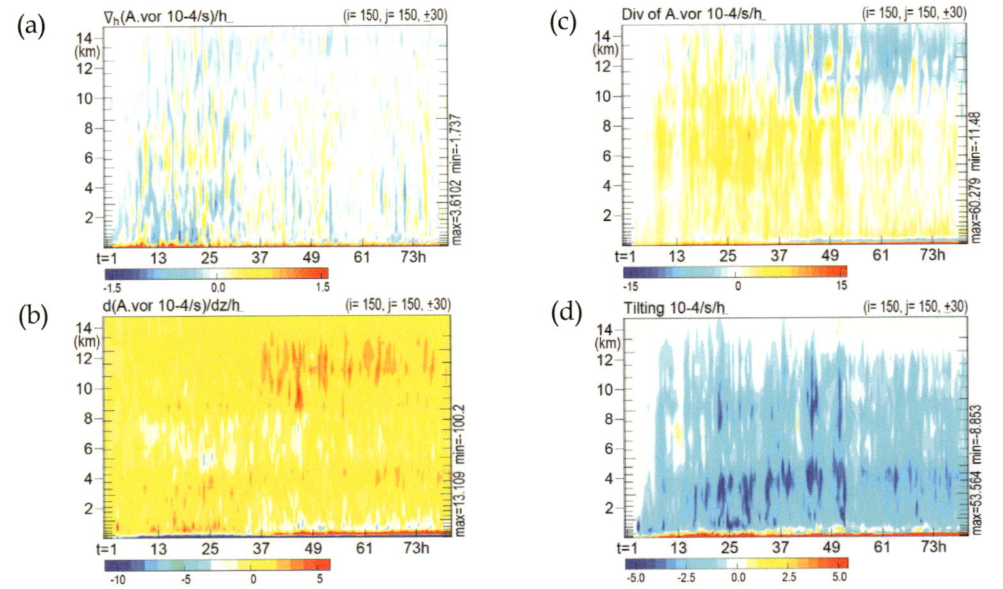

Fig. 14. As in Fig. 12, but for experiment OC with $m_d = 17.5$.

suppression of the intensification process. The negative processes thus lead to weakening of adiabatic heating in the upper troposphere and contribute to vortex Rossby wave activity.

4.2 Numerical simulations of Typhoon Choi-Wan in 2009

The discussion in this subsection focuses on the mature phase of Typhoon Choi-Wan in 2009, which passed near the moored buoy of the NOAA Kuroshio Extension Observatory (KEO) located at 32.3°N, 144.5°E in the North Pacific recirculation gyre south of the Kuroshio Extension (Bond et al., 2011). The JMA global spectral model with a grid spacing of 20 km predicted the intensity of Choi-Wan to be much stronger in its mature phase than indicated by JMA best-track analysis, although the JMA track prediction was accurate.

This study investigated the effect of NHM parameters (SSC, atmospheric and oceanic environmental conditions, horizontal resolution, and cloud physics) on the intensity and track of simulated Choi-Wan. TC intensity is measured in terms of MCP throughout subsection 4.2.

4.2.1 The impact of atmospheric environment on simulations

CPs obtained using GA for initial atmospheric and lateral-boundary conditions are lower than those obtained using JCDAS (Fig. 15). Simulated CP deepens in the early part of all simulations, but the period of deepening for JCDAS conditions is longer than that for GA conditions. The minimum simulated CP is maintained for almost 24 h for GA conditions, whereas the trend of CP for JCDAS conditions shows a sudden reversal after reaching its minimum (Fig. 15).

Track predictions indicate slower speeds of passage of simulated Choi-Wan than that of JMA best-track analysis (Fig. 16). Interestingly, the simulated tracks using GA conditions are close to the JMA best track, whereas those using JCDAS conditions clearly show a westward bias north of 25°N (Fig. 16).

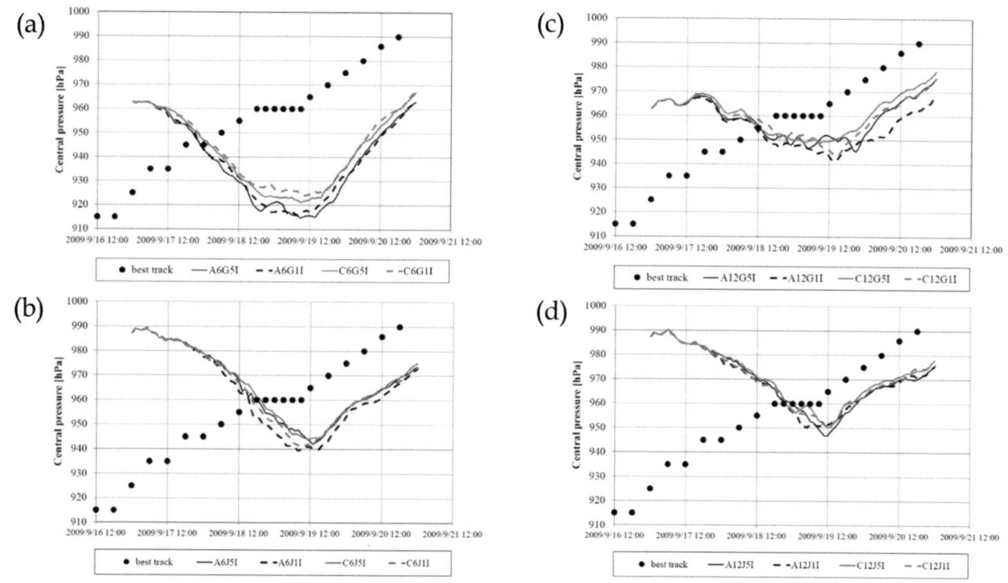

Fig. 15. Evolution of CPs for various simulations of Typhoon Choi-Wan in 2009 from 00:00 UTC 17 September and JMA best-track CP. Model specifications are (a) 6 km horizontal grid spacing and GA, (b) 6 km horizontal grid spacing and JCDAS, (c) 12 km horizontal grid spacing and GA and (d) 12 km horizontal grid spacing and JCDAS. Results shown include those obtained by both the NHM and NHM-wave-ocean coupled model in combination with the oceanic reanalysis dataset with a horizontal grid spacing of 0.1° or 0.5° (see Table 2).

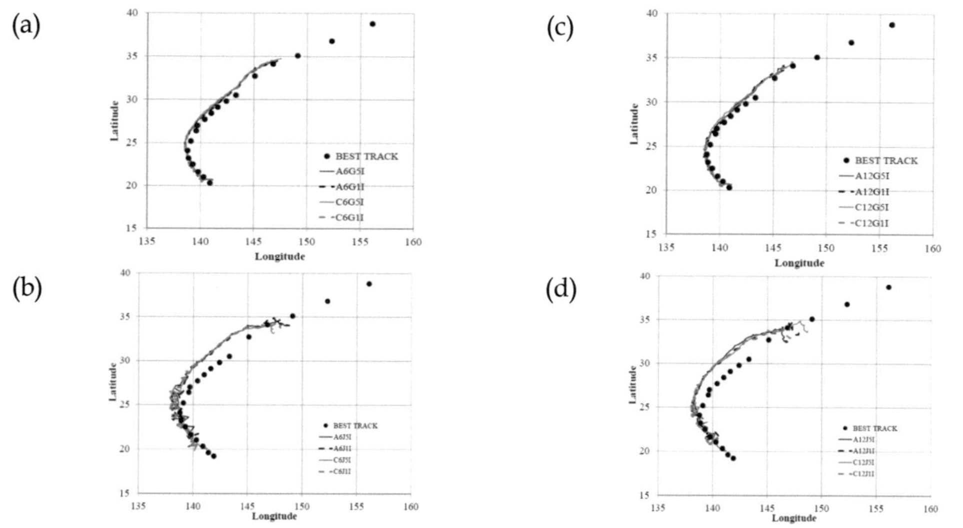

Fig. 16. Various simulated tracks for Typhoon Choi-Wan in 2009 from 00:00 UTC 17 September and the JMA best track. Model specifications are (a) 6 km horizontal grid spacing and GA, (b) 6 km horizontal grid spacing and JCDAS, (c) 12 km horizontal grid spacing and GA and (d) 12 km horizontal grid spacing and JCDAS. Results shown include those obtained by both the NHM and NHM-wave-ocean coupled model in combination with the oceanic reanalysis dataset with a horizontal grid spacing of 0.1° or 0.5° (see Table 2).

The horizontal distributions of initial atmospheric pressure at 5850 m altitude (near the 500 hPa level) for GA and JCDAS initial conditions show differences in both magnitude and contour patterns for Choi-Wan and the subtropical high to the northeast (Fig. 17). Moreover, the horizontal distributions of initial relative humidity at 5850 m altitude from GA and JCDAS data are clearly different, particularly around the center of Choi-Wan and along the cold front (Fig. 18). These differences of the initial atmospheric environment affect simulations of both the intensity and track of Choi-wan. These impacts on the simulations of intensity affect not only the value of CP but also its evolution.

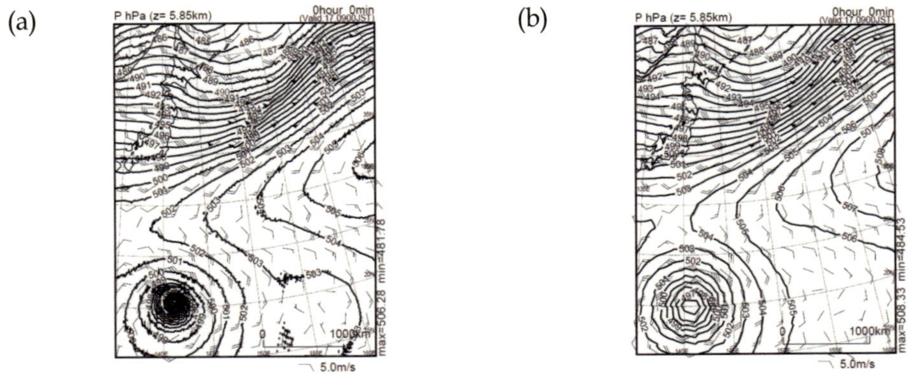

Fig. 17. Horizontal distributions of initial pressure at 5850 m altitude. Atmospheric initial and boundary conditions are from (a) GA and (b) JCDAS.

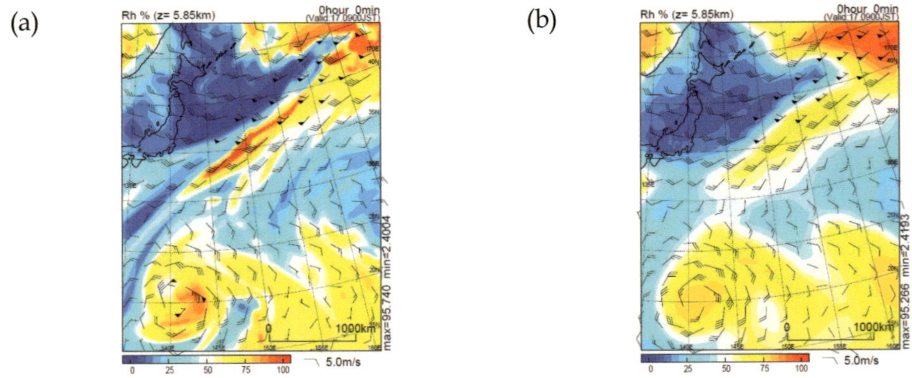

Fig. 18. As in Fig. 17, but for relative humidity.

4.2.2 The Impact of oceanic environment on simulations

The horizontal distributions of initial SST obtained from oceanic reanalysis data with horizontal grid spacings of 0.5 ° and 0.1° are clearly different around the center of Choi-Wan and in the Kuroshio Extension area. The data with a horizontal grid spacing of 0.1° resolve both the Kuroshio meander and warm-eddy structure around the Kuroshio Extension (Fig. 19). Both datasets reproduce the latitudinal SST front along 35°N, indicating that synoptic oceanic features are almost the same for both SST fields.

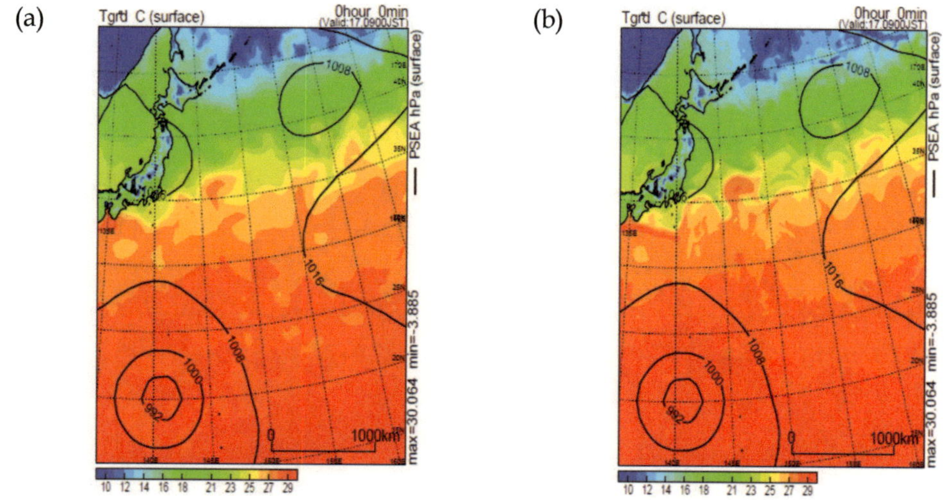

Fig. 19. Horizontal distributions of initial SST derived from oceanic reanalysis data with horizontal grid spacing of (a) 0.5° and (b) 0.1°. Contours show initial sea-level pressure.

These results indicate that the impacts of these two pre-existing oceanic environment datasets on the evolution of simulated CP are different, but the difference is considerably smaller than that obtained using GA and JCDAS atmospheric environment data (Fig. 15). The pre-existing oceanic environment dataset used has no noticeable effect on track predictions (Fig. 16).

4.2.3 The impact of horizontal resolution on simulations

Horizontal resolutions of 6 and 12 km show marked differences in the simulated evolution of CP when GA data are used for initial and lateral-boundary atmospheric conditions, but little difference when using JCDAS data (Fig. 15). Horizontal resolution has no impact on track prediction, as is also the case for choice of pre-existing oceanic environment (Fig. 16).

Differences of simulations of the structure of Choi-wan at the mature phase for horizontal resolutions of 6 and 12 km are investigated by considering the simulated horizontal distribution of PV at 1500 m altitude (Fig. 20) and that of hourly precipitation at the surface (Fig. 21). Each of the horizontal distributions of PV (Fig. 20) shows horizontal annular distributions about the center of the simulated Choi-Wan. PV is greatest for simulation C6G5I, but the PVs of simulations C6J5I and C12J5I are similar, indicating that PV near the center of the simulated Choi-Wan is not necessarily dependent on horizontal resolution.

However, the PV distributions of simulations C6G5I and C6J5I show more detailed structure, such as mesovortices within the spiral bands, than those of simulations C12G5I and C12J5I. Differences are notable in the size of horizontal annular distributions about the center of the simulated Choi-Wan among the four simulations shown in Fig. 20. For example, simulation C6G5I produces a small, compact PV pattern, whereas that of simulation C12J5I is considerably larger. These results indicate that the size of the simulated Choi-Wan is influenced by horizontal resolution as well as by the pre-existing atmospheric environment.

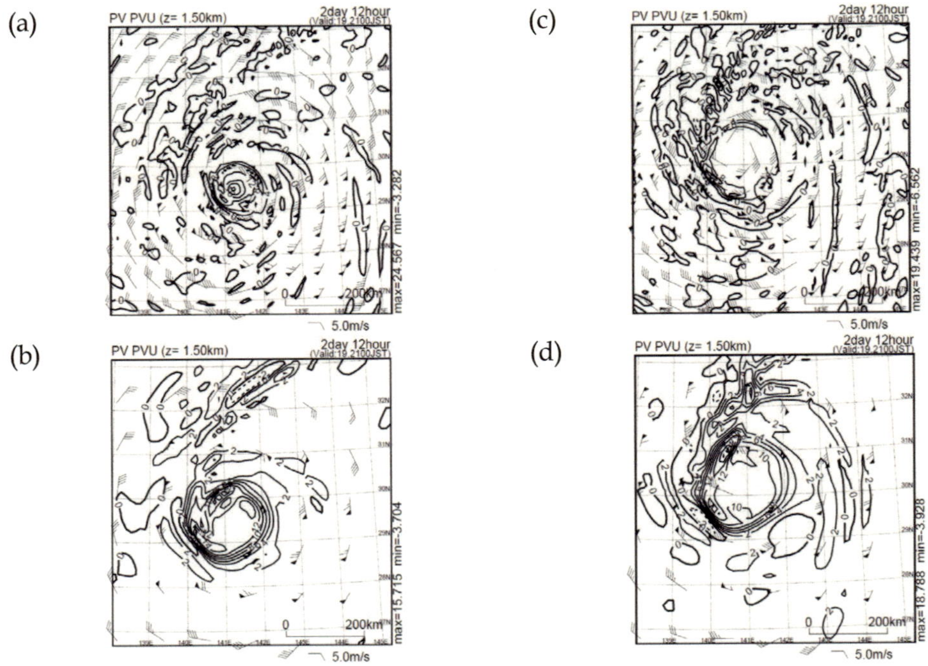

Fig. 20. Horizontal distributions of PV at 1500 m altitude at 60 h (12:00 UTC on 19 September 2009) for simulations (a) C6G5I, (b) C6J5I, (c) C12G5I, and (d) C12J5I.

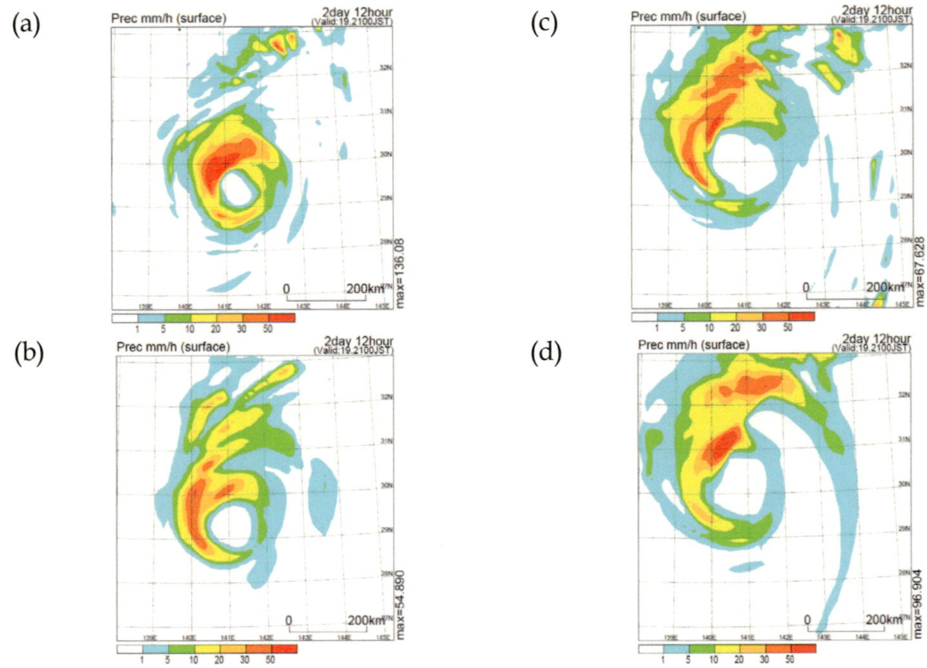

Fig. 21. Horizontal distribution of precipitation (mm h^{-1}) at the surface at 60 h (12:00 UTC on 19 September 2009) for simulations (a) C6G5I, (b) C6J5I, (c) C12G5I, and (d) C12J5I.

The simulations of the horizontal distributions of hourly precipitation each show a wave-number-1 pattern, except for simulation C6G5I, which shows a concentric eyewall (Fig. 21). Formation of the concentric eyewall is dependent on the intensity of the simulated Choi-Wan (Fig. 15). Indeed, the eye of Choi-Wan in simulation C6G5I is much smaller than in the other simulations (Figs. 20 and 21). The wave-number-1 patterns of simulated precipitation show heavy precipitation in the left upper quadrant of simulated Choi-Wan, where simulated PV is also high (Fig. 20). The National Aeronautics and Space Administration (NASA) Tropical Rainfall Measuring Mission Microwave Imager (TRMM/TMI) data at 4:43 UTC on 19 September 2009 clearly shows a wave-number-1 pattern (Fig. 22), although the heavy rain to the north of Choi-Wan does not match the simulations shown in Fig. 21. The eye of Choi-Wan in TRMM/TMI data is nearly 1° in diameter, which is comparable with the C12G5I simulation (Figs. 20c and 21c), although TRMM/TMI data show more debris of the inner eyewall east of the eye. It should be noted that TRMM/TMI data come with certain uncertainty so that the horizontal distributions of hourly precipitation should be validated using observed *in situ* rain gauge data. However, it is difficult to obtain the data when a typhoon exists in the open ocean.

Fig. 22. Horizontal distribution of rain rate (inches h^{-1}) from TRMM/TMI data at 4:43 UTC on 19 September 2009 (http://www.nrlmry.navy.mil/TC.html). The center position is 29.0°N, 141.4°E, with a central pressure of 967 hPa and maximum wind speed of 75 knots.

4.2.4 The impact of cloud physics on simulations

The impacts of ice cloud, snow and graupel on the intensity and track predictions of simulated Choi-wan are investigated using the NHM and NHM-wave-ocean coupled model, but without the cold-rain cloud physics associated with ice cloud, snow and graupel (Fig. 23). Under these simulation conditions, total water around Choi-Wan is lower than that obtained in previous simulations, particularly in the rain shield on the northern side of the simulated Choi-Wan, and simulated CP deepens to some extent (Fig. 23). Interestingly, the simulated tracks are shifted eastward when compared with simulations that include cold-rain physics (Fig. 23). The simulated tracks agree better to JMA best-track positions, except that the simulated speed of passage of Choi-Wan along its track is slower than JMA estimates (Fig. 23). These simulations suggest that cloud physics has considerable impact on simulations of the intensity and track of Choi-Wan.

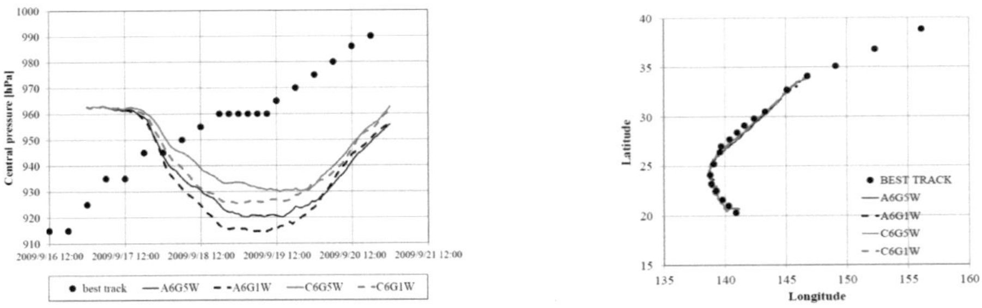

Fig. 23. Left panel is the same as Fig. 15 and right panel is the same as Fig. 16, except that the results shown here were obtained by using the NHM or NHM-wave-ocean coupled model, but without the physics of ice cloud, snow and graupel.

5. Discussion

The results of this study show that the effect of the ocean on TCs differs according to the phase of the TC, particularly during the intensification and mature or decaying phases. To accurately reproduce the intensification of a TC, the numerical model must resolve inner-core dynamics, such as the development of mesovortices and the vortex merger events. The results obtained here are consistent with those of previous studies that show that 1 to 2 km horizontal resolution is required (e.g. Wang & Wu, 2004; Chen et al., 2007). This study also shows that vortex-induced SSC is effective in suppressing the intensification of a TC. The effect of SSC is not instantaneous but accumulated over time within the inner core of the vortex (Wada & Usui, 2007; Wada, 2009).

However, the use of fine horizontal resolution for model parameters produces excessively strong TC intensity during the mature and decaying phases. Numerical simulations obtained using the JMANHM-ocean coupled model with 24-km horizontal resolution produce reasonable simulations of CP (not shown) that are comparable with JMA best-track estimates around the KEO moored buoy. During the mature phase, the effects of both vortex-induced SSC and pre-existing oceanic environment on a TC are smaller than the effects of pre-existing atmospheric environment and cloud physics. What is clear from this study is that the final TC intensity at the mature phase is not dependent simply on SST directly below the vortex and horizontal resolution, but is determined by the background effects of the evolution of CP and SST. Therefore, for accurate predictions of a TC, careful attention must be paid to model specifications other than SST and horizontal resolution.

6. Conclusion

This chapter describes the interactions between TCs and the ocean on the basis of numerical experiments using a nonhydrostatic atmosphere model coupled with a third generation wave model and a multi-layer ocean model developed jointly by NPD and MRI in JMA. The results of idealized numerical experiments and numerical simulations of Typhoon Choi-Wan in 2009 allow the following conclusion to be drawn.

The evolution of CP of a vortex can be divided into four phases: the spin-up, slow intensification, rapid intensification, and mature phases. Vortex-induced SSC is effective in suppressing TC intensification from the spin-up phase to the rapid intensification phase. CP evolution is thus sensitive to vertical turbulent mixing in the oceanic mixed layer caused by breaking surface waves. In contrast, the effects of processes in the surface-boundary and planetary-boundary layers are found in the relationship between MCP and MWS, although the impact of these on TC intensity and intensification is smaller than that of vortex-induced SSC. After formation of the annular ring within the inner core of the vortex, TC intensity is less sensitive to vortex-induced SSC. Budget analysis of AAM averaged over a 120 km x 120 km square at the center of the computational domain shows that vortex-induced SSC is effective in decreasing AAM due to vertical advection and stretching. This decrease leads to decay of the acceleration of vortex intensification and affects adiabatic heating processes in the upper troposphere.

Numerical simulations of Choi-Wan show that the effects of Choi-Wan-induced SSC and pre-existing oceanic conditions on simulations of TC track and intensity are smaller than those of pre-existing atmospheric conditions and cloud physics. Fine horizontal resolution of model parameters provides excessive simulations of TC intensity for the mature and decaying phases. Even though the excessive simulations of TC intensity are reduced by the effect of Choi-Wan-induced SSC, the improvement may have led to the simulation of excessive SSC induced by the passage of Choi-Wan. These results suggest that a model parameter, horizontal resolution of 1 to 2 km, is not always needed for TC simulations at the mature and decaying phases, or that both the atmosphere-wave-ocean coupled model and atmospheric reanalysis data will require further development.

7. Acknowledgment

This work was supported by the Japan Society for the Promotion of Science (JSPS) through the Grant-in-Aid for Scientific Research (C) (22540454) and by the Japanese Ministry of Education, Culture, Sports, Science and Technology (MEXT) under Grant-in-Aid for Scientific Research on Innovative Areas #2205 (in proposed research area 23106505).

8. References

Bender, M. & Ginis, I. (2000). Real-time simulation of hurricane-ocean interaction. *Monthly Weather Review*, Vol.128, pp. 917-946.

Bond, N. A.; Cronin, M. F.; Sabine, C.; Kawai, Y.; Ichikawa, H.; Freitag, P. & Ronnholm, K. (2011). Upper ocean response to Typhoon Choi-Wan as measured by the Kuroshio Extension Observatory mooring. *Journal of Geophysical Research*, Vol. 116, C02031.

Chen, S. S.; Price, J. F.; Zhao, W.; Donelan, D. A. & Walsh, E. J. (2007). The CBLAST-Hurricane program and the next-generation fully coupled atmosphere-wave-ocean models for hurricane research and prediction. *Bulletin of American Meteorological Society*, Vol. 88, pp.311-317.

Deardorff, J. W. (1980). Stratocumulus-capped mixed layers derived from a three-dimensional model. *Boundary-Layer Meteorology*, Vol. 18, pp.495-527.

Deardorff, J. W. (1983). A multi-limit mixed-layer entrainment formulation. *Journal of Physical Oceanography*, Vol.13, pp. 988-1002.

Donelan, M. A.; Haus, B. K.; Reul, N.; Plant, W. J.; Stiassnie, M.; Graber, H. C.; Brown, O. B. & Saltzman, E. S. (2004). On the limiting aerodynamic roughness of the ocean in very high winds. *Geophysical Research Letters*, Vol.31, L18306.

Emanuel, K. A., (1986). An air-sea interaction theory for tropical cyclones. Part I: Steady-state maintenance. *Journal of the Atmospheric Sciences*. Vol.43, pp. 585-604.

Ginis, I., (1995). Ocean response to tropical cyclone. In: *Global Perspective on Tropical Cyclones*, Elsberry, R. L., pp. 198-260. World Meteorological Organization, Geneva, Switzerland.

Klemp, J. B. & Wilhelmson R. B. (1978). The simulation of three-dimensional convective storm dynamics. *Journal of the Atmospheric Sciences*. Vol.35, pp. 1070-1096.

Kondo, J. (1975). Air-sea bulk transfer coefficients in diabatic conditions. *Boundary-Layer Meteorology*, Vol. 9, pp.91-112.

Lin, Y. G.; Farley, R. D. & Orville, H. D. (1983). Bulk parameterization of the snow field in a cloud model. *Journal of Climate and Applied Meteorology*, Vol.22, pp.1065-1092.

Palmén, E. (1948). On the formation and structure of tropical cyclones. *Geophysica*, Vol. 3, pp. 26-38.

Powell, M. D.; Vickery, P. J. & Reinhold, T. A. (2003). Reduced drag coefficient for high wind speeds in tropical cyclones. *Nature*, Vol. 422, pp. 279-283.

Smith, R. K.; Montgomery, M. T. & Vogl, S. (2008). A critique of Emanuel's hurricane model and potential intensity theory. *Quarterly Journal of Royal Meteorological Society*, Vol. 134, pp. 551-561.

Sugi, M.; Kuma, K.; Tada, K.; Tamita, K.; Hasegawa, N.; Iwasaki, T.; Yamada, S. & Kitade, T. (1990). Description and performance of the JMA operational global spectral model (JMA-GSM88). *Geophysical Magazine*, Vol. 43, pp.105-130.

Taylor, P. K. & Yelland, M. J. (2001). The dependence of sea surface roughness on the height and steepness of the waves. *Journal of Physical Oceanography*, Vol.31, pp. 572-590.

Usui, N.; Ishizaki S.; Fujii Y.; Tsujino H.; Yasuda T. & Kamachi M. (2006). Meteorological Research Institute multivariate ocean variational estimation (MOVE) system: Some early results. *Advances in Space Research*, Vol. 37, pp.896-822.

Wada, A. (2007). Numerical problems associated with tropical cyclone intensity prediction using a sophisticated coupled typhoon-ocean model. *Papers in Meteorology and Geophysics*, Vol. 58, pp.103-126.

Wada, A. & Usui, N. (2007). Importance of tropical cyclone heat potential for tropical cyclone intensity and intensification in the Western North Pacific. *Journal of Oceanography*, Vol.63, pp.427-447.

Wada, A. (2009). Idealized numerical experiments associated with the intensity and rapid intensification of stationary tropical cyclone-like vortex and its relation to initial sea-surface temperature and vortex-induced sea-surface cooling. *Journal of Geophysical Research*, Vol. 114, D18111.

Wada, A.; Niino, H. & Nakano, H. (2009). Roles of vertical turbulent mixing in the ocean response to Typhoon Rex (1998). *Journal of Oceanography*, Vol.65, pp. 373-396.

Wada, A. (2010). Tropical-cyclone-ocean interaction: Climatology. In: *Advances in Energy Research. Volume 1*, pp. 99-132, Acosta M. J., NOVA Publishers, ISBN:978-1-61668-994-0.

Wada, A. & Usui, N. (2010). Impacts of Oceanic preexisting conditions on predictions of Typhoon Hai-Tang in 2005. *Advances in Meteorology*, Vol. 2010, 756071.

Wada, A.; Kohno, N. & Kawai, Y. (2010). Impact of wave-ocean interaction on Typhoon Hai-Tang in 2005. *Scientific Online Letters on the Atmosphere*, Vol. 6A, pp.13-16.

Wang, Y. & Wu, C.-C. (2004). Current understanding of tropical cyclone structure and intensity change – a review. *Meteorological and Atmospheric Physics*, Vol. 87, pp.257-278.

4

Mean State and the MJO in a High Resolution Nested Regional Climate Model

Pallav Ray
International Pacific Research Center (IPRC), University of Hawaii
USA

1. Introduction

The Madden-Julian oscillation (MJO, Madden and Julian, 1971, 1972) is a dominant feature of intraseasonal (20-90 day) variability in the tropics. According to the classic view, the MJO begins as a positive convective anomaly in the equatorial western Indian Ocean. It then propagates eastward toward the maritime continent where convection weakens until the MJO reaches the west Pacific where the convection strengthens again. Convective coupling diminishes in the eastern Pacific in the presence of cooler sea surface temperature (SST), but the wind component in the upper troposphere may propagate eastward as free waves at about 12-15 m s^{-1}, much faster than the MJO propagation speed of 5 m s^{-1} (Knutson et al., 1986). Global circumnavigation associated with the MJO can also be noticed in the upper-tropospheric divergent wind (e.g., Krishnamurti et al., 1985; Knutson and Weickman, 1987) and moisture fields (Kikuchi and Takayabu, 2003), but is difficult to detect in parameters closer to the surface.

The MJO has been found to influence a number of features in the tropics including the Indian summer monsoon (e.g., Yasunari, 1979), Australian monsoon (e.g., Hendon and Liebmann, 1990), tropical storms (e.g., Liebmann et al., 1994), and the initiation of El Nino events (e.g., Lau and Chan, 1985). However, the influence of the MJO is not limited to the tropics. The MJO affects the global medium and extended range weather forecasts (e.g., Jones and Schemn, 2000) and modulates the global angular momentum (e.g., Weickmann et al., 1997). This tropics-extratropics interaction produced by the MJO affects the skill of the northern hemisphere weather forecasts (Ferranti et al., 1990). The long periodicity of the MJO convection relates it with the predictability on seasonal time scales. As a result, longer-range forecasts could be improved if the MJO can be predicted.

There have been considerable advancements in understanding the different aspects of the MJO using observation, theory and numerical modeling. However, an accurate MJO simulation using numerical models remains an extremely difficult task due to a number of model deficiencies (Lin et al., 2006; Zhang et al., 2006; Kim et al., 2009). One such deficiency is the model's inability in capturing the correct mean state. The role of the mean state on the MJO was previously explored using GCMs (e.g., Slingo et al., 1996; Inness et al., 2003; Maloney and Hartmann, 2001; Ajayamohan and Goswami, 2007; Maloney, 2009), observations (Zhang and Dong, 2004), and model-observation comparison (Zhang et al., 2006). It is found that the realistic distributions of precipitation, lower-tropospheric zonal wind and specific humidity, and boundary-layer moisture convergence in models are essential for them to reproduce realistic statistics of the intraseasonal variability. On the other hand, MJO events that are initiated by the extratropical influences may have less dependence on the mean state (e.g., Ray et al., 2011). A review of our present understanding of the MJO can be found in Zhang (2005).

The objective of this chapter is to further explore the role of the mean state on the MJO using a high-resolution nested regional climate model (NRCM). We use the NRCM (http://www.nrcm.ucar.edu), based on the Weather Research and Forecasting Model (WRF). The domain of this NRCM is global (periodic) in the zonal direction and is bounded in the meridional direction. The main advantage of the NRCM compared to a regular regional model is that, without the east-west boundaries, it isolates the external influences arriving solely from the extratropics. The added constraint provided by the lateral boundary conditions is expected to improve the simulated MJO statistics. Also, compared to a GCM, the NRCM has higher resolution and sophisticated physics that may be helpful to better capture the multi-scale organized convection associated with the MJO (Chen et al., 1996; Houze, 2004; Moncrieff, 2010).

The strategy of this study is to integrate the NRCM for several years and evaluate the role of the mean state on the MJO statistics. Our goal is to provide unique perspectives to the MJO dynamics and mean state.

Section 2 describes the configuration of the model, method and data. Section 3 explores the atmospheric mean state and its role on the MJO with an emphasis on the roles played by the mean precipitation and zonal winds at the 850 hPa (U850). Section 4 summarizes the results along with the implications and limitations of this study.

2. Model and data

2.1 Model

We use the NRCM based on the WRF model that was developed at the National Center for Atmospheric Research (NCAR). This is also known as a tropical channel model (TCM), since the model's computational domain is global (periodic) in the zonal direction. Conceptually, the configuration is similar to the TCM developed at the University of Miami based on the fifth-generation Pennsylvania State University-NCAR Mesoscale Model (MM5, Dudhia, 1993; Grell et al., 1995), known as the Tropical MM5 (TMM5, Ray et al., 2009; Ray and Zhang, 2010). The NRCM is atmosphere only and employs Mercator projection centered at the equator with open boundaries in the North-South direction. Global reanalyses data are used to provide the initial and boundary conditions for the model (see section 2.3).

The horizontal resolution of the NRCM is 36 km, and the meridional boundaries are placed at 30°S and 45°N. The model top is at 50 hPa, and 35 vertical levels are used. Output is taken every 3 hours. Based on a series of tests, the suite of parameterizations used for this study are: Kain-Fritsch cumulus parameterization (KF, Kain, 2004), WSM6 cloud microphysics (Hong et al., 2004), CAM 3.0 radiation scheme (Collins et al., 2006), YSU boundary layer scheme (Hong et al., 2006), and Noah land surface model (Chen and Dudhia, 2001). The model was integrated for 5 years from January 1, 1996 to January 1, 2000.

2.2 Method

The NRCM simulation is used to document the mean state and the MJO statistics. The MJO is defined as a planetary scale (zonal wavenumber 1 to 5), eastward propagating, intraseasonal (20-90 day) components in the U850 coupled with precipitation (P). To extract the coupled MJO signal, a singular vector decomposition (SVD) method (Wallace et al., 1992) is applied to U850 and P. This method is similar to EOF analysis, but with one advantage: it considers the wind-precipitation coupling associated with the MJO. The leading modes are selected based on North et al. (1982) rule. Three leading modes are found for both observations and model, and they explain 41% and 31% of the covariance for the observation and the NRCM, respectively. These selected modes represent the intraseasonal coupled components between U850 and P. Time series of U850 and P reconstructed through linear regression of intraseasonal bandpass filtered U850 and P upon their selected leading SVD modes, are considered to represent the MJO. Hereafter, they are referred to as U850* and P*.

2.3 Data

Model validation uses observations and reanalyses data. They include: National Centers for Environmental Prediction-National Center for Atmospheric Research (NCEP-NCAR) Re-analysis (Kalnay et al., 1996) winds and the merged analysis of precipitation (CMAP; Xie and Arkin, 1997).

The initial and boundary conditions of the NRCM are from the NCEP-NCAR reanalysis. The SSTs are from Atmospheric Model Intercomparison Project (AMIP; 1° x 1°, 6-hourly; Taylor et al., 2000). For brevity, both reanalysis and CMAP precipitation will be referred to as "observations".

3. Results

The simulated mean state is described first, followed by the MJO and how it has been affected by the model mean state.

3.1 Mean state

The mean state of the model is compared with the observation with respect to P and U850 (Fig. 1). The main error in the model precipitation is over the equatorial Indian and west Pacific Ocean and over the South Pacific Convergence Zone (SPCZ), where the variance of the MJO related precipitation is maximum (Zhang and Dong, 2005). This is the first indication that the simulated MJO may be affected by the mean state. The model precipitation seems to move further from the equator with much higher values over the southern Indian Ocean and north of maritime continent. Most of this error comes during the northern winter. On the other hand, simulated U850 is somewhat stronger than those of reanalysis over the Indian Ocean and the eastern and central Pacific. The model overestimates winds at 200 hPa in the equatorial Indian and west Pacific Ocean also (not shown). The simulation captures the winds quite well over the west African monsoon region, where the lack of precipitation in the model is obvious. Easterlies at 850 hPa are stronger over the southern Indian Ocean, where there is error in precipitation as well.

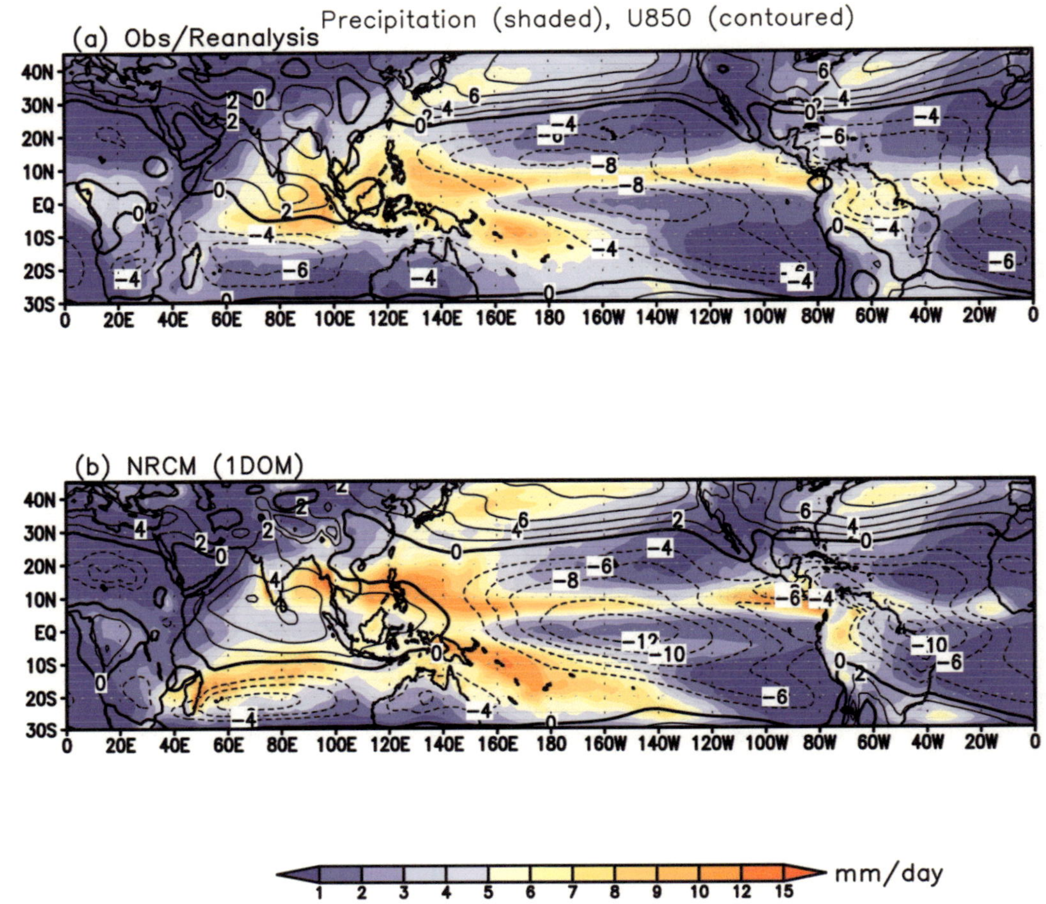

Fig. 1. Annual mean rainfall (shaded, mm/day) and U850 (contoured, m/s) during 1996-2000 from the (a) observation/reanalysis, and (b) NRCM. Zero contours are thickened.

Overall, the precipitation is underestimated over the equatorial (10°S-10°N) Indian Ocean, and is slightly overestimated over the west Pacific (Fig. 2a). However, U850 is overestimated over the Indian Ocean and is slightly underestimated over the west Pacific (Fig. 2b). The results indicate a possible lack of coupling between the winds and precipitation in the model compared to observations.

To further explore the simulated mean state, we show latitudinal distributions of P and U850 in Fig. 3. Precipitation is underestimated close to the equator (10°S-10°N), however, the model overestimates precipitation between 10°-30° latitudes, particularly in the southern hemisphere. The northern ITCZ is shifted towards the higher latitude. As expected, simulated winds are almost same as that of the reanalysis near the boundaries (Fig. 3b). Although there are large differences between the simulated and observed U850 over the equatorial Indian and west Pacific (Fig. 2b), the zonally averaged U850 match well due to the cancellation of errors (Fig. 3b).

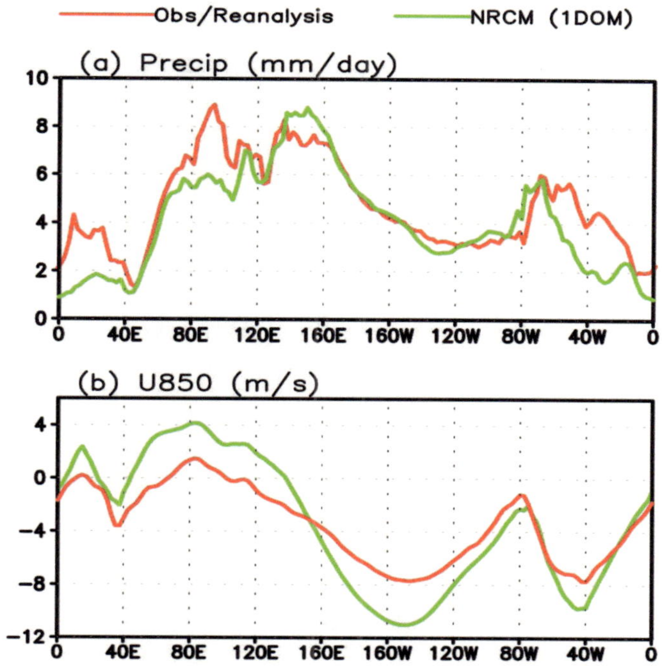

Fig. 2. Longitudinal distribution of (a) Precipitation (mm/day) and (b) U850 (m/s), averaged over 10°S-10°N from the observation/reanalysis (red) and NRCM (green).

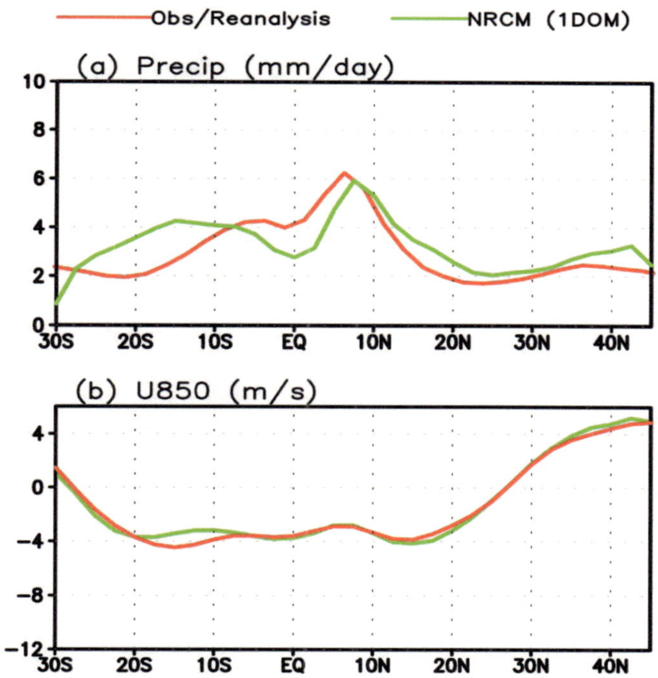

Fig. 3. Latitudinal distribution of (a) Precipitation (mm/day) and (b) U850 (m/s), averaged over 0°-360° from the observation/reanalysis (red) and NRCM (green).

3.2 MJO

A space-time spectrum analysis is performed on the filtered time series of U850 and P to compare the eastward and westward propagating intraseasonal (20-90 day) signal (Fig. 4). A necessary criterion for the MJO is the dominance of the eastward propagating power over its westward propagating counterpart at the intraseasonal and planetary scales. In the observations (Fig. 4, left), the eastward spectral power dominates its westward counterpart at the MJO space and time scales, but not quite so in the simulation (Fig. 4, right), particularly for P (Fig. 4d). The simulated MJO signal in P (Fig. 4d) is much weaker than that in U850 (Fig. 4c) in comparison to the observation. This discrepancy indicates a lack of physical-dynamical coherence in the NRCM simulation. This is consistent with the mean U850 and P in Fig. 1. The results are similar using other variables.

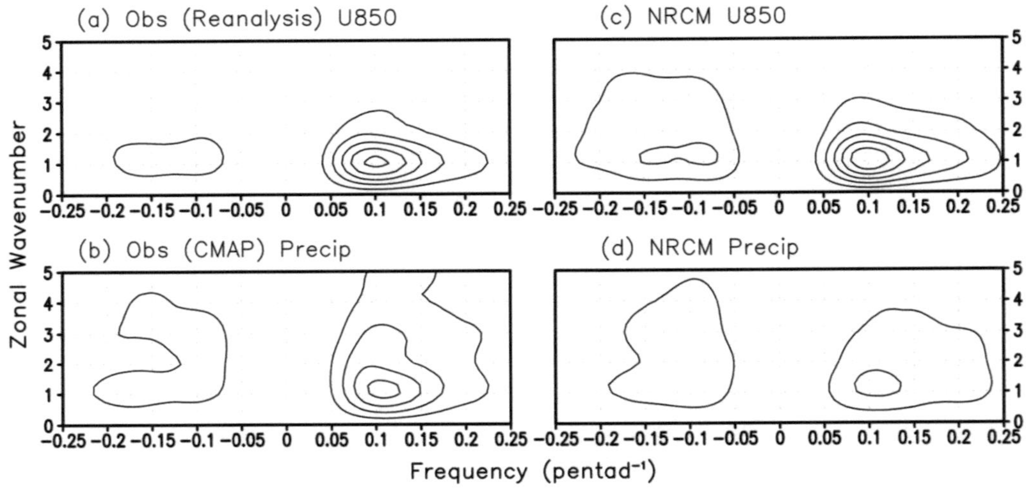

Fig. 4. Time-Space spectra for (a) U850 and (b) precipitation from the observation. The right panels are for the model. Zonal wavenumber 1, and frequency 0.1 (50 days), represent the dominant MJO scales. All are averaged over 10°S-10°N.

To further explore the MJO in the NRCM, the longitudinal variation of the MJO variance of U850 and P are shown in Fig. 5. For U850, over the Indian Ocean, the variance is underestimated, particularly near the equator (5°N-5°S, Fig. 5, right). However, when a larger area is considered (15°S-15°N, Fig. 5, left), the differences between the observation and the model become smaller. Over the west Pacific, however, the model overestimates the MJO variance in U850. For P, the MJO variance is greatly underestimated over the Indian Ocean (Fig. 6), particularly over the western Indian Ocean, where most MJO initiation occurs. This is consistent with the lack of precipitation over the equatorial Indian and west Pacific Ocean as shown in Fig. 1.

It is natural to enquire how the MJO simulation in the NRCM compares with those in GCM simulations. A quick comparison with GCM simulations reveals that the MJO in the NRCM is not better than those in GCMs. This is less than satisfactory considering that the model is forced by time-varying reanalysis boundary conditions. As a result, we further diagnoss the role of the mean state in the simulated MJO statistics.

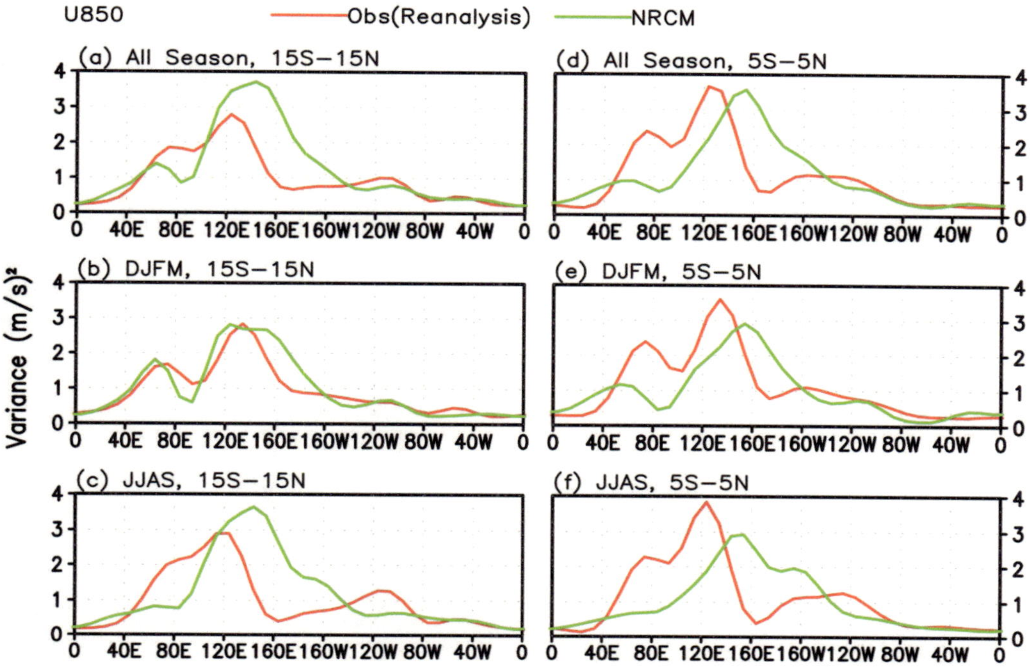

Fig. 5. (left) Variance of U850 averaged over 15°S-15°N during (a) all season, (b) boreal winter (DJFM), and (c) boreal summer (JJAS). Right panels are averaged over 5°S-5°N.

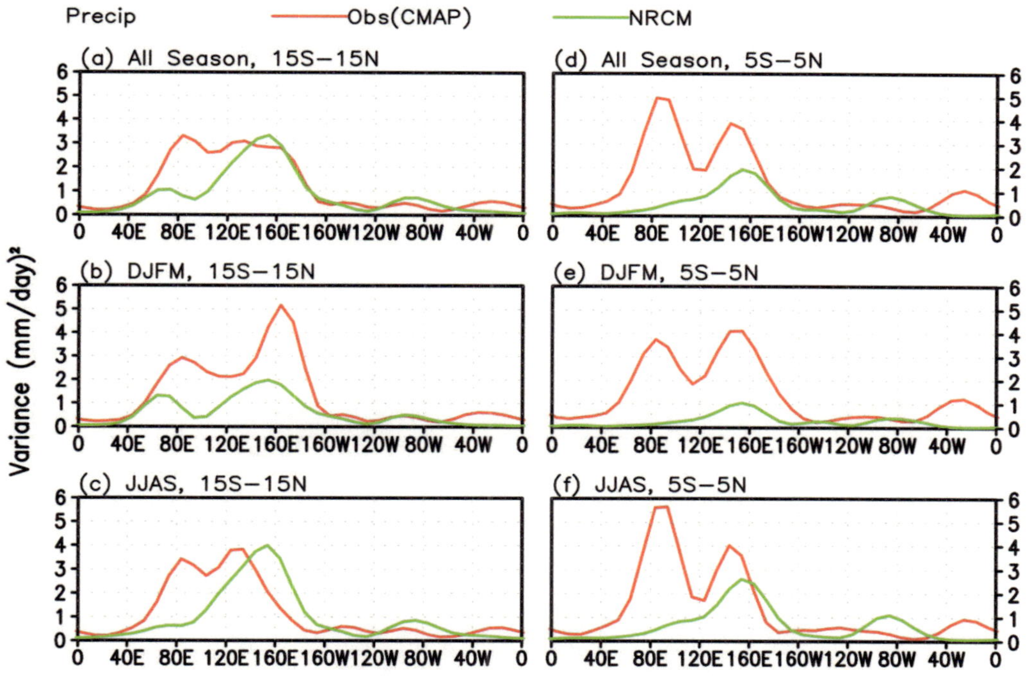

Fig. 6. (left) Variance of P averaged over 15°S-15°N during (a) all season, (b) boreal winter, and (c) boreal summer. Right panels are averaged over 5°S-5°N.

3.3 Role of the mean state on the MJO

Role of the mean state on the simulated MJO is described with respect to U850 and Precipitation. The MJO is represented by the variances of U850* and P*. Figs. 7 and 8 show the role of mean U850 on the U850* variance from the observation and model, respectively. The MJO variance (contoured) and the westerlies (yellow hues) are reasonably collocated in the reanalysis (Fig. 7), but not quite as well in the NRCM (Fig. 8), particularly over the equatorial Indian Ocean. During the boreal winter, simulated westerlies and the MJO variance (Fig. 8a) are stronger and located further from the equator compared to the reanalysis (Fig. 7a). This is the season when the MJO is strongest (Zhang and Dong, 2004). During the boreal summer, the observed variance of the MJO is located north of the equator (Fig. 7b). The simulated variance during the summer in the northern Indian Ocean is greatly reduced in the simulation (Fig. 8b).

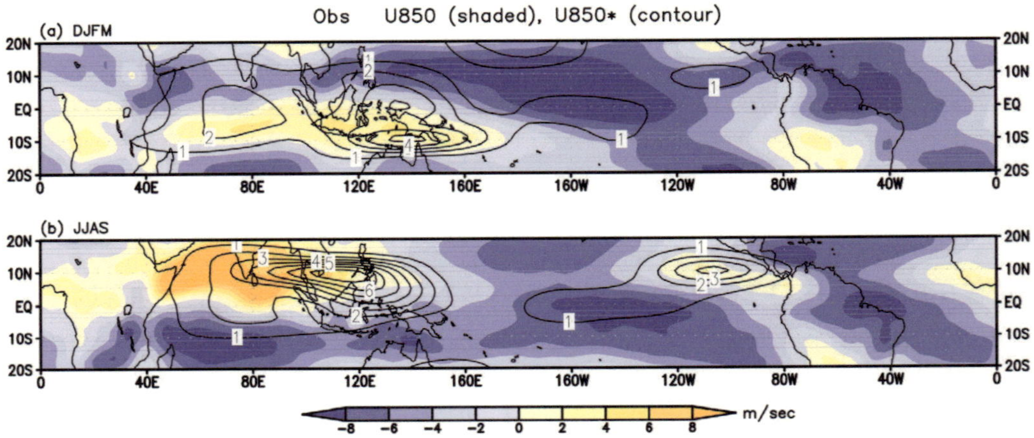

Fig. 7. Mean U850 (m s^{-1}, shaded) and variance of U850* (m^2 s^{-2}, contoured) from the NCEP-NCAR reanalysis during the (a) boreal winter (DJFM), and (b) boreal spring (JJAS). Contour intervals are 1 m^2 s^{-2}.

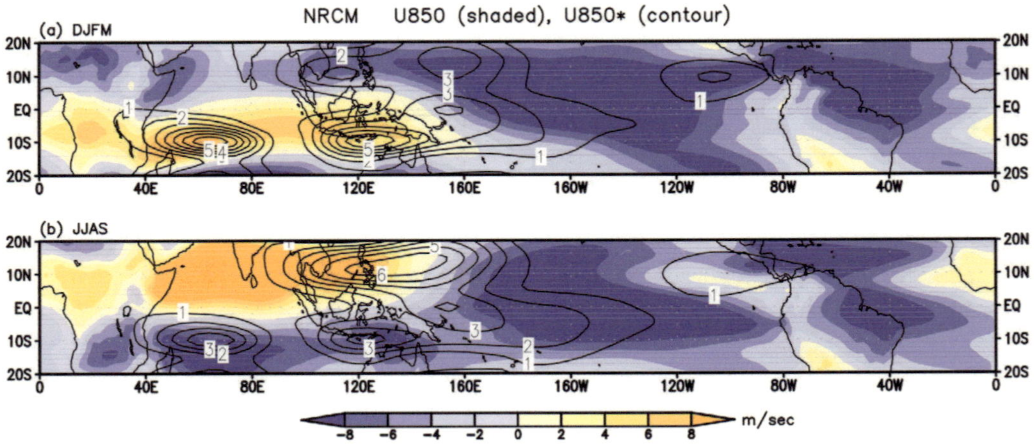

Fig. 8. Same as Fig. 7, but for the model.

Fig. 9 shows the role of the observed mean precipitation on the P* variance. During the boreal winter (Fig. 9a), the P* variance is over the southern hemisphere with three peaks, one over the Indian Ocean, and the other two over the west Pacific. The P* variance is always very well collocated with the stronger mean precipitation. This cannot be said for the NRCM simulation (Fig. 10a), in particular, the P* variance seems to avoid the equator. During the summer, the observed P* variance is in the northern hemisphere (Fig. 9b), however the model produces spurious variance in the SPCZ region and the eastern Pacific. Note that the P* variance is very small over the equatorial Indian Ocean due to the lack of precipitation in that region in the model. This is consistent with the mean annual precipitation (Fig. 1) and the spectrum (Fig. 4) indicating the role of the mean state on the simulated MJO. Next, we describe how the P* variance is affected by the mean distribution of U850.

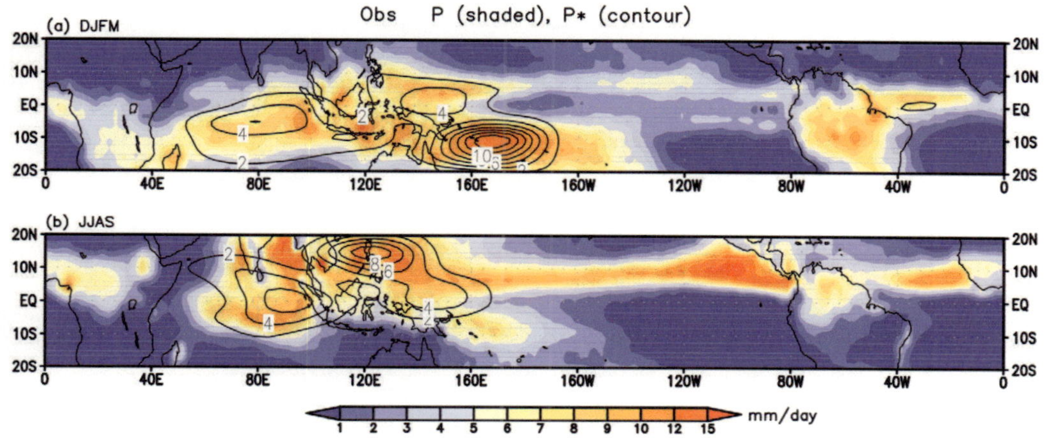

Fig. 9. Mean P (mm day^{-1}, shaded) and variance of P* (mm^2 day^{-2}, contoured) from the observation (CMAP) during the (a) boreal winter (DJFM), and (b) boreal spring (JJAS). Contour intervals are 2 mm^2 day^{-2}.

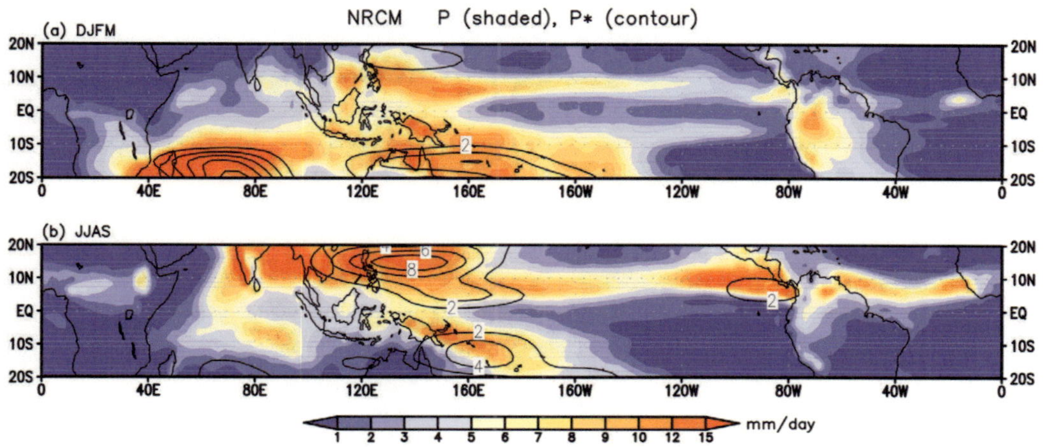

Fig. 10. Same as Fig. 9, but for the model.

Fig. 11 shows the distribution of observed P* variance (contoured) and the mean U850 (shaded). The observed P* maxima always follow the positive U850 or very weak zonal flow in both seasons. Latitudinal migration of mean U850 and P* are more prominent over the west Pacific than over the Indian Ocean. The amplitude of variance is also larger over the west Pacific. The model, however, does not reproduce the observation well (Fig. 12). Variance of P* seems to avoid the westerlies in both seasons. This is one of the most disturbing aspects of the simulated MJO in the NRCM. The larger values of P* variance avoids the equatorial region in the simulation. During the boreal summer, the model reproduces spurious P* variance over the eastern Pacific and in the SPCZ region (Fig. 12b) that is absent in the observation (Fig.11b). It seems that P* variance follows the mean precipitation (Fig. 10), and not the mean westerlies. This indicates a lack of coupling between the convection and circulation in the model.

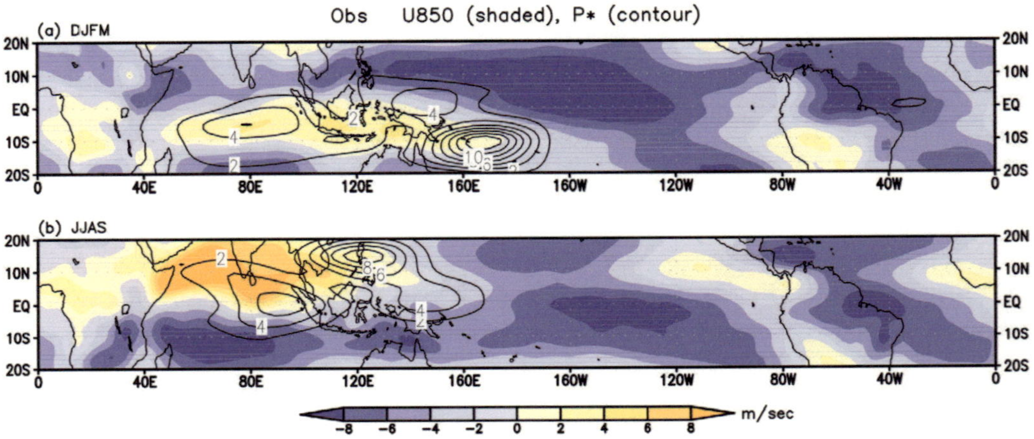

Fig. 11. Mean U850 (m sec^{-1}, shaded) and variance of P* (mm^2 day^{-2}, contoured) from the observation during the (a) boreal winter (DJFM), and (b) boreal spring (JJAS). Contour intervals are 2 mm^2 day^{-2}.

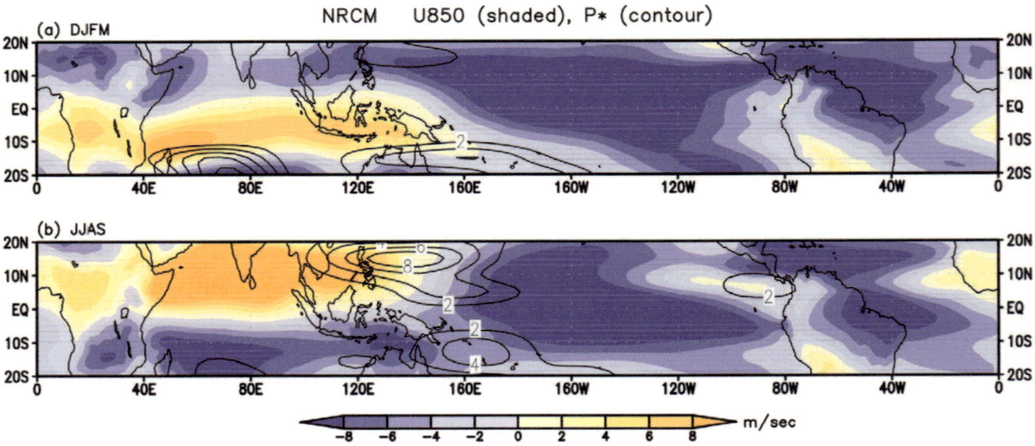

Fig. 12. Same as Fig. 11, but for the model.

4. Conclusion

A nested regional climate model (NRCM) is constructed at the NCAR based on Weather Research and Forecasting (WRF) model. This is also known as a tropical channel model (TCM), and is conceptually similar to the TCM developed at the University of Miami based on MM5. Both TCMs are useful tools to study the MJO dynamics and its initiation.

With the initial and lateral boundary conditions provided by a global reanalysis, the NRCM is integrated for several years. The simulated MJO statistics in the NRCM are not better than those found in the GCMs. This is less than satisfactory considering that the model is forced by time-varying reanalysis boundary conditions. Further diagnoses reveal that the error in the mean state is a reason for the poor MJO statistics in the simulation. For example, the MJO variance and the westerlies in the lower-troposphere are well collocated in the reanalysis, but not quite as well in the NRCM, particularly over the equatorial Indian Ocean where the initiation of the MJO events usually occur. The model also lacks precipitation in the equatorial Indian Ocean. The large error in the precipitation (through modifying the latent heating) must have inhibited any dynamical effects from the lateral boundaries from reaching the interior of the domain. Thus, the lateral boundary conditions couldn't participate effectively in simulating the mean conditions.

However, the multi-year simulation with large error in the mean state was able to capture two individual MJO events that were initiated by the extratropical influences (Ray et al., 2011a). In other words, the negative effect of mean state error can be overcome if there are extra dynamical influences, either from the meridional boundary conditions or initial conditions. Note that, it is not known to what extent the error in the mean state inhibits tropical variability, although it is likely to be model dependent.

The large error in the precipitation over the southern Indian Ocean was thought to be due to the interactions between tropical cyclones and the southern boundaries. To rectify this problem, southern boundaries were further moved to 45°S in another experiment. This simulation also has more vertical levels (55 levels instead of 35) and higher model top at 10 hPa level (instead of 50 hPa). However, this did not improve the result significantly, indicating potential problems with the model physics (Tulich et al., 2011; Murthi et al., 2011). Use of nested domains inside the model also did not improve the mean state (Ray et al., 2011).

In a regular regional model, the domain size is vital for the model mean state through the influence of boundary conditions. For example, a small domain may lead to very little "climate error" because the model is fundamentally controlled by its boundary conditions. On the other hand, the mean state in a global model would be less constrained. The NRCM lies between the regular regional model and the global model. Thus, climate drift in the NRCM simulation would not be noticeable in the smaller regional domains used by Gustafson and Weare (2004a, b) and Monier et al. (2009). How much error in the mean state is sufficient to prevent the initiation of an MJO in the model is not known; arguably, it is event dependent. Thus a systematic study for multiple MJO events including several "primary" (no prior MJO, Matthews, 2008) and "successive" (with prior MJO) events is needed to have a better idea of the effect of mean state on the MJO.

Is the poor skill of the NRCM to simulate MJO due to shortcomings from the cumulus parameterization (Park et al. 1990; Raymond and Torres, 1998; Wang and Schlesinger, 1999;

Maloney and Hartmann, 2001)? Or do we need further increase in the model resolution (Hayashi and Golder, 1986; Gualdi et al., 1997; Grabowski and Moncrieff, 2001; Inness et al., 2001; Liess and Bengtsson, 2004)? The use the Betts-Miller-Janjic (Janjic, 1994) scheme did not improve the MJO simulation. Similarly, higher resolution nested domains inside the NRCM made minor differences. Further works need to be done to investigate the roles of cumulus parameterization and horizontal resolution on the simulated MJO.

The lack of MJO in the NRCM does not necessarily imply a lack of tropical-extratropical interaction. For example, if the observed source of perturbations that eventually initiate an MJO event is located inside the model domain, then the lateral boundary conditions may not be effective beyond the MJO predictability limit. As a result, the locations of the meridional boundaries of the NRCM are crucial for capturing the extratropical influences, if any, on MJO dynamics.

The NRCM is an atmosphere only model forced by the SST without true oceanic feedback. Therefore, it is difficult to rule out the role of coupled air-sea feedbacks in modulating the mean state (Hendon, 2000; Zheng et al. 2004; Vitart et al. 2007; Woolnough et al. 2007). Pegion and Kirtman (2008a, b) found that air-sea coupling was responsible for differences in the simulation of the MJO between the coupled and uncoupled models, specifically in terms of organization and propagation in the western Pacific. The role of intraseasonally varying SST was found to be important to the amplitude and propagation of the oscillation beyond the Maritime continent in their model. After removing the intraseasonally varying component in the SST and lateral boundary conditions in MM5, Gustafson and Weare (2004b) found only minor differences in the MJO simulation compared to the simulation forced with observed SST. Ray et al. (2009) also reported that use of constant SST did not influence the MJO initiation in the Indian Ocean. These results indicate that the MJO amplitude and propagation are influenced by the air-sea interactions whose effect is dominant over the Pacific.

In short, we have shown that the erroneous mean state may be responsible for poor MJO simulation in the model. Our results call for further research attention towards using the untapped potential of high-resolution models in the MJO simulation and forecasting.

5. Acknowledgment

Acknowledgment is made to the NCAR, which is sponsored by the National Science Foundation, for making the NRCM model output available. The NCEP-NCAR reanalysis data were taken from the NOAA/CDC.

6. References

Ajayamohan, R.S. & Goswami, B.N. (2007). Dependence of simulation of boreal summer tropical intraseasonal oscillations on the simulation of seasonal mean. *J. Atmos. Sci.*, Vol.64, 460-478

Chen, F. & Dudhia, J. (2001). Coupling an advanced land surface-hydrology model with the Penn State-NCAR MM5 modeling system, Part I: Model implementation and sensitivity. *Mon. Wea. Rev.*, Vol.129, pp. 569-585

Chen, S.S.; Houze, R.A. & Mapes, B.E. (1996). Mult-scale variability of deep convection in relation to large-scale circulation in TOGA-COARE. *J. Atmos. Sci.*, Vol.53, pp. 1380-1409

Collins, W.D.; Bitz, C.M.; Blackmon, M.L.; Bonan, G.B.; Bretherton, C.S.; Carton, J.A.; Chang, P.; Doney, S.C.; Hack, J.J.; Henderson, T.B.; Kiehl, J.T.; Large, W.G.; McKenna, D.S.; Santer, B.D. & Smith, R.D. (2006). The Community Climate System Model: CCSM3. *J. Clim.*, Vol.19, pp. 2122-2143

Done, J.M.; Holland, G.J. & Webster, P.J. (2011). The role of wave energy accumulation in tropical cyclogenesis over the tropical north Atlantic. *Clim. Dyn.*, Vol.36, pp. 753-767, doi 10.1007/s00382-010-0880-5

Dudhia, J. (1993). A nonhydrostatic version of the Penn State-NCAR Mesoscale Model: Validation tests and simulation of an Atlantic cyclone and cold front. *Mon. Wea. Rev.*, Vol.121, pp. 1493-1513

Ferranti, L.; Palmer, T.N.; Molteni, F. & Klinker, K. (1990). Tropical-extratropical interaction associated with the 30-60 day oscillation and its impact on medium and extended range prediction. *J. Atmos. Sci.*, Vol.47, pp. 2177-2199

Grell, G.A.; Dudhia, J. & Stauffer, D.R. (1995). A description of the fifth-generation Penn-State/NCAR Mesoscale Model (MM5). *NCAR/TN-398.*

Grabowski, W.W. & Moncrieff, M.W. (2001). Large-scale organization of tropical convection in two-dimensional explicit numerical simulations. *Quart. J. Roy. Meteor. Soc.*, Vol.127, pp. 445-468

Gualdi, S. ; Navarra, A. & Von Starch, H. (1997). Tropical intraseasonal oscillation appearing in operational analyses and in a family of general circulation models. *J. Atmos. Sci.*, Vol.54, pp. 1185-1202

Gustafson, W.I. & Weare, B.C. (2004a). MM5 modeling of the Madden-Julian oscillation in the Indian and west Pacific Oceans: Model description and control run results. *J. Clim.*, Vol.17, pp. 1320-1337

Gustafson, W.I. & Weare, B.C. (2004b). MM5 modeling of the Madden-Julian oscillation in the Indian and west Pacific Oceans: Implications of 30-70 day boundary effects on MJO development. *J. Clim.*, Vol.17, pp. 338-1351

Hayashi, Y. & Golder, D.G. (1986). Tropical intraseasonal oscillation appearing in the GFDL general circulation model and FGGE data. Part I : Phase propagation. *J. Atmos. Sci.*, Vol.43, pp. 3058-3067

Hendon, H.H. & Liebmann, B. (1990). The intraseasonal (30-50 day) oscillation of the Australian summer monsoon. *J.* Atmos. *Sci.*, Vol.47, pp. 2909-2923

Hendon, H.H. (2000). Impact of air-sea coupling on the Madden-Julian Oscillation in a general circulation model. *J. Atmos. Sci.*, Vol.57, pp. 3939-3952

Hong, S.-Y.; Dudhia, J. & Chen, S.-H. (2004). A revised approach to ice microphysical processes for the bulk parameterization of clouds and precipitation. *Mon. Wea. Rev.*, Vol.132, pp. 103-120

Hong, S.-Y.; Noh, Y. & Dudhia, J. (2006). A new diffusion package with an explicit treatment of entrainment processes. *Mon. Wea. Rev.*, Vol.134, pp. 2318-2341

Houze, R.A. (2004). Mesoscale convective systems. *Rev. Geophys.*, Vol.42, RG4003, doi:10.1029/2004RG000150

Inness, P.M.; Slingo, J.M.; Woolnough, S.J.; Neale R.B. & Pope, V.D. (2001). Organization of tropical convection in a GCM with varying vertical resolution: Implications for the simulation of the Madden-Julian oscillation. *Clim. Dyn.*, Vol.17, pp. 777-793

Inness, P.M.; Slingo, J.M.; Guilyardi, E. & Cole, J. (2003). Simulation of the Madden-Julian oscillation in a coupled general circulation model. Part II: The role of the basic state. *J. Clim.*, Vol.16, 365-382

Janjic, Z.I. (1994). The step-mountain coordinate model: Further development of the convection, viscous sublayer, and turbulence closure schèmes. *Mon. Wea. Rev.*, Vol.122, pp. 927-945

Jones, C., & Schemn, J.-K.E. (2000). The influence of intraseasonal variations on medium-range weather forecasts over South America. *Mon. Wea. Rev.*, Vol.128, pp. 486-494

Kain, J.S. (2004). The Kain-Fritsch convective parameterization: An update. *J. Appl. Meteorol.*, Vol.43, pp. 170-181

Kalnay, E. & Coauthors (1996). The NCEP-NCAR 40-year reanalysis project. *Bull. Amer. Meteor. Soc.*, Vol.77, pp. 437-471

Kikuchi K. & Takayabu, Y.N. (2003). Equatorial circumnavigation of moisture signal associated with the Madden-Julian Oscillation (MJO) during boreal winter. *J. Met. Soc. Jpn.*, Vol.81, No.4, pp. 851-869

Kim, D. & Coauthors (2009). Application of MJO simulation diagnostics to climate models. *J. Clim.*, Vol.22, pp. 6413-6436

Knutson, R.R.; Weickmann, K.M. & Kutzbach, J.E. (1986). Global-scale intraseasonal oscillations of outgoing longwave radiation and 250 mb zonal wind during northern hemisphere summer, *Mon. Wea. Rev.*, Vol.114, pp. 605-623

Knutson, R.R. & Weickmann, K.M. (1987). 30-60 day atmospheric oscillations: Composite life-cycles of convection and circulation anomalies. *Mon. Wea. Rev.*, Vol.115, pp. 1407-1436

Krishnamurti, T.N.; Jayakumar, P.K.; Sheng, J.; Surgi, N. & Kumar, A. (1985). Divergent circulations on the 30 to 50 day time scale, *J. Atmos. Sci.*, Vol.42, pp. 364-375

Lau, K.-M. & Chan, P.H. (1985). Aspects of the 40-50 day oscillation during the northern winter as inferred from outgoing longwave radiation, *Mon. Wea. Rev.*, Vol.113, pp. 1889-1909

Liebmann, B.; Hendon, H.H. & Glick, J.D. (1994). The relationship between tropical cyclones of the western Pacific and Indian Oceans and the Madden-Julian Oscillation, *J. Meteor. Soc. Jpn.*, Vol.72, pp. 401-411

Liess, S. & Bengtsson, L. (2004). The intraseasonal oscillation in ECHAM4 part II: Sensitivity studies. *Clim. Dyn.*, Vol.22, pp. 671-688, doi :10.1007/s00382-004-0407-z

Lin, J.-L., & Coauthors (2006). Tropical intraseasonal variability in 14 IPCC AR4 climate models. Part I: Convective signals. *J. Clim.*, Vol.19, pp. 2665-2690

Madden, R.A. & Julian, P.R. (1971). Detection of a 40-50 day oscillation in the zonal wind in the tropical Pacific. *J. Atmos. Sci.*, Vol.28, pp. 702-708

Madden, R.A. & Julian, P.R. (1972). Description of global-scale circulation cells in the tropics with a 40-50 day period. *J. Atmos. Sci.*, Vol.29, pp. 1109-1123

Maloney, E.D. (2009). The moist static energy budget of a composite tropical intraseasonal oscillation in a climate model. *J. Clim.*, Vol.22, pp. 711-729

Maloney, E.D. & Hartman, D.L. (2001). The sensitivity of intraseasonal variability in the NCAR CCM3 to changes in convective parameterization. *J. Clim.*, Vol.14, pp. 2015-2034

Matthews, A.J. (2008). Primary and successive events in the Madden-Julian oscillation. *Quart. J. Roy. Meteor. Soc.*, Vol.134, pp. 439-453

Moncrieff, M.W. (2010). The multi-scale organization of moist convection and the interaction of weather and climate, In D.-Z. Sun and F. Bryan (Eds.), Climate Dynamics: Why Does Climate Vary? *Geophysical monograph series*, Vol.189, American Geophysical Union, Washington DC, 3-26, doi:10.1029/2008GM000838

Monier, E.; Weare, B.C. & Gustafson, W.I. (2009). The Madden-Julian oscillation wind-convection coupling and the role of moisture processes in the MM5 model. *Clim. Dyn.*, doi:10.1007/s00382-009-0626-4

Murthi, A.; Bowmann, K.P. & Leung, L.R. (2011). Simulations of precipitation using NRCM and comparisons with satellite observations and CAM: annual cycle. *Clim. Dyn.* (in press)

North, G.R.; Bell, T.L.; Cahalan, R.F. & Moeng, F.J. (1982). Sampling errors in the estimation of empirical orthogonal functions. *Mon. Wea. Rev.*, Vol.110, pp. 699-710

Park, C.-K.; Strauss, D.M. & Lau, K.-M. (1990). An evaluation of the structure of tropical intraseasonal oscillations in three general circulation models. *J. Meteorol. Soc. Jpn.*, Vol.68, pp. 403-417

Pegion, K. & Kirtman, B.P. (2008a). The impact of air-sea interactions on the simulation of tropical intraseasonal variability. *J. Clim.*, Vol.21, pp. 6616-6635

Pegion, K. & Kirtman, B.P. (2008b). The impact of air-sea interactions on the predictability of tropical intraseasonal variability. *J. Clim.*, Vol.21, pp. 5870-5886

Ray, P.; Zhang, C.; Dudhia, J. & Chen S.S. (2009). A numerical case study on the initiation of the Madden-Julian oscillation. *J. Atmos. Sci.*, Vol.66, pp. 310-331

Ray, P. & Zhang, C. (2010). A case study of the mechanics of extratropical influence on the initiation of the Madden-Julian oscillation. *J. Atmos. Sci.*, Vol.67, pp. 515-528

Ray, P.; Zhang, C.; Moncrieff, M.W.; Dudhia, J., Caron, J., Leung, R. & Bruyere, C. (2011). Role of the atmospheric mean state on the initiation of the Madden-Julian oscillation in a tropical channel model. *Clim. Dyn.*, Vol.36, pp. 161-184, doi 10.1007/s00382-010-0859-2

Ray, P., Zhang, C. ; Dudhia, J. ; Li, T. & Moncrieff, M.W. (2012) Tropical channel model, InTech publication, *Climate Models*, ISBN 979-953-307-338-4

Raymond, D.J. & Torres, D.J. (1998). Fundamental moist modes of the equatorial troposphere. *J. Atmos. Sci.*, Vol.55, pp. 1771-1790

Slingo, J.M. & Coauthors (1996). Intraseasonal oscillations in 15 atmospheric général circulation models: Results from an AMIP diagnostic subproject. *Clim. Dyn.*, Vol.12, pp. 325-357

Taylor, K.E.; Williamson, D. & Zwiers, F. (2000). The sea surface temperature and sea-ice concentration boundary conditions for AMIP II simulations. PCMDI Report No. 60 and UCRL-MI-125597, 25 pp

Tulich, S.N.; Kiladis, G.N. & Suzuki-Parker, A. (2011). Convectively coupled Kelvin and easterly waves in a regional climate simulation of the tropics. *Clim. Dyn.*, Vol.36, pp. 185-203, doi 10.1007/s00382-009-0697-2

Vitart, F.; Woolnough, S.J.; Balmaseda, M.A. & Tompkins, A.M. (2007). Monthly forecast of the Madden-Julian oscillation using a coupled GCM. *Mon. Wea. Rev.* Vol.135, pp. 2700-2715

Wallace, J.M. ; Smith, C. & Bretherton, C.S. (1992). Singular value decomposition of wintertime sea surface température and 500-mb height anomalies. *J. Clim.*, Vol.5, pp. 561-576

Wang, W.Q. & Schlesinger, M.E. (1999). The dependence on convective parameterization of the tropical intraseasonal oscillation simulated by the Uiuc 11-layer atmospheric GCM. *J. Clim.*, Vol.12, pp. 1423-1457

Weickmann, K.; Kiladis G.N., & Sardeshmukh, P. (1997). The dynamics of intraseasonal atmospheric angular momentum oscillations. *J. Atmos. Sci.*, Vol.54, 1445-1461

Woolnough, S.J.; Vitart, F. & Balmaseda, M. (2007). The role of the ocean in the Madden-Julian oscillation: sensitivity of an MJO forecast to ocean coupling. *Quart. J. Royal. Meteor. Soc.*, Vol.133, pp. 117-128

Xie, P. & Arkin, P.A. (1997). Global precipitation: A 17-year monthly analysis based on gauge observations, satellite estimates, and numerical model outputs. *Bull. Amer. Meteor. Soc.*, Vol.78, pp. 2539-2558

Yasunari, T., (1979). Cloudiness fluctuations associated with the Northern Hemisphere summer monsoon, *J. Meteor. Soc. Jpn.*, 57, 227-242

Zhang, C. & Dong, M. (2004). Seasonality of the Madden-Julian oscillation. J. *Clim.*, Vol.17, pp. 3169-3180

Zhang, C. (2005). Madden-Julian oscillation. *Rev. Geophys.*, Vol.43, RG2003, doi:10.1029/2004RG000158.

Zhang, C.; Dong, M.; Gualdi, S.; Hendon, H.H.; Maloney, E.D.; Marshall, A.; Sperber, K.R. & Wang, W. (2006). Simulations of the Madden-Julian oscillation by four pairs of coupled and uncoupled global models. *Clim. Dyn.*, DOI: 10.1007/s00382-006-0148-2

Zheng, Y.; Waliser, D.E.; Stern, W.F. & Jones, C. (2004). The role of coupled sea surface temperatures in the simulation of the tropical intraseasonal oscillation. *J. Clim*, Vol.17, pp. 4109-4134

5

The JMA Nonhydrostatic Model and Its Applications to Operation and Research

Kazuo Saito
Meteorological Research Institute
Japan

1. Introduction

Nonhydrostatic models were initially developed as research tools for small scale meteorological phenomena. Today, several nonhydrostatic models have been developed and applied to numerical simulations and operational numerical weather prediction (NWP). In this chapter, we review the Japan Meteorological Agency (JMA) nonhydrostatic model (JMA-NHM, hereafter referred to as NHM) and its applications to operational forecasts and research fields. Section 2 presents a brief history of the model development from a research tool to a full-scale NWP model. In section 3, we review applications of the model to several research fields of various time/spatial scales from tornado to regional climate modelling, and mesoscale data assimilation and ensemble prediction studies. Section 4 introduces on-going relevant topics including the Japanese next generation supercomputer project.

2. The JMA nonhydrostatic model

2.1 Development of NHM at MRI

The JMA nonhydrostatic model (NHM) was first developed as a research tool at the Meteorological Research Institute (MRI). Ikawa (1988) developed a nonhydrostatic model with orography and compared computational schemes with a 2-dimensional numerical experiment. Following Gal-Chen and Somerville (1975), the terrain-following vertical coordinates

$$z^* = \frac{H(z - z_s)}{H - z_s}, \quad (1)$$

and the metric tensors for the coordinate transformations were employed:

$$G^{\frac{1}{2}} = 1 - \frac{z_s}{H}, \qquad G^{\frac{1}{2}}G^{13} = (\frac{z^*}{H} - 1)\frac{\partial z_s}{\partial x}, \qquad G^{\frac{1}{2}}G^{23} = (\frac{z^*}{H} - 1)\frac{\partial z_s}{\partial y}, \quad (2)$$

where z_s is the surface height and H is the model top height.

2.1.1 Anelastic equation model

The first version of NHM used the anelastic (AE) scheme to solve the Navier-Stokes momentum equations for a fluid. The AE model removes sound waves from solutions in the equation system by scale analysis (Ogura and Phillips, 1962). Field variables are divided into the time independent horizontal uniform reference state $\phi\,(z)$ and its perturbation $\phi'\,(x, y, z, t)$ as

$$p = \bar{p} + p', \quad \rho = \bar{\rho} + \rho', \quad \theta = \bar{\theta} + \theta' \ . \tag{3}$$

Following Clark (1977), the continuity equation was given by

$$DIVT = \frac{\partial U}{\partial x} + \frac{\partial V}{\partial y} + \frac{\partial W^*}{\partial z^*} = 0, \tag{4}$$

where U, V, and W are momentum multiplied by the metric tensor as the prognostic variables in the model:

$$U = \bar{\rho} G^{\frac{1}{2}} u, \quad V = \bar{\rho} G^{\frac{1}{2}} v, \quad W = \bar{\rho} G^{\frac{1}{2}} w, \tag{5}$$

$$W^* = \frac{1}{G^{\frac{1}{2}}} (W + G^{\frac{1}{2}} G^{13} U + G^{\frac{1}{2}} G^{23} V) \ . \tag{6}$$

Taking the total divergence of the momentum equation yields a 3-dimensional Poisson-type pressure diagnostic equation, which was solved by the Dimension Reduction Method.

This model was further evolved by including a bulk cloud microphysics scheme based on Lin et al. (1983), a turbulent closure model based on Deardorff (1980), and treatment surface processes for sea and land. Comprehensive documentation was published in the Technical Report of MRI (Ikawa and Saito, 1991) as a nonhydrostatic model developed at the Forecast Research Department of MRI.

2.1.2 Nested model

Ikawa and Saito's (1991) model was modified to a nested model (MRI-NHM) to realistically simulate mesoscale phenomena (Saito, 1994). For the dynamical core, the AE scheme was adopted. Variational calculus (Sherman, 1978) was implemented to obtain a non-divergent mass consistent initial field, where the continuity equation (4) was used as the strong constraint to modify the interpolated wind field. Orlanski's (1976) radiation condition was employed with the time-dependent lateral boundary condition, and mass fluxes through the lateral boundaries were adjusted to maintain mass conservation.

A hydrostatic version of MRI-NHM was developed by Kato and Saito (1995) and was used to examine the applicability of hydrostatic approximation to a high-resolution simulation of moist convection.

2.1.3 Fully compressible version with a map factor

Saito (1997) developed a semi-implicit, fully compressible version of MRI-NHM including a map factor, where linearization using the reference atmosphere was removed. Density was defined by the sum of masses of moist air and the water substances per unit volume as

$$\rho \equiv \rho_d + \rho_v + \rho_c + \rho_r + \rho_i + \rho_s + \rho_g \ , \tag{7}$$

where subscripts *c, r, i, s* and *g* stand for the cloud water, rain, cloud ice, snow, and graupel, respectively. ρ_d is the density of dry air and ρ_v that of water vapor.

Introducing a map factor *m*, the continuity equation is given by

$$G^{\frac{1}{2}} \frac{\partial \rho}{\partial t} + DIVT = PRC, \tag{8}$$

where *DIVT* is the total divergence in z* coordinate and *U, V,* and *W** are defined by

$$DIVT = m^2 \left(\frac{\partial U}{\partial x} + \frac{\partial V}{\partial y}\right) + m \frac{\partial W^*}{\partial z^*}, \tag{9}$$

$$U = \frac{\rho G^{\frac{1}{2}} u}{m}, \quad V = \frac{\rho G^{\frac{1}{2}} v}{m}, \quad W = \frac{\rho G^{\frac{1}{2}} w}{m}, \tag{10}$$

$$W^* = \frac{1}{G^{\frac{1}{2}}} \{W + m(G^{\frac{1}{2}} G^{13} U + G^{\frac{1}{2}} G^{23} V)\}. \tag{11}$$

PRC in (8) is the fall-out of water substances written in *z** coordinate:

$$PRC = \frac{\partial}{\partial z^*} (\rho_a V_r q_r + \rho_a V_s q_s + \rho_a V_g q_g), \tag{12}$$

where ρ_a is the density of moist air, and Vr, Vs *and* Vg the terminal velocity of precipitable water substances (rain, snow and graupel, respectively). The state equation is given as the diagnostic equation for density:

$$\rho = \frac{p_0}{R\theta_m} \left(\frac{p}{p_0}\right)^{C_v/C_p}, \tag{13}$$

where θ_m is the mass-virtual potential temperature defined by

$$\theta_m = \theta(1 + 0.608 q_v)(1 - q_c - q_r - q_i - q_s - q_g). \tag{14}$$

In the AE model, the buoyancy term was computed from the perturbations of potential temperature and pressure, but in the fully compressible model, it is directly computed from the density perturbation.

The pressure equation is obtained from (8) and (13) as

$$\frac{\partial P}{\partial t} = C_m{}^2 (PFT - DIVT + PRC), \tag{15}$$

where C_m is the sound wave speed and *PFT* is the local time tendency of the mass-virtual potential temperature.

To stabilize the acoustic mode, the HI-VI (Horizontally Implicit-Vertically Implicit) scheme, which treats sound waves implicitly for both vertical and horizontal directions, was used. This scheme, often referred as the *semi-implicit* method, was first implemented by Tapp and White (1976). A 3-dimensional Helmholtz pressure equation, which is formally similar to the Poisson equation in the AE model, is obtained by HI-VI treatment of sound waves. For details, see Saito (1997) and Saito et al. (2007).

2.1.4 MRI/NPD-NHM

In 1999, a cooperative effort to develop a community model for NWP and research started between the Numerical Prediction Division (NPD) of JMA and MRI. The HE-VI (Horizontally Explicit-Vertically Implicit) scheme, which solves a vertically 1-dimensional Helmholtz-type pressure equation, was re-implemented into MRI-NHM by Muroi et al. (2000). Code parallelization of the model was also performed to cope with the distributed memory parallel computers. A flux limiter advection correction scheme (Kato, 1998) was implemented to reduce numerical errors due to the finite difference approximation and to assure monotonicity. In the cloud physics, the box Lagrangian scheme (Kato 1995) was introduced to assure computational stability for sedimentation of rain.

A comprehensive description of the unified model (MRI/NPD-NHM) was published in the Technical Report of MRI (Saito et al., 2001a).

2.2 Operational applications

2.2.1 Development at NPD/JMA

In 2001, full-scale development of an operational nonhydrostatic mesoscale model at JMA (NHM) was started at NPD in collaboration with MRI. Several modifications were added to enhance computational efficiency, robustness, and accuracy for an operational NWP model.

Higher-order (third to fifth) advection schemes that consider a staggered grid configuration were implemented by Fujita (2003). The fourth-order scheme was chosen for the operational forecasting, considering computational cost and matching with the advection correction scheme. A time-splitting scheme of gravity waves and advection terms was implemented by Saito (2003). In this scheme, higher-order advection terms with advection correction were fully evaluated at the center of the leapfrog time step; the lower-order (second-order) components at each short time step were then adjusted in the latter half of the leapfrog time integration. In the continuity equation, the diffusion of water vapor in unit time, which includes sub-grid scale turbulent mixing and computational diffusion, was considered by Saito (2004) as

$$G^{\frac{1}{2}}\frac{\partial \rho}{\partial t}+DIVT=PRC+\rho DIFq_v. \tag{16}$$

This term was implemented to consider the surface evaporation of water vapor, which offsets the loss of mass by precipitation in total mass conservation.

As for physical processes, the Kain-Fritsch convective parameterization scheme (K-F scheme; Kain and Fritsch, 1993) was implemented with modification by Yamada (2003).

Several points in the K-F scheme were revised to improve its performance as a mesoscale NWP model in Japan, where a moist and unstable maritime air mass prevails in summer. The targeted moisture diffusion (TMD) was implemented to attenuate the grid-point storms and the associated intense grid-scale precipitation (Saito and Ishida, 2005), where an artificial second-order horizontal diffusion is applied to water vapor when strong upward motions exist.

2.2.2 Operational MSM

On 1 September 2004, JMA replaced the former hydrostatic mesoscale model (hydrostatic MSM) with NHM (a nonhydrostatic MSM). Eighteen-hour forecasts were run four times a day to support disaster prevention and the very short-range forecast of precipitation at JMA. A domain of 3600 km x 2880 km that covers Japan and its surrounding areas was taken. Vertically, 40 levels with variable grid intervals were employed, where the model top was located at 22 km and the lowest level was 20 m.

Initial conditions of horizontal wind, temperature, water vapor, and surface pressure were given by the JMA Meso 4D-Var (Koizumi et al., 2005) six-hourly analyses, as in the former hydrostatic MSM. Initial conditions of cloud microphysical quantities were given by the six-hourly forecast-forecast cycle. Details of the 10 km nonhydrostatic MSM are given in Saito et al. (2006).

2.2.3 Upgrade of MSM after 2006

In March 2006, horizontal and vertical resolutions of the nonhydrostatic MSM were enhanced from 10 km L40 to 5 km L50. The model operation was also increased from four times a day (six hourly) to eight times (three hourly), to provide more frequent forecasts. Several modifications were added to physical processes. In the atmospheric radiation scheme of the 10 km MSM, a cloud was assumed as a black body and cloud optical properties were given by empirical constants. In the new radiation scheme, cloud optical properties were determined by cloud water/ice contents and their effective radius, which reduced the negative bias of the upper air temperature. Convective parameterization was still required at 5 km, because without convective parameterization the model overestimated intense rain and underestimated weak to moderate rains. In the K-F scheme, the following revisions were made (Ohmori and Yamada 2006): (1) conversion from convective condensate to rain was reduced, (2) time scales of deep and shallow convection were reduced, and (3) threshold values for conversion from cloud water/ice to precipitation were increased.

In May 2007, JMA extended the forecast time of MSM from 15 to 33 hours at the initial times of 03, 09, 15 and 21 UTC to provide information for disaster prevention up to 24 hours (Hara et al., 2007). To modify the dynamics, the generalized hybrid vertical coordinate,

$$\zeta = z - z_s f(\zeta) \,, \tag{17}$$

was introduced, which approaches the z^* coordinate near the surface and the z coordinate near the model top (Ishida, 2007). Here, $f(0) = 1$, $f(H) = 0$, *and* $f(\zeta)$ is determined so that $\partial z/\partial\zeta$ is positive and the second derivative of $f(\zeta)$ is differentiable. In the cloud microphysics, fall-out of cloud ice was considered in addition to rain, snow, and graupel in order to prevent excessive accumulation of cloud ice in the upper model atmosphere. In the K-F scheme for convection, perturbation depending on relative humidity was added in the trigger function to reduce the overestimation of convective rain induced by orography. For the turbulent model, a Mellor and Yamada level-3 closure model (MYNN3; Nakanishi and Niino, 2004; 2006) was implemented first as the operational NWP model to reduce model bias in the boundary layer (Hara, 2007). In addition to the prognostic turbulent kinetic energy (*TKE*), fluctuations of liquid water potential temperature ($\theta_l'^2$), total water content ($q_w'^2$), and their correlation ($\theta_l' q_w'$) were treated as prognostic variables. To evaluate the degree of cloudiness, partial condensation computed by the probability density function in MYNN3 was considered. Results of these modifications are given in Saito et al. (2007)

A nonhydrostatic 4D-Var data assimilation system (JNoVA; JMA Nonhydrostatic model based Variational data assimilation system) (Honda et al., 2005) was implemented in April 2009 to supply MSM more accurate initial conditions (Honda and Sawada, 2008). The horizontal resolution of the 4D-Var inner-loop model was enhanced from 20 km of Meso 4D-Var to 15 km in JNoVA.

2.2.4 QPF performance of MSM

Figure 1 plots the quantitative precipitation forecast (QPF) performance of MSM since it began actual operation (March 2001) to November 2011. In this figure, threat scores averaged for FT=3 to 15 for moderate rain with a threshold value of 5 mm in 3 hours are indicated. The verification grid size is 20 km. The averaged score in 2001 was about 0.2, but the score improved year by year, and the latest score approaches 0.4. Given the fact that statistical PQF performance of high resolution regional models is sometimes notoriously bad due to the difficulty of predicting mesoscale precipitation and the *double penalty* problem, this threat score improvement is remarkable.

Table 1 lists the main modifications added to the operational mesoscale NWP at JMA from 2001 to 2011. In addition to the modifications discussed in the former subsections, the global positioning system (GPS)-derived total precipitable water vapor (TPWV) data (Ishikawa, 2010) has been assimilated since October 2009, and the 1D-Var retrieved water vapor data from radar reflectivity (Ikuta and Honda, 2011) has been used since June 2011. These modifications have contributed to the recent improvement of the QPF performance of MSM through improving water vapor analysis.

2.2.5 Mesoscale tracer transport model

The GPV of the operational MSM is used as input to the atmospheric transport model at MRI and JMA. This mesoscale ATM takes a Lagrangian scheme (Seino et al., 2004) with many tracer particles that follow advection, horizontal and vertical diffusion, fallout, and dry and wet deposition processes. JMA incorporated photochemical oxidant information in June 2007 (Takano et al., 2007) and the tephra fall forecast in March 2008 (Shimbori et al., 2010).

3. Applications to research

3.1 Mountain flow

As described in section 2.1, NHM was first developed as a research tool at MRI. Ikawa and Nagasawa (1989) conducted a numerical experiment for a dynamically induced foehn event

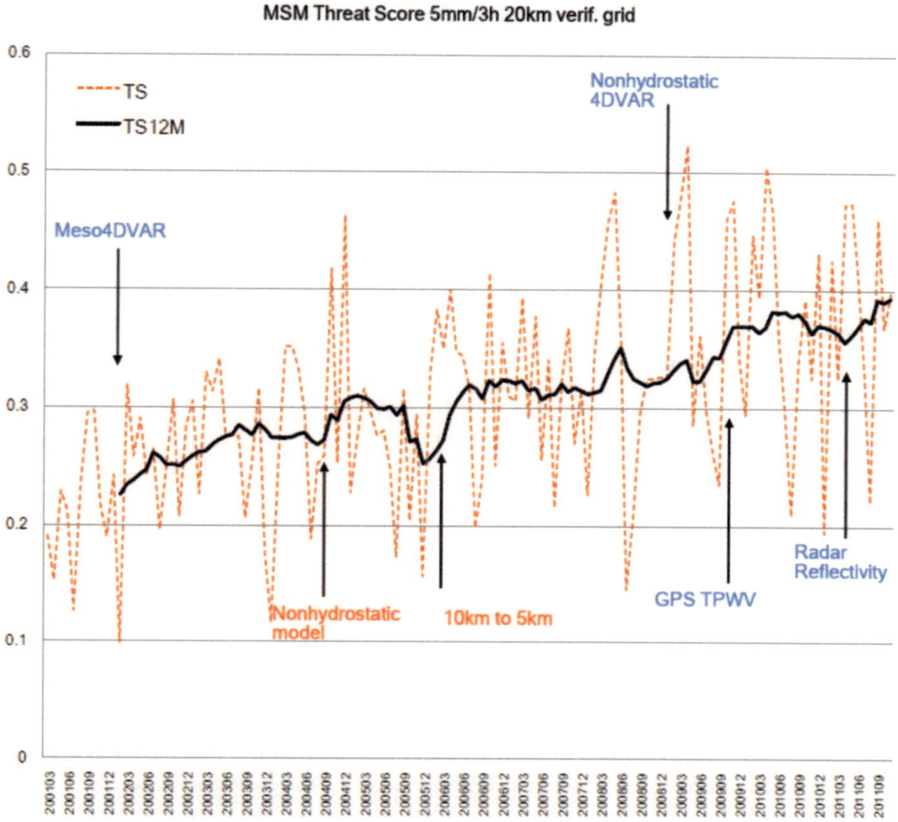

Fig. 1. Threat score of MSM for three-hour precipitation averaged for FT = 3 to 15 with a threshold value of 5 mm/3 hour from March 2001 to November 2011. The red broken line denotes the monthly value, while the black solid line indicates the 12-month running mean. Courtesy of NPD/JMA.

Year. Month	Modification
2001. 3	Start of Mesoscale NWP (10kmL40+OI)
2001. 6	Wind profiler data
2002. 3	Meso 4D-Var
2003. 10	SSM/I microwave radiometer data
2004. 7	QuikSCAT Seawinds data
2004. 9	Nonhydrostatic model
2005. 3	Doppler radar radial winds data
2006. 3	Enhancement of model resolution (5kmL50)
2007. 5	Upgrade of physical processes
2009. 4	Nonhydrostatic 4D-Var
2009. 10	GPS total precipitable water vapor (TPWV) data
2011. 6	Water vapor data retirieved from radar reflectivity

Table 1. Modifications for operational mesoscale NWP at JMA.

observed in Hokkaido, northern Japan, using the 2-dimensional AE model. Inspired by their works, Saito and Ikawa (1991) conducted 2-dimensional simulation of the local downslope wind *Yamaji-kaze* in Shikoku, western Japan. The averaged orography of Shikoku Island in the east-west direction was regarded as the typical orography, and the development and propagation of an internal hydraulic jump were simulated by a numerical experiment using the observed thermal stratification and time-changing wind profile.

Saito (1993) conducted numerical experiments using the real orography of Shikoku Island with the surface friction, and studied the geographical characteristics of the Yamaji-kaze. Smith's (1980) linear analytic solution of the mountain flow over an isolated mountain was extended to the flow over a mountain range with a col, and compared with the non-linear aspect of the simulated flow.

Saito (1994) developed a double-nested model to reproduce the Yamaji-kaze of the 27 September 1991 windstorm. A realistic simulation of the observed phenomena was first conducted in Japan using a nested nonhydrostatic model with a horizontal resolution of 2.5 km. Good agreement between the simulation and the observed time evolution of surface wind was obtained when a strong typhoon approached western Japan.

The Kii Peninsula, central Japan, is famous for its abundant rainfall which reaches 3000 to 4000 mm a year. Airflow over the Kii Peninsula and its relation to the orographic enhancement of rainfall has been studied by Saito et al. (1994) and Murata (2009). Saito et al. (1998) compared the Deutscher Wetterdienst nonhydrostatic regional model (DWD LM) and MRI-NHM for numerical solutions of the 3-dimensional mountain waves, focusing on the computational efficiency of HI-VI and HE-VI schemes. Fujibe et al. (1999) studied diurnal wind variation in the lee of a mountain range using MRI-NHM and demonstrated agreement with the daytime advance of downslope wind in the Canterbury Plains in New Zealand.

A model intercomparison of mountain flow over a steep mountain was conducted by Satomura et al. (2003) as the Steep Mountain Model Intercomparison Project (St-MIP). To examine the accuracy of the terrain-following coordinates, mountain waves over two-dimensional bell-shaped mountains with various half-widths and heights were compared with theoretical calculations and among models, including NHM.

3.2 Tropical meteorology

3.2.1 Tropical convection and heavy rainfall

A model intercomparison for a tropical squall line observed during the Ttropical Ocean/Global Atmosphere Coupled Ocean-Atmosphere Response Experiment (TOGA COARE) was conducted in the GEWEX Cloud Systems Study (GCSS). Redelsperger et al. (2000) compared results from eight cloud-resolving models including NHM. Most of the models were able to predict similar rainfall and integrated water content evolutions and agreed quantitatively, but the apparent heat and moisture sources indicated some quantitative differences.

The diurnal evolution of tropical island convection observed during the Maritime Continent Thunderstorm Experiment (MCTEX; Keenan et al., 2000) was simulated by Saito et al (2001b) using NHM nested with the Australian Bureau of Meteorology Research Centre's

Limited-Area Assimilation and Prediction System (BMRC LAPS). The left panel of Fig. 2 presents the visible satellite images over Melville and Bathurst Islands, Northern Territory of Australia, on 27 November 1995. At 1200 CST, shallow convective clouds corresponding to Rayleigh–Benard convection covered the interior of the islands. Along the southern coasts of the two islands, line-shaped clouds suggest organization associated with the sea breeze front (SBF). At 1300 CST, the clouds merged and organized at the central part of the islands in the form of an east–west line. An hour later (1400 CST), deep convection (*Hector*) developed at the southwestern part of Melville Island and along the southern coast of Bathurst Island. The right panel of Fig. 2 indicates the corresponding simulation with a 1 km NHM. Details of the observed evolution of the clouds on this day (Rayleigh–Benard convection, cloud merger along the convergence zone between the two SBFs, and succeeding explosive growth of deep convection) were very well reproduced.

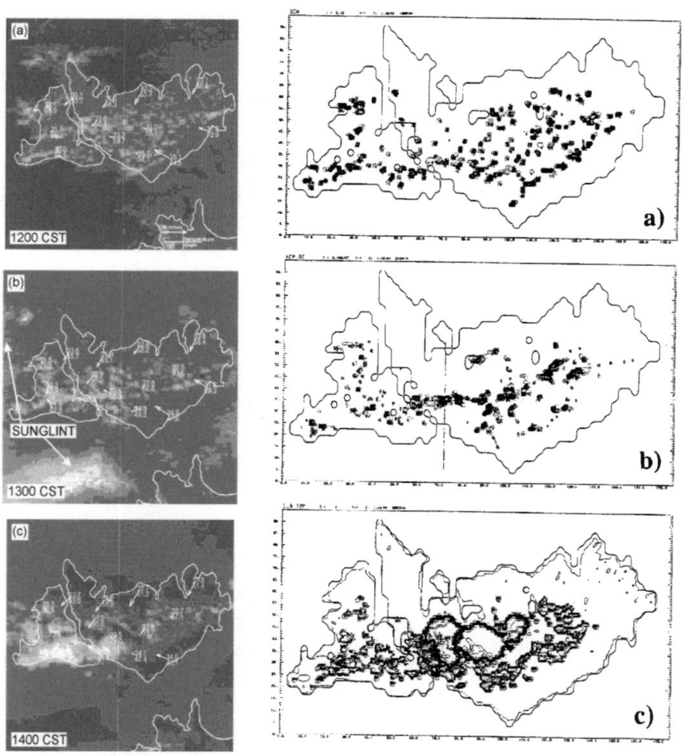

Fig. 2. Left: Visible image on 27 Nov 1995: (a) at 0230 UTC (1200 CST), (b) at 0330 UTC (1300 CST), and (c) at 0430 UTC (1400 CST). Right: Fields derived from the 1 km NHM simulation. (a) Cloud water mixing ratio at z = 1.46 km and t = 180 min (1300 CST). Contour interval is 0.1 g Kg^{-1}. (b) Vertically accumulated cloud water at t = 240 min (1400 CST). Contour interval is 0.1 Kg m^{-2}. (c) Simulated cloudtop temperature at t = 300 min (1500 CST). Contour interval is 5 K. After Saito et al. (2001b).

On 26 July 2005, an intense rainfall system caused heavy rain in excess of 900 mm at Mumbai, on the west coast of India. This system was simulated by Seko et al. (2008) using the global analysis data of JMA as the initial condition of NHM. A maximum rainfall exceeding 1,100 mm in 17 hours was reproduced by the simulation with a horizontal resolution of 5 km. The detailed structure of the rainfall system at the intense rain stage was also investigated by NHM with a horizontal resolution of 1 km and sensitivity experiments.

A flood was caused by heavy rainfall that lasted for several days from late January to early February 2007 in Jakarta and its vicinity. Trilaksono et al. (2012) investigated spatiotemporal modulation of precipitation using time-lagged ensemble. The National Centers for Environmental Prediction (NCEP) Global Analyses with a horizontal resolution of 1° × 1° was used for the initial and boundary conditions of NHM with a horizontal resolution of 20 km. They demonstrated that the event associated with a cold surge was preceded by a Borneo vortex event. Trilaksono et al. (2011) conducted downscaling experiments for the Jakarta Flood, and studied the dependence of heavy precipitation simulated by the model on the horizontal resolution. The downscale runs with higher resolutions (2, 4, and 5 km) demonstrated the ability to reproduce a region of strong convective activity to the north of Java Island during the flood.

Seko et al. (2012) investigated generation mechanisms of convection cells in tropical regions. Convection cells near Sumatra Island were reproduced using the reanalysis data of the JMA Climate Data Assimilation System (JCDAS) and downscale experiments with a horizontal grid interval of 1 km. Updrafts of gravity waves that were trapped in the lower atmosphere triggered new convection cells.

3.2.2 Tropical cyclone and tornado

Murata et al. (2003) was the first to apply MRI-NHM to a typhoon. They numerically simulated the major spiral rainband in typhoon Flo (T9019), which was the subject of an international model intercomparison (COMPARE III; Nagata et al., 2001), with several horizontal resolutions (20, 14, 10, 7 and 5 km). The effects of precipitation schemes and horizontal resolution on the representation of the simulated rainband were studied.

Effects of ice-phase processes and evaporation from raindrops on the development and structures of tropical cyclones were studied by Sawada and Iwasaki (2007; 2010). Wada (2009; 2012) and Wada et al. (2010) used an atmosphere-wave-ocean coupled version of NHM to examine the effect of the ocean on tropical-cyclone intensity and structural change. Wong et al. (2010) developed a new parameterization scheme of air-sea momentum, heat, and moisture fluxes, considering the saturation properties of bulk transfer coefficients under a high winds regime, and tested its impact on tropical cyclone intensity. Kanada et al. (2012) compared PBL schemes and examined their impact on the development of intense tropical cyclones.

In May 2008, Cyclone Nargis hit southern Myanmar and claimed more than 100,000 lives there in one of the largest meteorological disasters in Southeast Asia, mainly due to the storm surge. Numerical modeling studies of this event were performed as part of the research project "International Research for Prevention and Mitigation of Meteorological Disasters in Southeast Asia" (see 3.2.3). Kuroda et al. (2010) conducted a forecast experiment of Nargis and its associated storm surge using NHM with a horizontal resolution of 10 km and the Princeton Ocean Model (POM) with a horizontal resolution of 3.5 km. The impact of SST in the Bay of Bengal and ice phase on Nargis' rapid development was also examined.

The greatest typhoon damage in Japan was caused by 'Isewan Typhoon' (Vera) and its associated storm surge in 1959. Kawabata et al. (2012) performed a reanalysis experiment of the typhoon using JNoVA, assimilating drop sonde observations taken by the US Air Force.

Mashiko et al. (2009) performed a numerical simulation of tornadogenesis in an outer-rainband minisupercell of a typhoon using NHM with a grid spacing of 50 m.

3.2.3 International research for prevention and mitigation of meteorological disasters in Southeast Asia

In 2007, Kyoto and MRI started the project "International Research for Prevention and Mitigation of Meteorological Disasters in Southeast Asia" (http://www-mete. kugi.kyoto-u.ac.jp/project/MEXT/) with institutions in southeast Asia. This project was supported for fiscal years 2007 through 2009 by the Asia S & T Strategic Cooperation Program of the Ministry of Education, Culture, Sports, Science and Technology of Japan (MEXT) Special Coordination Funds for Promoting Science and Technology. Numerical experiments were performed at the Institut Technologi Bandung (ITB) in Indonesia, Hong Kong Observatory, Nanyang Technological University in Singapore, National Center for Hydro-Meteorological Forecasting of Vietnam, and CSIR Centre for Mathematical Modelling and Computer Simulation in India.

MRI was a major participating institution in Japan, and was responsible for NWP model development and application. Case studies with downscale prediction and statistical verifications of forecast accuracy around Japan and Southeast Asia were performed. Hayashi et al. (2008) compared NHM and the Weather Research Forecasting (WRF) model for two seasons (July 2007 and January 2008) with the same initial and boundary conditions and the same domain size and resolutions. Tools were prepared for numerical experimentations with NHM using the JMA global analysis, the global model forecast, and the JMA one-week global ensemble forecast. Saito et al. (2010a) conducted ensemble predictions of Nargis and the associated storm surge employing perturbations from the JMA one-week global ensemble forecast. Ensemble prediction results were used as input data for the decision support system developed by Kyoto University (Otsuka and Yoden, 2011).

Data assimilation experiments were conducted by modifying the JMA Meso 4D-VAR system to apply to tropical areas. A tropical cyclone (TC) bogus procedure was developed for the Bay of Bengal, and the impact on TC forecasts was investigated by Kunii et al. (2010b). Near-real-time (NRT) analysis of precipitable water vapor using the international ground based GPS network around the Bay of Bengal was performed to demonstrate its positive impact on the Nargis forecast (Shoji et al., 2011).

A comprehensive description of this international project including newsletters, the users' guide to the decision support tool, and links to related published papers has been published as a technical report of MRI (Saito et al., 2011b).

3.3 Cloud simulations

3.3.1 Winter monsoon snow clouds over the Sea of Japan

Ikawa et al. (1991) implemented a bulk parameterization scheme of cloud microphysics to MRI-NHM, and performed a numerical simulation of the convective snow cloud over the Sea of Japan. Sensitivity experiments to ice nucleation rates were conducted to simulate the cloud more realistically and to examine the effects of an increase in the number concentration of ice crystals on the formation of the convective snow cloud.

Saito et al. (1996) conducted sensitivity experiments on the orographic snowfall over the mountainous region of northern Japan, and investigated the orographic effect on the snowfall from cloud microphysical aspects. A 2-dimensional model with a horizontal resolution of 2 km was employed. In the experiments with full cloud microphysics, the precipitation amount over the land increased significantly with mountain heights exceeding the height of the cloud base. However, in the experiments with a warm-rain process, the precipitation amount was only 1/3 that of the experiments with an ice phase. A seeding experiment in which ice nucleation rates were enhanced over a specified zone in the Sea of Japan demonstrated the possibility of artificial modification of the snowfall. A further seeding experiment was conducted by Hashimoto et al. (2008), and seedability of the winter orographic snow clouds over northern Japan was assessed using a nested NHM with a horizontal resolution of 1 km.

Saito (2001) conducted a numerical simulation of cloud bands during a cold air outbreak over the Sea of Japan with a horizontal resolution of 3 km, and showed outstanding similarity between the satellite image and simulated clouds. Four nodes of the HITAC SR8000 of MRI were employed. A higher resolution (1 km) simulation of the snow clouds over the southern coastal area of the Japan Sea was conducted by Eito et al. (2005). Eito et al. (2010) extended the simulation area to (2000 km)^2 and examined the structure and formation mechanism of transversal cloud bands using the Earth Simulator.

Yanase et al. (2002) conducted high-resolution simulation of an observed polar low over the Japan Sea on 21 January 1997 with the 2-km horizontal resolution MRI-NHM. The simulation successfully reproduced the observed features of the polar low (*e.g.*, its horizontal scale, movement, spiral bands, and a cloud-free eye). Detailed three-dimensional structures of the simulated polar low were clarified, and its development mechanism was investigated. Futher siumulations of the polar low were conducted by Yanase et al. (2004) and Yanase and Niino (2007).

3.3.2 Maritime boundary layer clouds and spectral bin methods

Nagasawa et al. (2006) conducted numerical simulation of *Yamase* clouds, typical maritime boundary-layer clouds over the sea off the east coast of northern Japan. NHM with a high resolution of 100 m was employed to simulate convective structures observed from satellite images.

Iguchi et al. (2008) implemented a bin-based microphysics scheme for cloud into NHM to reproduce realistic and inhomogeneous condensation nuclei (CN) fields. Nested simulations were performed for two precipitation events over an area of the East China Sea, where the general features of the horizontal distributions of variables (*e.g.*, effective droplet radius derived from satellite data retrieval) were reproduced. Iguchi et al. (2012) evaluated the bin-based cloud microphysical scheme through comparison with observation data by shipborne Doppler and spaceborne cloud profiling radars.

3.4 Regional climate modelling and urban simulation

A regional climate model version of NHM employing spectral boundary coupling (SBC) was developed by Yasunaga et al. (2005). Forty-day simulations with a horizontal grid interval of 5 km were performed using the Earth Simulator in the MEXT Research Revolution 2002 project. Kanada et al. (2005) used this version of NHM and examined structure of mesoscale convective systems (MCSs) during the late Baiu season in the global warming climate. Sasaki et al. (2008) performed five-year integrations of the non-hydrostatic regional climate model (NHRCM) with a 4 km grid interval, and successfully reproduced monthly precipitation, seasonal change, and regional features in Japan. The relationship between precipitation and elevation in the present climate by NHRCM was compared with observation by Sasaki and Kurihara (2008). The climatological features of warming events over Toyama Plain, central Japan, were investigated with NHRCM by Ishizaki and Takayabu (2009).

A single-layered square prism urban canopy (SPUC) scheme for NHM was developed by Aoyagi and Seino (2011). The SPUC run more accurately reproduced the expected behavior of the urban canopy effect than the slab run did.

3.5 Mesoscale data assimilation and ensemble forecast

As shown in Fig. 1, QPF performance of operational mesoscale NWP at JMA has been remarkably improved in these ten years for weak to moderate rains. However, there are still many difficulties in producing predictions of severe mesoscale phenomena that specify their intensity, location, and timing. Convective rains without strong synoptic or orographic forcing are still very hard to predict due to the smallness of their spatiotemporal scales and their sensitivity to the small perturbations in the initial condition. To overcome these difficulties, research and development on mesoscale data assimilation and ensemble forecast have been conducted.

3.5.1 Mesoscale assimilation of GPS data

Water vapor is one of the most important parameters in weather forecasting, and GPS is a powerful tool for retrieving accurate water vapor information. In Japan, the Geospatial Information Authority of Japan has been operating a nationwide permanent dense GPS array, the GPS Earth Observation Network (GEONET), since 1994. Shoji (2009) developed a near real-time (NRT) analysis system of TPWV derived from GEONET data to contribute to water vapor monitoring and NWP. Ishikawa (2010) applied this system to the operational GPS TPWV data assimilation at JMA using JNoVA and contributed to improve the MSM's QPF performance (see 2.2.4).

On 28 July 2008, a local heavy rainfall occurred over the Hokuriku and Kinki districts, central Japan. Shoji et al. (2009) used Meso 4DVAR to perform data assimilation experiments of GEONET TPWV data for this event, and demonstrated that the rainfall forecast was improved. Further improvements were obtained by adding TPWV derived from GPS stations of the International GNSS Service in Korea and China.

Seko et al. (2010) used the Meso 4D-Var system for a heavy rainfall event in northern Japan on 16 July 2004 to investigate the impacts of three kinds of GPS-derived water vapor data: TPWV, slant water vapor along the path from the GPS satellite to the receiver (SWV), and radio occultation (RO) data along the path from the GPS satellite to the CHAllenging Minisatellite Payload (CHAMP) satellite. When SWV and RO data were assimilated simultaneously, both the rainfall region and rainfall amount were similar to the observed ones.

3.5.2 Storm scale 4DVAR

Although JNoVA contributed to improve the QPF performance of operational MSM, its horizontal grid spacing in the inner loop model is 15 km and only large-scale condensation is considered in the adjoint model. Kawabata et al. (2007) developed a cloud-resolving 4DVAR system based on the JMA-NHM (NHM-4DVAR), and applied it to reproduce the deep convection observed in Tokyo on 21 July 1999. The cost function to be minimized was formulated as

$$J(x_0) = \frac{1}{2}(x_0 - x_0{}^b)^{\mathrm{T}} \mathbf{B}^{-1}(x_0 - x_0{}^b) + \frac{1}{2}(Hx - y^o)^{\mathrm{T}} \mathbf{R}^{-1}(Hx - y^o), \tag{18}$$

where x denotes the model prognostic variables, subscript 0 means those at the beginning time of the assimilation window, and $x_0{}^b$ is the first guess of x_0. **B** represents the background error covariance matrix. H is the observation operator, y^o is the observations, and **R** the observation error covariance matrix. Lateral boundary conditions were also included in the analysis. The adjoint model included only dry dynamics and advection of water vapor, but they successfully reproduced observed cumulonimbi by assimilating Doppler radar radial winds and GPS TPWV data. This study was the first to demonstrate the feasibility of short-range forecasting of local heavy rainfall brought about by deep convection, using a full-scale numerical model and a dense observation network.

Kawabata et al. (2011) implemented the warm rain process into NHM-4DVAR and applied it to a data assimilation experiment of a heavy rainfall event in Tokyo on 5 September 2005 with a horizontal resolution of 2 km. GPS-TPWV data was assimilated at 5-min intervals within the 30-min assimilation windows, and surface in-situ data and wind profiler data were assimilated at 10-min intervals. Doppler radial winds and radar-reflectivity data were assimilated at 1-min intervals. The 4DVAR assimilation reproduced a line-shaped mesoscale convective system (MCS) with a shape and intensity consistent with the observation (Fig. 3). Assimilation of radar-reflectivity data intensified the MCS and suppressed false convection.

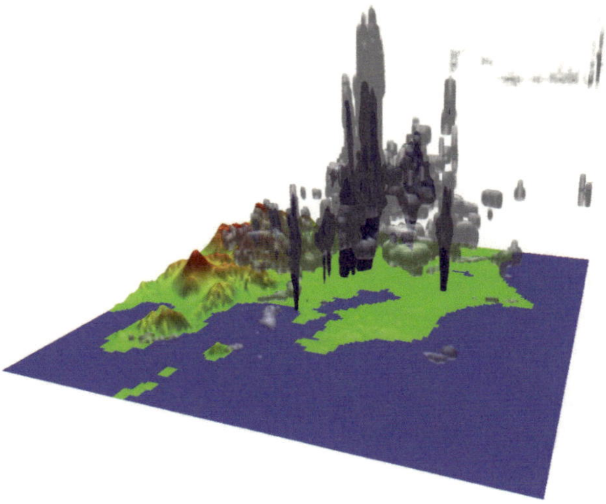

Fig. 3. Three-dimensional image of the MCS in Tokyo on 5 September 2005 reproduced by NHM-4DVAR.

3.5.3 Mesoscale ensemble prediction

Another reason of difficulty in predicting local heavy rainfall is the inherent low predictability of severe small-scale phenomena, that occur under convectively unstable atmospheric conditions. To consider forecast errors due to uncertainties in initial conditions and numerical models, ensemble prediction systems (EPSs) are widely used for medium range NWP, and there is now an increasing need to develop mesoscale EPSs.

Seko et al. (2009) conducted 11-member mesoscale ensemble experiments using NHM with a horizontal resolution of 15 km for two tornado events in Japan. Initial conditions were produced by adding the normalized perturbation of operational one-week ensemble forecast of JMA. Probabilities that the Energy Helicity Index (EHI) exceeds some criteria were examined, and feasibility of the probability forecast of tornadoes using the EPS-derived potential parameters was indicated.

As discussed in 3.2.3, Saito et al. (2010a) conducted ensemble predictions of Nargis and the associated storm surge. NHM with a horizontal scale of 10 km was used for the 21 member ensemble prediction. In addition to the initial perturbations, the effect of lateral boundary perturbations (LBPs) on tropical cyclone ensemble prediction was examined. When LBPs were implemented, the ensemble spread increased by 50%, and root mean square errors (RMSEs) of the ensemble mean forecast became smaller than without LBPs.

A mesoscale singular vector (MSV) method was developed by Kunii (2010a). The tangent linear model and the adjoint model of NHM developed for JNoVA (Honda et al., 2005) were used in the Lanczos method with Gram-Schmidt re-orthogonalization to obtain the singular vector. Kunii (2010b) tested the MSV for a heavy rainfall event to assess MSV performance as the initial perturbation method. A torrential rainfall that occurred over central Japan on 29 August 2008 (Fig. 3a) was chosen. Ensemble forecast with a horizontal resolution of 15 km was conducted, where probabilistic values were determined by the proportion of members that predicted precipitation above a certain threshold. A greater-than-30% probability of precipitation was estimated even with the large threshold of 50 mm/3 hours (Fig. 3c), whereas the control forecast (Fig. 3b) predicted rainfall of only about 20 mm/3 hours.

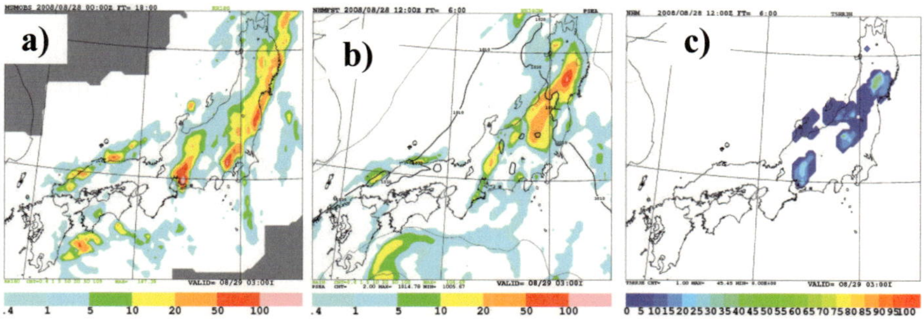

Fig. 4. a) 3-hour accumulated precipitation at 1800 UTC 28 August 2008 (Radar AMeDAS analyzed rainfall). b) Rainfall predicted by a control forecast (FT = 06). c) Probability forecasts of 3-hour accumulated precipitation with a threshold of 50 mm. After Kunii (2010b).

Kunii and Saito (2009) conducted a sensitivity analysis experiment using MSV to support the THORPEX Pacific Asian Regional Campaign (T-PARC). For TY0813 (SINLAKU), MSVs-based sensitivity areas for the typhoon were located on the right side of the moving direction of the typhoon, which was dominated by potential energy components in the mid-lower troposphere. Compared with global model singular vectors (GSVs), MSVs reflected small scale structures that affect mesoscale disturbances.

A mesoscale EPS system using MSV and GSV is under development at JMA (Ono et al., 2011), assuming operational application.

3.5.4 WWRP Beijing Olympics 2008 RDP project

An international research project of the World Weather Research Programme (WWRP), Beijing 2008 Olympics Research and Development Project (B08RDP; Duan et al. 2012) was conducted in conjunction with the Beijing 2008 Olympic Games. The main component of B08RDP was short-range forecasting of up to 36 h based on mesoscale EPSs with a horizontal resolution of 15 km. Six institutions from five countries [MRI, NCEP, Meteorological Service of Canada (MSC), Central Institute for Meteorology and Geodynamics (ZAMG) of Austria, National Meteorological Center (NMC) of CMA, and Chinese Academy of Meteorological Sciences (CAMS)] participated in the project, and were requested to run their EPSs for a forecast time of up to 36 h, starting every day at 1200 UTC.

Prior to the 2008 intercomparison period (one month, from 25 July to 23 August 2008), MRI developed five initial perturbation methods: (1) a downscaling of JMA's operational one-week EPS (WEP; Saito et al., 2010a), (2) a targeted global model singular vector (GSV; Yamaguchi et al., 2009; Hara, 2010) method, (3) MSV method based on the adjoint model of NHM (Kunii, 2010a), (4) a mesoscale breeding of growing modes (MBD) method based on the NHM forecast (Saito, 2007), and (5) a local ensemble transform Kalman filter using NHM (LETKF; Miyoshi and Aranami, 2006; Seko, 2010). Saito et al. (2011a) objectively compared the results of the ensemble forecasts made with these five methods by evaluating the evolution of the ensemble spreads, the RMSE of the ensemble mean. GSV was selected as the initial perturbation method, considering its performance for weak to moderate rain QPF and the RMSE characteristics. The initial condition of the MRI/JMA system was prepared by applying Meso 4DVAR to the B08RDP area (Kunii et al., 2010a).

Kunii et al. (2011) reported the results of the international EPS intercomparison. Verification was performed using the MEP outputs interpolated into a common verification domain. For all systems, the ensemble spreads grew as the forecast time increased, and the ensemble mean improved the RMSEs compared with individual control forecasts in the verification against the analysis fields. MRI/JMA's EPS and the control run had the best performance among the six EPS systems for predicting surface conditions (2m temperature and relative humidity) and weak to moderate rains, in terms of RSMEs against the initial condition and the threat scores.

Details of MRI and JMA's activities in B08FDP/RDP have been published as an MRI Technical Report (Saito et al., 2010b).

3.5.5 Ensemble Kalman filter

The ensemble Kalman filter (EnKF) technique is a new method of data assimilation that employs ensemble prediction to estimate forecast error. Miyoshi and Aranami (2006) applied four-dimensional expansion of LETKF to NHM and performed data assimilation experiments in a perfect model scenario with 5-km grid spacing. The analysis equation for LETKF is:

$$\mathbf{X}^a = \overline{x}^f e + \delta\mathbf{X}^f \tilde{\mathbf{P}}^a (\mathbf{H}\delta\mathbf{X})^T \mathbf{R}^{-1} (y^o - \overline{H(x^f)}) e + \delta\mathbf{X}^f \mathbf{T}. \qquad (19)$$

Here, the overbar means the ensemble mean, and $\delta\mathbf{X}^f$ the ensemble perturbation matrix. $\mathbf{H}$ is the tangent linear of the observation operator, and e is an m-dimensional row vector. $\mathbf{P}^a$ with tilde is the analysis error covariance matrix in the space spanned by forecast ensemble perturbations obtained by eigenvalue decomposition.

This system, NHM-LETKF, was modified by Seko (2010) and was tested as the initial perturbation generator for the mesoscale ensemble prediction in the B08RDP project.

Saito et al. (2012) examined the effects of LBPs on the MBD method and LETKF for mesoscale ensemble prediction. Introducing LBPs in the data assimilation cycles of LETKF improved the ensemble spread, the ensemble mean accuracy, and the performance of precipitation forecast. The accuracy of the LETKF analyses was compared with those of the Meso 4D-VAR analyses. With LBPs in the LETKF cycles, the RMSEs of the forecasts from the LETKF analyses improved, and some became comparable to those of the Meso 4D-VAR analyses.

Seko et al. (2011) performed data assimilation experiments with NHM-LETKF for an intense local rainfall event near the city of Kobe, western Japan, on 28 July 2008. They assimilated GEONET GPS TPWV data with conventional observation data. Adding GPS TPWV data tended to increase low-level water vapor and improve the precipitation forecast. The experiment with 5-km resolution generated a rainfall band in western Japan that was not reproduced using conventional data alone. The experiment with 1.6-km resolution effectively reproduced the observed band of intense rainfall.

4. Ongoing plans

4.1 Next generation supercomputer project in Japan

The next-generation supercomputer project, "Strategic Programs for Innovative Research (SPIRE)", is being carried out under a MEXT initiative. A supercomputer center was built in the city of Kobe by the RIKEN Advanced Institute for Computational Science (AICS). The supercomputer 'K' achieved the benchmark performance of 10.51 petaflops in November 2011 with a total of 88,128 CPUs of the FUJITSU SAPARC64 processor.

The SPIRE project consists of five strategic research fields (life science and medicine, new material and energy, disaster prevention, engineering, and matter and universe). A five-year research plan of high performance NWP with cloud resolving ensemble data assimilation has been endorsed as a sub-subject of Field 3 in SPIRE (Saito et al., 2011c). The sub-subject on mesoscale NWP has three goals:

a. Development of a cloud resolving 4 dimensional data assimilation system,
b. Development and validation of a cloud resolving ensemble analysis forecast system, and
c. Basic research using very high resolution atmospheric models.

The goal of a) is to dynamically predict local heavy rainfalls with deep convection by assimilating dense observation data. As described in the previous section, high resolution data assimilation methods (*e.g.*, 4D-VAR and LETKF) have been developed and applied to case studies of cloud resolving forecast experiments of precipitation. In addition, a displaced ensemble variational assimilation method has also been developed and tested for a data assimilation experiment on satellite microwave imager data (Aonashi and Eito, 2011).

The goal of b) is to demonstrate the plausibility of the probabilistic quantitative forecast of heavy rainfalls for disaster prevention by cloud resolving ensemble NWP. A NHM-LETKF system using incremental approach has been developed by Fujita et al. (2011) and has been tested at MRI in anticipation of application to the K computer (Kuroda et al., 2011). A cloud resolving (2 km) ensemble forecast experiment was performed for the summer of 2010 as a test trial. The JNoVA 4DVAR analyses were used as the initial conditions, and the JMA one-week ensemble prediction was used as the boundary perturbations. Duc et al. (2012) evaluated the ensemble prediction with the fraction skill score extended to temporal spaces. Results of the ensemble prediction will also be used for input data of river and flood models for risk management applications at Kyoto University.

In c), parameterizations in the cloud resolving model (*e.g.*, the bulk cloud microphysics and the PBL scheme) are assesed and modified using the spectral BIN method and the LES model. Mechanisms of typhoons and tornados are also examined by very high resolution simulations.

4.2 Tokyo metropolitan area convection study (TOMACS)

Observation data are critical in high resolution data assimilation experiments. A field campaign in the Tokyo metropolitan area with a dense observation network is being conducted by MRI, the National Research Institute for Earth Science and Disaster Prevention (NIED), and twelve national institutions and universities in Japan. This field experiment is part of the research program 'Tokyo Metropolitan Area Convection Study for Extreme Weather Resilient Cities (TOMACS)'. The intensive operational periods are set in the summers of 2011 to 2013. The research project consists of the following three subjects.

1. Studies of extreme weather with dense meteorological observations.
2. Development of an extreme weather early detection and prediction system.
3. Social experiments on extreme weather resilient cities.

Unprecedented dense observations including fourteen Doppler radars, six Doppler lidars, a high resolution (3 km) AWS network, a KU-band fast scan MP radar, and five additional GPS stations are deployed as a possible international test-bed for deep convection (Fig. 5).

4.3 Local forecast model at JMA

JMA is planning to run a high resolution (2 km horizontally and 60 layers vertically) local forecast model in 2012 for the aviation weather forecast and the very short range forecast of precipitation. A rapid update cycle with a 3D-VAR version of JNoVA (Fukuda et al., 2011) will be used to prepare initial conditions hourly. Performance of NHM with a horizontal resolution of 2 km as a regional NWP model has been verified (Hirahara et al., 2011), while a new dynamical core 'asuca' is under development considering better computational efficiency in the new parallel computer system at JMA (Ishida et al. 2010).

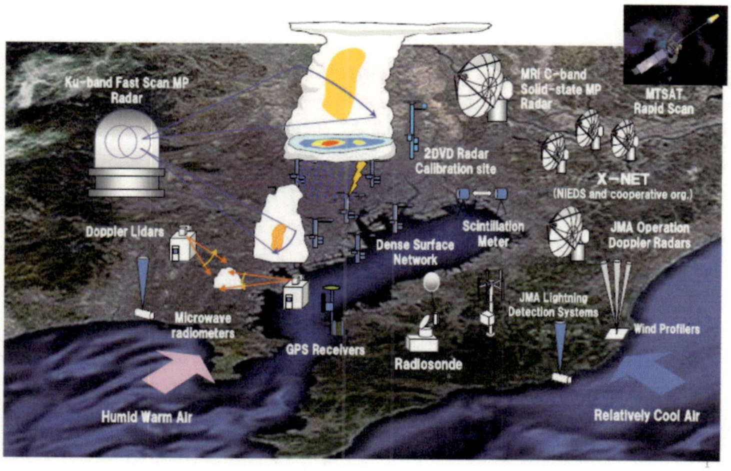

Fig. 5. Field campaign in TOMACS. Courtesy of NIED and the Meteorological Satellite and Observing System Research Department of MRI.

5. Acknowledgments

The author thanks T. Kato, C. Muroi, H. Eito, T. Hara, Y. Honda, Ts. Fujita, Ta. Fujita, J. Ishida and many scientists of NPD/JMA for their significant contributions towards developing NHM and its data assimilation systems. Figure 1 was quoted by courtesy of T. Hara and H. Kusabiraki of NPD. The author also thanks H. Seko, T. Kawabata, M. Kunii, Y. Shoji, S. Origuchi, K. Aonashi, T. Tsuyuki, Y. Yamada, M. Murakami, and several researchers of MRI, as well as T. Kuroda and L. Duc of JAMSTEC, for their research and development. Thanks are extended to T. Tokioka and F. Kimura of JAMSTEC and M. Maki of NIED for their leadership in conducting the research projects SPIRE and TOMACS. This study was partly supported by MEXT through a Grant-in-Aid for Scientific Research (21244074) "Study of advanced data assimilation and cloud resolving ensemble technique for prediction of local heavy rainfall".

6. References

Aonashi, K. and Eito, H. (2011). Displaced ensemble variational assimilation method to incorporate microwave imager data into a cloud-resolving model. *J. Meteor. Soc. Japan*, Vol.89, pp.175-194.

Aoyagi, T., and Seino, N. (2011). A square prism urban canopy scheme for the NHM and its evaluation on summer conditions in the Tokyo metropolitan area, Japan. *J. Appl. Meteor. Climatol.*, Vol.50, pp.1476-1496.

Clark, T.L. (1977). A small scale numerical model using a terrain following coordinate system. *J. Comp. Phys.*, Vol.24, pp.186-215.

Deardorff, J.W. (1980). Stratocumulus-capped mixed layers derived from a three-dimensional model. *Boundary-Layer Meteorol.*, Vol.18, pp.495-527.

Duan, Y., Gong, J., Du, J., Charron, M., Chen, J., Deng, G., DiMego, G., Hara, M., Kunii, M., Li, X., , Li, Y., Saito, K., Seko, H., Wang, Y., and Wittmann, C. (2012). An overview of Beijing 2008 Olympics Research and Development Project (B08RDP). *Bull. Amer. Meteor. Soc.*, 381-403.

Duc, L., Saito, K. and Seko, H. (2012). Application of spatial-temporal fractions skill score to high-resolution ensemble forecast verification. *Weather and Forecasting*. (submitted)

Eito, H., Kato, T., Yoshizaki, M. and Adachi A. (2005). Numerical simulation of the quasi-stationary snowband observed over the southern coastal area of the Sea of Japan on 16 January 2001. *J.Meteor.Soc.Japan*, Vol. 83, pp. 551-576.

Eito, H., Murakami, M., Muroi, M., Kato, T., Hayashi, S. Kuroiwa, H., and Yoshizaki, M. (2010). The structure and formation mechanism of transversal cloud bands associated with the Japan-Sea polar-airmass convergence zone. *J. Meteor. Soc. Japan*, Vol.88, pp.625-648.

Fujibe, F., Saito, K., Wratt, D. S. and Bradley, S. G. (1999). A numerical study on the diurnal variation of low-level wind in the lee of a two-dimensional mountain. *J. Meteor. Soc. Japan*, Vol.77, pp.827-843.

Fujita, Ts. (2003). Higher order finite difference schemes for advection of NHM. *Proceedings, CAS/JSC WGNE Res. Act. Atmos. Ocea. Model.*, Vol.33, pp.3.09-3.10.

Fujita, Ta., Kuroda, T. Seko, H. and Saito, K. (2011). Development of a meso ensemble data assimilation system. *Presentation, 2011 Meeting on the Study of Advanced Data Assimilation and Cloud Resolving Ensemble Technique for Prediction of Local Heavy Rainfall.*

Fukuda, J., Fujita, T., Ikuta, Y., Ishikawa, Y. and Yoshimoto, K. (2011). Development of JMA local analysis. *CAS/JSC WGNE Res. Act. Atmos. Ocea. Model.*, Vol.41, pp.1.07-1.08.

Gal-Chen, T. & Somerville, R. C. J. (1975). On the use of a coordinate transform for the solution of the Navier-Stokes equation. *J. Comp. Phys.*, Vol.17, pp. 209-228.

Hara, T. (2007). Implementation of improved Mellor-Yamada Level 3 scheme and partial condensation scheme to JMANHM and their performance. *CAS/JSC WGNE Res. Act. in Atmos. and Ocea. Modelling*, 37, 0407-0408.

Hara, T., Aranami, K., Nagasawa, R., Narita, M., Segawa, T., Miura, D., Honda Y., Nakayama, H. and Takenouchi, K. (2007). Upgrade of the operational JMA non-hydrostatic mesoscale model. *CAS/JSC WGNE Res. Act. Atmos. Ocea. Model.*, Vol.37, pp.5.11-5.12.

Hara, M. (2010). Global singular vector method. *Tech. Rep. MRI.* Vol.62, pp.61–72.

Hashimoto, A., Kato, T. Hayashi, S. and Murakami, M. (2008). Seedability assessment for winter orographic snow clouds over the Echigo Mountains. *SOLA*, Vol.4, pp.69-72.

Hayashi S., Aranami, K. and Saito, K. (2008). Statistical verification of short term NWP by NHM and WRF-ARW with 20 km horizontal resolution around Japan and Southeast Asia., *SOLA*, Vol.4, pp.133-136.

Hirahara, Y., Ishida, J. and Ishimizu, T. (2011). Trial operation of the Local Forecast Model at JMA. *CAS/JSC WGNE Res. Act. Atmos. Ocea. Model.*, Vol.41, pp.5.11-5.12.

Honda, Y., Nishijima, M., Koizumi, K., Ohta, Y., Tamiya, K., Kawabata, T. and Tsuyuki, T. (2005). A pre-operational variational data assimilation system for a nonhydrostatic model at Japan Meteorological Agency: Formulation and preliminary results. *Q. J. R. Meteorol. Soc.*, Vol.131, pp. 3465-3475.

Honda, Y., Y. and Sawada, K. (2008). A new 4D-Var for mesoscale analysis at the Japan Meteorological Agency. *CAS/JSC WGNE Res. Act. Atmos. Ocea. Model.*, Vol.38, pp.01.7-01.8

Ikawa, M. (1988). Comparison of some schemes for nonhydrostatic models with orography. *J. Meteor. Soc. Japan*, Vol.66, pp.753-776.

Ikawa, M., and Nagasawa, Y., (1989). A numerical study of a dynamically induced foehn observed in the Abashiri-Ohmu area. *J. Meteor. Soc. Japan*, Vol.67, pp.429-458.

Ikawa, M., and Saito, K. (1991). Description of a nonhydrostatic model developed at the Forecast Research Department of the MRI. *Tech. Rep. MRI*, Vol.28, 238pp.

Ikawa, M., Mizuno, H., Murakami, M., Yamada, Y. and Saito, K. (1991). Numerical modeling of the convective snow cloud over the Sea of Japan. --Precipitation mechanism and sensitivity to ice crystal nucleation rates--. *J. Meteor. Soc. Japan*, Vol.69, pp.641-667.

Ikuta, Y. and Honda, Y. (2011). Development of 1D+4DVAR data assimilation of radar reflectivity in JNoVA. *CAS/JSC WGNE Res. Act. Atmos. Ocea. Model.*, Vol.41, pp. 1.09-1.10.

Iguchi, T., Nakajima, T., Khain, A. P., Saito, K., Takemura, T. and Suzuki, K. (2008). A study of the cloud microphysical properties in an East Asia region by a bin-type cloud model coupled with a meso-scale model. *J. Geophys. Res.*, Vol.113, D14215, doi:10.1029/2007JD009774.

Iguchi, T., Nakajima, T., Khain, A. P., Saito, K., Takemura, Okamoto, H., Nishizawa, T. and Tao, W. (2012). Evaluation of cloud microphysics simulated using a meso-scale model coupled with bin and bulk microphysical schemes through comparison with data observed by cloud radars *J. Atmos. Sci.* (in press)

Ishida, J. (2007). Development of a hybrid terrain-following vertical coordinate for JMA nonhydrostatic model. *CAS/JSC WGNE Res. Act. Atmos. Ocea. Model.*, Vol.37, pp.3.09-3.10.

Ishida, J., Muroi, C., Kawano, K., and Kitamura, Y. (2010). Development of a new nonhydrostatic model ASUCA at JMA. *CAS/JSC WGNE Res. Act. Atmos. Ocea. Model.* , Vol.40, pp.5.11-5.12.

Ishikawa, Y. (2010). Data assimilation of GPS precipitable water vapor into the JMA mesoscale numerical weather prediction model. *CAS/JSC WGNE Res. Act. Atmos. Ocea. Model.*, Vol.40, pp.1.13-1.14.

Ishizaki, N and Takayabu, I. (2009). On the warming events over Toyama Plain by using NHRCM. *SOLA*, Vol.5: pp.129–132.

Kain, J. and Fritsch, J. (1993). Convective parameterization for mesoscale models: The Kain-Fritsch scheme. *The Representation of Cumulus Convection in Numerical Models, Meteor. Monogr.*, Vol.24, pp.165-170.

Kanada, S., Muroi, C., Wakazuki, Y., Yasunaga, K., Hashimoto, A., Kato, T., Kurihara, Yoshizaki M. and Noda, A. (2005). Structure of mesoscale convective systems during the late Baiu season in the global warming climate simulated by a non-hydrostatic regional model. *SOLA*, Vol.1, pp.117-120.

Kanada, S., S., Wada, A., Nakano M., and Kato, T. (2012). Effect of PBL schemes on the development of intense tropical cyclones using a cloud resolving model. *J. Geophys. Res.* (in press)

Kato, T., (1995). Box-Lagrangian rain-drop scheme. *J. Meteor. Soc. Japan*, Vol.73, pp. 241-245

Kato, T., and Saito, K. (1995). Hydrostatic and nonhydrostatic simulation of moist convection: Applicability of hydrostatic approximation to a high-resolution model. *J. Meteor. Soc. Japan*, Vol.73, pp.59-77.

Kato, T., (1998). Numerical simulation of the band-shaped torrential rain observed southern Kyushu, Japan on 1 August 1993. *J. Meteor. Soc. Japan*, Vol.76, pp. 97-128.

Kawabata, T., Seko, H., Saito, K., Kuroda, T., Tamiya, K., Tsuyuki, T., Honda, Y. and Wakazuki, Y. (2007). An assimilation experiment of the Nerima heavy rainfall with a cloud-resolving nonhydrostatic 4-dimensional variational data assimilation system. *J. Meteor. Soc. Japan*, Vol.85, pp.255-276.

Kawabata, T., Kuroda, T., Seko, H. and Saito, K. (2011). A cloud-resolving 4D-Var assimilation experiment for a local heavy rainfall event in the Tokyo metropolitan area. *Mon. Wea. Rev.*, Vol.139, pp.1911-1931.

Kawabata, T., Kunii, M., Bessho, K., Nakazawa, T., Kohno, N., Honda, Y. and Sawada, K. (2012). Reanalysis and reforecast of typhoon Vera (1959) using a mesoscale four dimensional variational assimilation system, *J. Meteor. Soc. Japan*, (submitted).

Keenan, T., Rutledge, S., Carbone, R., Wilson, J., Takahashi, T., May, P., Tapper, N., Platt, M., Hacker, J., Sekelsky, S., Moncrieff, M., Saito, K., Holland, G., Crook, A. and Gage, K. (2000). The Maritime Continent Thunderstorm Experiment (MCTEX): Overview and some results. *Bull. Amer. Meteor. Soc.*, Vol.81, pp.2433–2455.

Koizumi, K., Ishikawa, Y. and Tsuyuki, T. (2005). Assimilation of precipitation data to JMA mesoscale model with a four-dimensional variational method and its impact on precipitation forecasts. *SOLA*, Vol.1, pp.45-48.

Kunii, M. and Saito, K. (2009). Sensitivity Analysis using the Mesoscale Singular Vector. *CAS/JSC WGNE Res. Act. Atmos. Ocea. Model.*, Vol.39, pp.1.27-1.28.

Kunii, M., (2010a). MSV method. *Tech. Rep. MRI*, Vol.62, pp.73 - 77.

Kunii, M., (2010b). Heavy rainfall experiments using mesoscale SV. *Tech. Rep. MRI*, Vol.62, pp.73 - 77. pp.186-187.

Kunii, M., Saito, K. and Seko, H. (2010a). Mesoscale data assimilation experiment in the WWRP B08RDP. *SOLA*, Vol.6, pp.33-36.

Kunii, M., Shoji, Y., Ueno M. and Saito, K. (2010b). Mesoscale data assimilation of Myanmar cyclone Nargis. *J. Meteor. Soc. Japan*, Vol.88, pp.455-474.

Kunii, M., Saito, K., Seko, H., Hara, M., Hara, T., Yamaguchi, M., Gong, J., Charron, M., Du, J., Wang, Y. and Chen, D. (2011). Verifications and intercomparisons of mesoscale ensemble prediction systems in B08RDP. *Tellus*, Vol.63A, pp.531-549.

Kuroda, T., Saito, K., Kunii, M. and Kohno, N. (2010). Numerical simulations of Myanmar cyclone Nargis and the associated storm surge Part 1 : Forecast experiment with NHM and simulation of storm surge. *J. Meteor. Soc. Japan*, Vol.88, pp.521-545.

Kuroda, T., Fujita, Ta., Seko H. and Saito, K. (2011). Development and near future utilization of incremental LETKF data assimilation system at MRI. *Presentation, 2011 Meeting on the Study of Advanced Data Assimilation and Cloud Resolving Ensemble Technique for Prediction of Local Heavy Rainfall.*

Lin, Y.H., Farley, R.D. and Orville, H.D. (1983). Bulk parameterization of the snow field in a cloud model. *J. Clim. Appl. Meteor.*, Vol.22, pp. 1065-1092.

Mashiko, W., Niino, H. and Kato, T. (2009). Numerical simulation of tornadogenesis in an outer-rainband minisupercell of typhoon Shanshan on 17 September 2006. *Mon. Wea. Rev.*, 137, 4238-4260.

Miyoshi, T. and Aranami, K. (2006). Applying a four-dimensional local ensemble transform Kalman filter (4D-LETKF) to the JMA nonhydrostatic model (NHM). *SOLA*, Vol.2, pp.128 - 131.

Murata, A., Saito, K. and Ueno, M., (2003). The effects of precipitation schemes and horizontal resolution on the major rainband in typhoon Flo (1990) predicted by the MRI mesoscale nonhydrostatic model. *Meteorol. Atmos. Phys.*, Vol. 82, pp. 55-73.

Murata, A., (2009). A mechanism for heavy precipitation over the Kii peninsula accompanying typhoon Meari (2004). *J. Meteor. Soc. Japan,* Vol.87, pp.101-117.

Muroi, C., Saito, K., Kato, T. and Eito, H. (2000). Development of the MRI/NPD nonhydrostatic model. *CAS/JSC WGNE Res. Act. Atmos. Ocea. Model.,* Vol.30, pp.5.25-5.26.

Nagasawa, R. T. Iwasaki, S. Asano, K. Saito and H. Okamoto, (2006). Resolution dependence of nonhydrostatic models in simulating the formation and evolution of low-level clouds during a "Yamase" event. *J. Meteor. Soc. Japan,* Vol.84, pp.969-987.

Nagata, M., Leslie, L., Kamahori, H., Nomura, R., Mino, H., Kurihara, Y., Rogers, E., Elsberry, R. L., Basu, B. K., Buzzi, A., Calvo, J., Desgagne, M., D'Isidoro, M., Hong, S.-Y., Katzfey, J., Majewski, D., Malguzzi, P., McGregor, J., Murata, A., Nachamkin, J., Roch, M. and Wilson, C. (2001). A mesoscale model intercomparison: A case of explosive development of a tropical cyclone (COMPARE III), *J. Meteor. Soc. Japan,* Vol.79, pp.999–1033.

Nakanishi, M. and Niino, H. (2004). An improved Mellor-Yamada level 3 model with condensation physics: Its design and verification. *Bound.-Layer Meteor.,* Vol.112, pp.1-31.

Nakanishi, M., and Niino, H. (2006). An improved Mellor-Yamada level-3 model: Its numerical stability and application to a regional prediction of advection fog. *Bound.-Layer Meteor.,* Vol.119, pp.397-407.

Ogura, Y. and Phillips, N.A. (1962). Scale analysis of deep and shallow water convection in the atmosphere. *J. Atmos. Sci.,* Vol.19, pp.173-179.

Ohmori, S. and Yamada, Y. (2006). Development of cumulus parameterization scheme in the nonhydrostatic mesoscale model at the Japan Meteorological Agency. *CAS/JSC WGNE Res. Act. Atmos. Ocea. Model.* Vol.35, pp.4.21-4.22.

Ono, K., Honda, Y. and Kunii, M. (2011). A mesoscale ensemble prediction system using singular vector methods. *CAS/JSC WGNE Res. Act. Atmos. Ocea. Model.* Vol.40, pp.5.15-5.16.

Orlanski, I. (1976). A simple boundary condition for unbounded hyperbolic flows. *J. Comp. Phys.,* Vol.21, pp.251-269.

Otsuka, S., and Yoden, S. (2011). Experimental development of a decision support system for prevention and mitigation of meteorological disasters based on ensemble NWP Data. *Tech. Rep. MRI,* Vol.65, pp.103-108.

Redelsperger, J.L., Brown, P., Guichard, F., Hoff, C., Kawasima, M., Lang, S., Montmerle, T., Nakamura, K., Saito, K., Seman, C. and Tao, W. K. (2000). A GCSS intercomparison of models for a tropical squall line observed during TOGA-COARE. Part 1: Cloud-resolving models. *Q. J. R. Met. Soc.,* Vol.126, pp.823-863.

Saito, K. and Ikawa, M. (1991). A numerical study of the local downslope wind "Yamaji-kaze" in Japan. *J. Meteor. Soc. Japan,* Vol. 69, pp.31-56.

Saito, K., (1993). A numerical study of the local downslope wind "Yamaji-kaze" in Japan. Part 2: Non-linear aspect of the 3-D flow over a mountain range with a col. *J. Meteor. Soc. Japan,* Vol.71, pp.247-271.

Saito, K., (1994). A numerical study of the local downslope wind "Yamaji-kaze" in Japan. Part 3: Numerical simulation of the 27 September 1991 windstorm with a non-hydrostatic multi-nested model. *J. Meteor. Soc. Japan,* Vol. 72, pp.301-329.

Saito, K., Thanh, L. and Takeda, T. (1994). Airflow over the Kii Peninsula and its relation to the orographic enhancement of rainfall. *Pap. Met. Geophys.,* Vol.45, pp.65–90.

Saito, K., Murakami, M., Matsuo, T. and Mizuno, H. (1996). Sensitivity experiments on the orographic snowfall over the mountainous region of northern Japan. *J. Meteor. Soc. Japan*, Vol.74, pp.797-813.

Saito, K., (1997). Semi-implicit fully compressible version of the MRI mesoscale nonhydrostatic model ---Forecast experiment of the 6 August 1993 Kagoshima torrential rain---. *Geophys. Mag. Ser. 2*, Vol.2, pp.109-137.

Saito, K., Doms, G., Schaetter, U. and Steppeler, J. (1998). 3-D mountain waves by the Lokal-Modell of DWD and the MRI-mesoscale nonhydrostatic model. *Pap. Met. Geophys.*, Vol. 49, pp.7-19.

Saito, K., (2001). Numerical simulation of clouds during the cold air outbreak. *Parity*, Vol.16, pp. 46-51. (in Japanese)

Saito, K., Kato, T., Eito H. and Muroi, C. (2001a). Documentation of the Meteorological Research Institute/Numerical Prediction Division unified nonhydrostatic model. *Tech. Rep. MRI*, Vol.42, 133pp.

Saito, K., Keenan, T., Holland, G. and Puri, K. (2001b). Numerical simulation of tropical diurnal thunderstorms over the Maritime Continent. *Mon. Wea. Rev.*, Vol.129, pp.378-400.

Saito, K., (2003). Time-splitting of advection in the JMA Nonhydrostatic Model. *CAS/JSC WGNE Res. Act. Atmos. Ocea. Model.* Vol.33, pp.3.15-3.16.

Saito, K., (2004). Direct evaluation of the buoyancy and consideration of moisture diffusion in the continuity equation in the JMA nonhydrostatic model. *CAS/JSC WGNE Res. Act. Atmos. Ocea. Model.* Vol.34, pp.5.25-5.26.

Saito, K. and Ishida, J. (2005). Implementation of the targeted moisture diffusion to JMA-NHM. *CAS/JSC WGNE Res. Act. Atmos. Ocea. Model.* Vol.35, pp.5.17-5.18.

Saito, K., Fujita, Ta., Yamada, Y., Ishida, J., Kumagai, Y., Aranami, K., Ohmori, S., Nagasawa, R., Kumagai, S., Muroi, C., Kato, T., Eito H. and Yamazaki, Y. (2006). The operational JMA nonhydrostatic model. *Mon. Wea. Rev.*, Vol.134, pp.1266-1298.

Saito, K., (2007). Development of a BGM method with the JMA nonhydrostatic mesoscale model. *CAS/JSC WGNE Res. Act. Atmos. Ocea. Model.* Vol.37, pp.5.27-5.28.

Saito, K., Ishida, J., Aranami, K., Hara, T., Segawa, T., Narita, M. and Honda, Y. (2007). Nonhydrostatic atmospheric models and operational development at JMA. *J. Meteor. Soc. Japan*, Vol.85B, pp.271-304.

Saito, K., Kuroda, T., Kunii, M. and Kohno, N. (2010a). Numerical Simulations of Myanmar Cyclone Nargis and the Associated Storm Surge Part 2: Ensemble prediction. *J. Meteor. Soc. Japan.* Vol.88, pp.547-570.

Saito, K., Kunii, M. Hara, M., Seko, H., Hara, T., Yamaguchi, M., Miyoshi, T. and Wong, W.K. (2010b). WWRP Beijing 2008 Olympics Forecast Demonstration / Research and Development Project (B08FDP/RDP). *Tech. Rep. MRI*, Vol.62, 214pp.

Saito, K., Hara, M., Kunii, M., Seko, H. and Yamaguchi, M. (2011a). Comparison of initial perturbation methods for the mesoscale ensemble prediction system of the Meteorological Research Institute for the WWRP Beijing 2008 Olympics Research and Development Project (B08RDP). *Tellus*, Vol.63A, pp.445-467.

Saito, K., Kuroda, T., Hayashi, S., Seko, H., Kunii, M., Shoji, Y., Ueno, M., Kawabata, T., Yoden, S. Otsuka, S., Trilaksono, N. J., Koh, T.Y., Koseki, S., Duc, L., Xin, X.K., Wong W.K. and Gouda, K.C. (2011b). International research for prevention and mitigation of meteorological disasters in Southeast Asia. *Tech. Rep. MRI.* Vol.65, 198pp.

Saito, K., H. Seko, T. Kuroda, T. Fujita, T. Kawabata, K. Aonashi, and T. Tsuyuki, (2011c). Next generation supercomputer project toward cloud resolving NWP. *CAS/JSC WGNE Res. Act. Atmos. Ocea. Model.*, Vol.41, pp.5.19-5.20.

Saito, K., Seko, H.,Kunii M. and Miyoshi, T. (2012). Effect of lateral boundary perturbations on the breeding method and the local ensemble transform Kalman filter for mesoscale ensemble prediction. *Tellus,* Vol.64, doi: 10.3402/tellusa.v64i0.11594.

Sasaki, H., Kurihara, K. Takayabu, I. and Uchiyama, T. (2008). Preliminary experiments of reproducing the present climate using the non-hydrostatic regional climate model, *SOLA*, Vol.4, pp.25-28.

Sasaki, H., and Kurihara, K. (2008). Relationship between precipitation and elevation in the present climate reproduced by the non-hydrostatic regional climate model. *SOLA*, Vol.4, pp.109–112.

Satomura, T., Iwasaki, T., Saito, K., Muroi, C. and Tsuboki, K. (2003). Accuracy of terrain following coordinates over isolated mountain: Steep mountain Model intercomparison project (St-MIP). *Annuals, Disas, Prev. Res. Inst., Kyoto University*, Vol. 46, pp.337-346.

Sawada, M. and Iwasaki, T. (2007). Impacts of ice phase processes on tropical cyclone development. *J. Meteor. Soc. Japan*, Vol.85, pp.479-494.

Sawada, M., and Iwasaki, T. (2010). Impacts of Evaporation from Raindrops on Tropical Cyclones. Part I: Evolution and Axisymmetric Structure. *J. Atmos. Sci.*, Vol.67, pp.71–83.

Seino, N., Sasaki, H., Sato, J., and Chiba, M. (2004). High-resolution simulation of volcanic sulphur dioxide dispersion over the Miyake Island. *Atmospheric Environment*, Vol.38, pp.7073-7081.

Seko, H., Hayashi, S., Kunii, M. and Saito, K. (2008). Structure of the regional heavy rainfall system that occurred in Mumbai, India, on 26 July 2005. *SOLA*, Vol.4, pp.129-132.

Seko, H., K. Saito, M. Kunii and M. Kyouda, (2009). Mesoscale ensemble experiments on potential parameters for tornado formation. *SOLA*, Vol.5, pp.57-60.

Seko, H., (2010). Local ensemble transform Kalman filter (LET) method. Tech. Rep. MRI, Vol.62, pp.80–84.

Seko, H., H., M. Kunii, Y. Shoji and K. Saito, (2010). Improvement of rainfall forecast by assimilations of ground-based GPS data and radio occultation data. *SOLA*. Vol.6, pp.81-84.

Seko, H., T. Miyoshi, Y. Shoji and K. Saito, (2011). A data assimilation experiment of PWV using the LETKF system -Intense rainfall event on 28 July 2008-. *Tellus*, Vol.63A, pp.402-414.

Seko, H., S. Hayashi and K. Saito, (2012). Generation mechanisms of convection cells near Sumatra Island in the Monsoon Season. *Pap. Met. Geophys.* (submitted)

Sherman, C.A., (1978). A mass-consistent model for wind fields over complex terrain. *J. Appli. Meteor.*, Vol.17, pp. 312-319.

Shimbori, T., Aikawa, Y., Fukui, K., Hashimoto, A., Seino, N., and Yamasato, H. (2010). Quantitative tephra fall prediction with the JMA mesoscale tracer transport model for volcanic ash: A case study of the eruption at Asama volcano in 2009. *Pap. Met. Geophys.*, Vol.61, pp.13-29. (in Japanese with English abstract and figure captions)

Shoji, Y. (2009). A study of near real-time water vapor analysis using a nationwide dense GPS network of Japan. *J. Meteor. Soc. Japan*, Vol.87, pp.1-18.

Shoji, Y., Kunii, M. and Saito, K. (2009). Assimilation of nationwide and global GPS PWV data for a heavy rain event on 28 July 2008 in Hokuriku and Kinki, Japan. *SOLA*, Vol.5, pp.45-48.

Shoji, Y., Kunii, M. and Saito, K. (2011). Mesoscale Data Assimilation of Myanmar Cyclone Nargis. Part 2 : Assitilation of GPS derived Precipitable Water Vapor. *J. Meteor. Soc. Japan*, Vol.89, pp.67-88.

Smith, R. B., (1980). Linear theory of stratified hydrostatic flow past an isolated mountain. *Tellus*, Vol. 32, pp. 348-364.

Takano, I., Aikawa, Y. and Gotoh, S. (2007). Improvement of photochemical oxidant information by applying transport model to oxidant forecast, *CAS/JSC WGNE Res. Act. Atmos. Ocea. Model.*, Vol.37, pp.5.35-5.36.

Tapp, M. C. & White, P. W., (1976). A non-hydrostatic mesoscale model. *Q. J. R. Meteorol. Soc.*, Vol.102, pp. 277-296.

Trilaksono, N. J., Otsuka, S., Yoden, S., Saito, K. and Hayashi, S. (2011). Dependence of model simulated heavy rainfall on the horizontal resolution during the Jakarta flood event in January-February 2007. *SOLA*, Vol.7, pp.193-196.

Trilaksono, N. J., Otsuka, S. and Yoden, S. (2012). A time-lagged ensemble simulation on the modulation of precipitation over West Java in January-February 2007. *Mon. Wea. Rev.*, Vol.140. 601-606.

Wada, A. (2009). Idealized numerical experiments associated with the intensity and rapid intensification of stationary tropical cyclone-like vortex and its relation to initial sea-surface temperature and vortex-induced sea-surface cooling. *Journal of Geophysical Research*, Vol.114, doi:10.1029/2009JD011993.

Wada, A., Kohno, N. and Kawai, Y. (2010). Impact of wave-ocean interaction on typhoon Hai-Tang in 2005. *SOLA*, Vol.6A, pp.13-16.

Wada, A., (2012). Numerical study on the effect of the ocean on tropical-cyclone intensity and structural change. *InTech*, (this volume).

Wong, W.K., Sumdin, S. and Lai, S.T. (2010). Development of Air-Sea Bulk Transfer Coefficients and Roughness Lengths in JMA Non-hydrostatic Model and Application in Prediction of an Intense Tropical Cyclone. *SOLA*, Vol. 6, pp.65-68.

Yamada, Y., (2003). Introduction to the Kain-Fritsch scheme. *Separate volume of annual report of NPD*, Vol.49, pp.84-89. (in Japanese)

Yamaguchi, M., Sakai, R., Kyoda, M., Komori, T. and Kadowaki, T. (2009). Typhoon ensemble prediction system developed at the Japan Meteorological Agency. *Mon. Wea. Rev.* Vol.137, pp.2592 - 2604.

Yanase, W., Niino, H., and Saito, K. (2002). High-resolution numerical simulation of a polar low. *Geophys. Res. Lett.* , Vol.29, 1658, doi:10.1029/2002GL014736.

Yanase, W., G. Fu, H. Niino, and T. Kato, (2004). A polar low over the Japan Sea on 21 January 1997. Part II: A numerical study. *Mon. Wea. Rev.*, Vol.132, pp. 1552–1574.

Yanase, W., and Niino, H., (2007). Dependence of polar low development on baroclinicity and physical processes: an idealized high-resolution numerical experiment. *J. Atmos. Sci.*, Vol.64, pp.3044–3067.

Yasunaga, K., Sasaki, H., Wakazuki, Y., Kato, T., Muroi, C., Hashimoto, A., Kanada, S., Kurihara, K., Yoshizaki M. and Sato, Y. (2005). Performance of the long-term integrations of the Japan Meteorological Agency nonhydrostatic model with use of the spectral boundary coupling method. *Weather and Forecasting*, Vol.20, pp.1061-1072.

6

Variability of Intertropical Convergence Zone (ITCZ) and Extreme Weather Events

Yehia Hafez
Cairo University, Faculty of Science Department of Astronomy, Space Science and Meteorology
Egypt

1. Introduction

For case one, the UK severed from abnormal severs cold winter season on 2009. The mean temperature for that winter was 3.2 °C, which was 0.5 °C below average of (1971-2000), provisionally making it the coldest winter since 1996/97. Whereas, Mean temperatures over the UK were 1.1 °C below the average during December 2008, 0.6 °C below average during January and 0.2 °C above average during February during that season. A generally cold first half to December was followed by a milder period, before turning very cold by the first of January see Figure (1). This very cold spell persisted for the first 10 days of January, with some severe frosts, followed by alternating milder and colder periods. Despite a cold (and snowy) first half of February, milder conditions later resulted in near-normal temperatures overall. Rainfall amounts over the UK were below the 1971-2000 average during December with 70%, January was close to average with 98% and February was drier than average with 63%. In December, parts of south-east England, East Anglia and Wales had less than 50% of the average rainfall and in February much of Wales, north-west England and western Scotland recorded less than 50% of average. Significant snowfalls occurred in the first half of February, particularly over England and Wales during the first week, when depths greater than 15 cm were recorded quite widely. The last time of that winter season a comparable snowy spell occurred was in February 1991 (MetOffice., UK, 2009). However, there are several scientific literatures challenge the abnormal weather conditions [e.g. (Cohen et al., 2001; Hafez 2007, 2008; and Rosting & Kristjansson 2008)]. In addition to that identification, oscillations, and influence of the ITCZ (Intertropical Convergence Zone) in the atmospheric cooling weather conditions had studied by (Bates 1970; Pike 1972; Citeau 1988b; Gadgil & Guruprasad 1990; Waliser 1992, 1994; Hess et al., 1993; Philander et al., 1996; Kraus 1997; Sultan & Janicot 2000; Hafez 2003a; Broccoli et al., 2006; and Raymond 2006). However, climate simulations, using models with different levels of complexity, indicated that the north-south position of the intertropical convergence zone (ITCZ) responds to changes in interhemispheric temperature contrast. The present work aims to investigate the relationship between the Atlantic Western Africa ITCZ variability and the surface air temperature over UK through months of the winter 2009. For case two, the intertropical convergence zone (ITCZ) is one of the most recognizable aspects of the global circulation that influence in the atmospheric weather. The ITCZ forms as a zonally elongated band of cloud at low latitudes nearness of the equator where the northeasterly and southeasterly

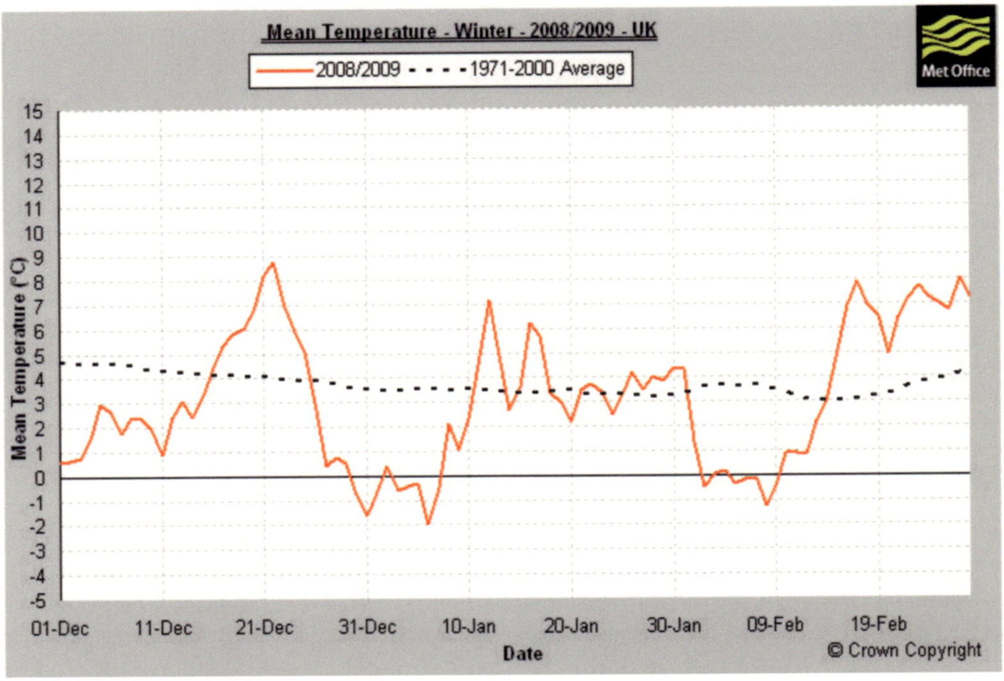

Fig. 1. Variation of UK daily mean temperature for Winter 2008/2009 [Source: MetOffice., UK, 2009: Winter Summary, 2009, ©Crow copyright]

trade winds converge. The focus of this study is to introduce the rule of ITCZ variability on the occurrence of unseasonably heavy rains, widespread flash floods, over Eastern Mediterranean (EM) in the period (17- 20) on January 2010. In fact, the topography in the EM such as high mountains, is essential factor for huge disasters of flash floods (Llasat, 2009). The flood disasters in EM through this period had been recorded and reported. Whereas, in Egypt, in the southern city of Aswan, floods and strong winds disrupted power in several neighborhoods. The floods were a surprise as North Sinai had not seen floods in 30 years. Sinai Peninsular was flash floods damage left more than 1000 homes totally destroyed, 1,076 submerged and the area suffered material losses of over US$25.3 million. The floods ruined 59km of roads, killed 1,838 animals and felled 27,820 (mostly olive) trees. Five Egyptians died in flooding in the southern Sinai desert. All 75 patients at the El-Arish general hospital in the Sinai had to be evacuated when the first floor was flooded. Flooding wiped out large sections of a major road in Egypt's south Sinai and destroyed two dozen homes in Ras Sudr. In Palestine, Heavy rain and flooding has forced hundreds of people from their homes in Khan Younis in the south of the Gaza Strip. Over 100 families had been made homeless.. Some 300 families were also displaced. The flooding along Egypt's Red Sea coast, the border with Israel and in the south left six people dead. It also damaged the roads leading to the resorts in the Sinai desert and brought down telephone and power lines. Israel temporarily closed its southern border crossings with Egypt and Jordan. Jordanians were warned off the streets after nearly a dozen accidents in one area. Rains of this magnitude, rains reached to 90 mm/day, which are rare in this largely arid region and where heavy precipitation can result in sudden and deadly flash floods. In Israel, a woman drowned when her car was caught in a flash flood in the south, where stormy weather also blocked the main road to the Red Sea resort of Eilat. A bridge also collapsed near a cargo crossing between Egypt and Israel. In addition to that, heavy rainfall recorded over south Turkey, Syria and Lebanon. One positive aspect of the flooding is that it helped replenish groundwater reserves. Whereas, the floods boosted groundwater reserves which are the main source of freshwater in this region. It also brought silt, which is very good for crops. Silt also reduces erosion of the coast when flood water reached the sea. The disasters information getting from Dartmouth Flood Observatory. Historical records of flash flood episodes over EM show that it was existed in autumn season of months (September, October and November) not in January. So that the present case study is outstanding extreme case. However, the flash floods problem in the EM was challenged several times in scientific literatures (e.g, Hafez, 2003b; Barnolas et al., 2007&2008, Papadopoulos & Katsafados, 2009; Houssos et al. 2009; Hatzaki et al., 2010;Michaelides et al., 2010 and Llasat et al., 2010). The previous studies referred the occurrence of these floods to deep of upper air trough of low pressure system with cold advection over the EM region. The present paper aims to uncover the rule played by ITCZ variability in the occurrence of extreme flash floods in EM in January 2010.

2. Data and methodology

For the first case, the daily NCEP/NCAR reanalysis data composites for mean surface air temperature, over the UK, [(49° N- 61° N) latitudes and (11° W- 2° E) longitudes] , for the period from 1 December 2008 to 28 February 2009 (Kalnay et al., 1996) are used in this study. The available meteorological data obtained from UK meteorological office are also used in the present study. In addition to that, the Atlantic - Western Africa [15° W - 10° E] ITCZ mean position data for summer months June, July and august of 2008 are used. The movement of the ITCZ over Atlantic-Western Africa has been monitored by plotting the daily location of the surface 15-degree C dew point temperature at 1200 UTC for every 5 degrees of longitude, (Ilesanmi, 1971). Over Atlantic-Western Africa, a mean position for each 10-day period is calculated for the area from 15 degrees west longitude to 10 degrees east longitude. However, the ITCZ data series begin in 1979 for Atlantic - Western Africa and the long-term means use 1979-2001 data. These data were obtained from website through the internet of the Climate Prediction Centre at http://www.cpc.ncep.noaa.gov/products/monitoring_data/. In the present work, these datasets are analyzed using the anomalies methodology and correlation coefficient technique. The formula for calculating the correlation coefficient was taken from (Spiegel, 1961). For the second case; The 6-hour and daily NCEP/NCAR reanalysis data composites for precipitation MSL pressure, gepotential height at 500 hpa level, surface vector and meridonal winds over the eastern Mediterranean region [(22° N- 40° N) latitudes and (24° E- 42° E) longitudes] , for the period (17 - 20) January 2010 (Kalnay et al., 1996) are used in the present study. The available floods disaster data obtained from dartmouth flood observatory are also used. In addition to that, 6-hour infrared (IR) satellite images are obtained to identify the ITCZ position by using of cloud clusters through the same period. In the present work, datasets are analyzed using the anomalies methodology. These data were obtained from websites through the internet of the Climatic centers, Climate Diagnostics Centre for supporting the data used throughout this study. Plots and images were provided by the NOAA-CIRES Climate Diagnostics Centre, Boulder, Colorado, USA from their Web site at http://www.cdc.noaa.gov. Available data of flash floods disasters obtained from dartmouth flood observatory through its website http://www.dartmouth.edu.

3. Results

3.1 For the first case

3.1.1 Analysis of the surface temperature anomalies over UK in winter 2009

In the present work, daily data for surface air temperature in UK through the period (1 December 2008 – 28 February 2009) are analyzed using of statistical anomalies methodology. Table (1) shows the anomalies in the 10-day mean of surface temperature (°C) over UK in winter 2008/2009. The results revealed that almost of UK severed from abnormal cooling whereas there are negative anomalies in the surface air temperature during December 2008. However, surface air temperature was less than its normal values by -0.5 °C. However, the normal value taken as average of the period of years 1968-1996. As shown in Figure 2a and Table 1. Meanwhile, for the month of January 2009 the temperature becomes around its normal values. Whereas, the anomalies in temperature values alternative around its normal between positive at the north of the UK, and negative values at the south of the UK, (Figure 2b and Table 1). For the month of February, the first half had a cooling and the last half had a warming rather than its normal values. In general the temperature remands around its normal values for that month (Figure 2c, and Table 1). In general, winter 2008/2009, from 1 December 2008 to 28 February 2009, had recorded a negative anomalies in surface air temperature with -0.5 °C. This cooling during that winter season occurred mainly at the central and southern parts of UK (Figure 2d).

Duration time (10-day interval)	Anomalies in the 10- day Mean surface air temperature (°C) over UK
1-10 Dec. 2008	-2.2
11-20 Dec. 2008	0.0
21-31 Dec. 2008	-0.5
1-10 Jan. 2009	-5.0
11-20 Jan. 2009	+1.2
21-31 Jan. 2009	+1.0
1-10 Feb. 2009	-3.5
11-20 Feb. 2009	+2.0
21-28 Feb. 2009	+3.5

Table 1. Anomalies in the 10-day mean of surface temperature (°C) over UK in winter 2008/2009.

3.1.2 Study variability of the Atlantic-Western Africa ITCZ during summer 2008

The movement of the ITCZ over Atlantic-Western Africa had been monitored by plotting the daily location of the surface 15-degree C dew point temperature at 1200 UTC for every 5 degrees of longitude, (Ilesanmi, 1971). Over Atlantic-Western Africa, a mean position for each 10-day period is calculated for the area from 15 degrees west longitude to 10 degrees east longitude. The data series begin in 1979 for Atlantic-Western Africa and the long-term means use 1979-2001 data. In the present study the changes of Atlantic-Western Africa ITCZ

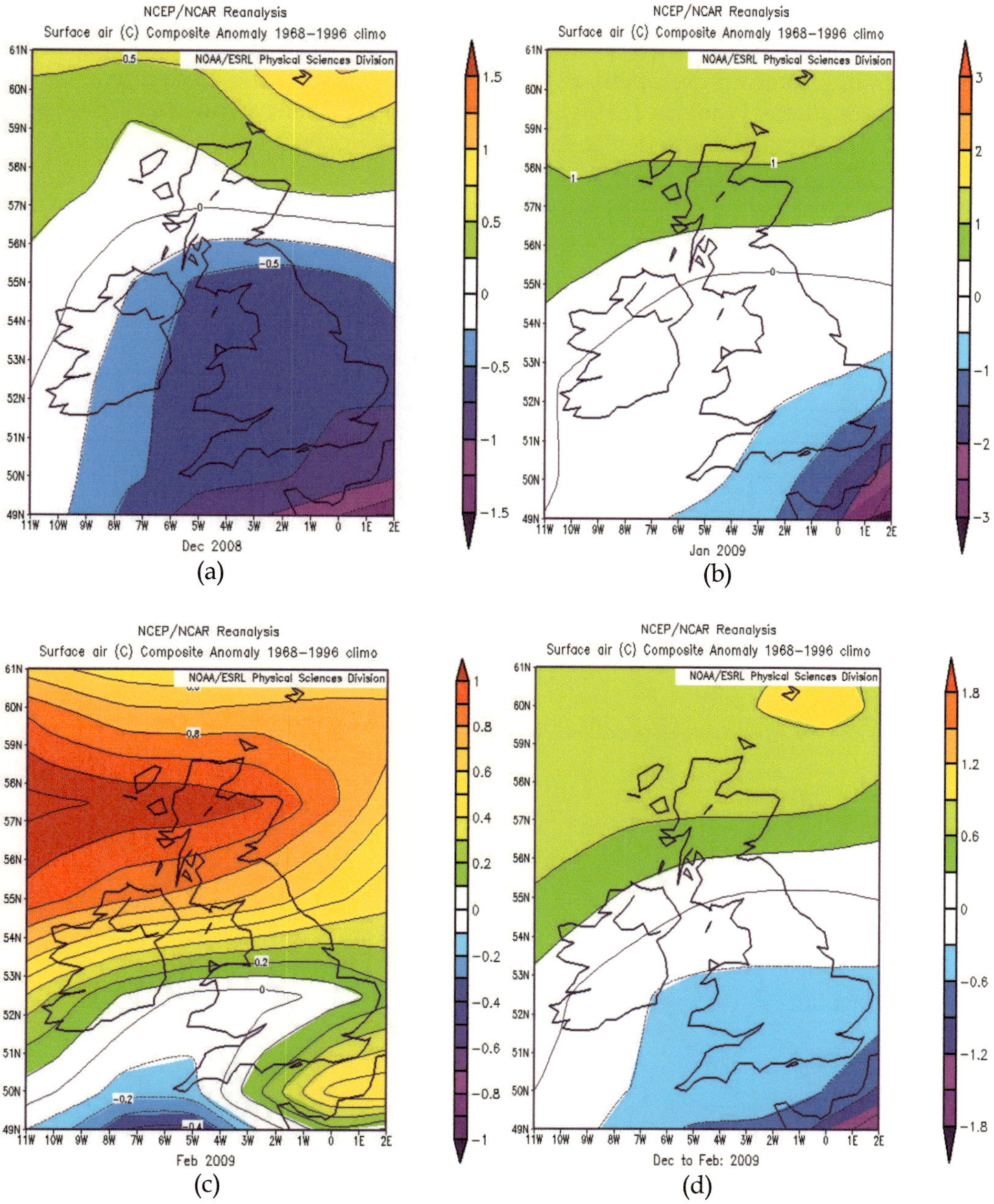

Fig. 2. The distribution of mean surface air temperature anomalies (°C) over UK for winter 2008/2009, (a) For month of December 2008, (b) For month of January 2009, (c) For month of February 2009 and (d) For months of winter season 2008/2009 (December, January and February).

variability through the period (1 June 2008 - 31 August 2008) are analyzed using of anomalies methodology. The result shows that the Atlantic-Western Africa ITCZ moved southward direction south of its average position from 1 June to 20 July with negative anomalies of its values. The maximum negative anomaly is recorded -2.351 latitudinal degrees at 15 W of ITCZ position through the 10 day interval (11-20 June 2008). In general, the outstanding southward changes of ITCZ variability existed over western part of the Greenwich longitude. Whereas, the significant negative anomalies occurred at the western part of the ITCZ through the period of study (Table 2 and Figure 3).

Duration time (10-day interval)	Anomalies in 10- day mean position of Atlantic-Western Africa ITCZ (Longitude degree)					
	15W	10W	5W	0	5E	10E
1-10 June 2008	-1.716	-1.719	-1.347	-0.625	-0.840	-0.488
11-20 June 2008	-2.351	-1.757	-1.250	-0.718	-0.892	-0.427
21-30 June 2008	-0.291	-0.481	-0.048	-0.090	0.059	0.071
1-10 July 2008	-1.374	-1.104	-1.565	-0.737	-0.295	-0.555
11-20 July 2008	-1.757	-2.058	-1.133	-0.808	-0.596	-0.374
21-31 July 2008	0.091	0.136	-0.213	-0.039	0.379	0.604
1-10 August 2008	-0.075	-0.846	-0.661	0.103	0.142	0.518
11-20 August 2008	0.033	0.371	0.759	1.023	1.194	1.379
21-31August 2008	0.200	0.433	0.562	0.954	0.571	0.035

Table 2. Anomalies in 10-day mean position of Atlantic-Western Africa ITCZ during summer 2008.

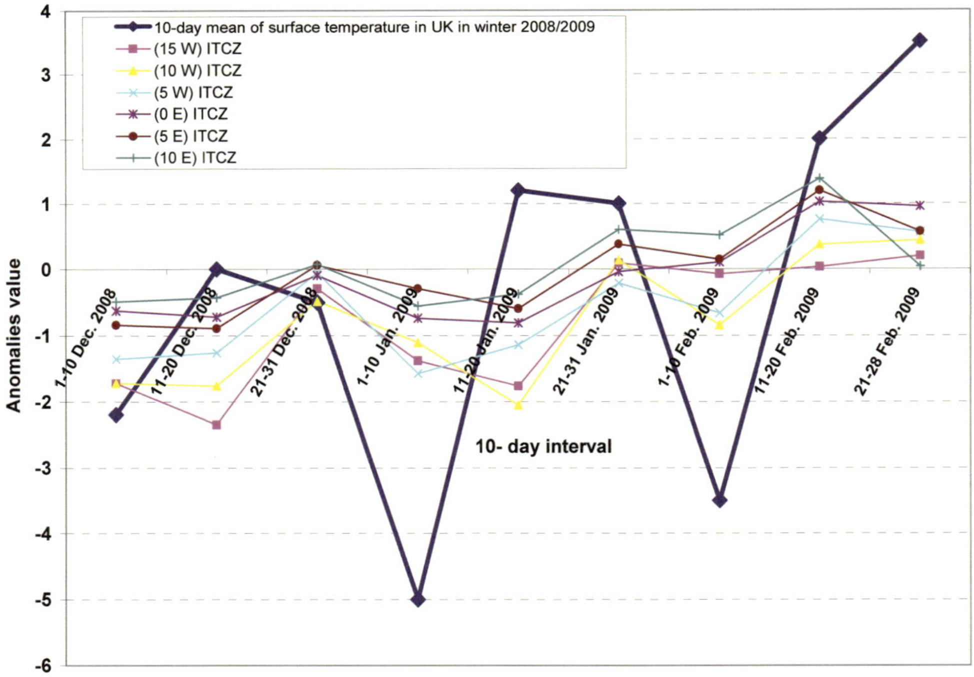

Fig. 3. The variation of 10-day Atlantic-Western Africa ITCZ mean position anomalies during summer 2008 and the variation of 10-day mean surface temperature anomalies over UK through winter 2008/2009.

3.1.3 Relationship between the Atlantic-Western Africa ITCZ and abnormal cold winter 2009 over UK

The relationship between the Atlantic-Western Africa ITCZ variability and abnormal cold winter 2009 over UK are studied in this section. Whereas, a 10-day time series analysis of anomalies in both of the variation of Atlantic-Western Africa ITCZ mean position during summer 2008 and the variation of mean surface temperature anomalies over UK through winter 2008/2009 are analyzed. The results revealed that there are outstanding relationship between the southward variations of Atlantic-Western Africa ITCZ and the occurrence of negative anomalies in surface air temperature over UK through winter 2009 (Figure 3). In addition to that a correlation coefficient technique analysis has been made to study this relationship. There are significant positive correlation coefficients between the southward variability of summer Atlantic-Western Africa ITCZ and the abnormal cooling weather that existed in UK through winter 2009. The highest correlation coefficient value is +0.7 at 5° W longitude of ITCZ (Figure 4).

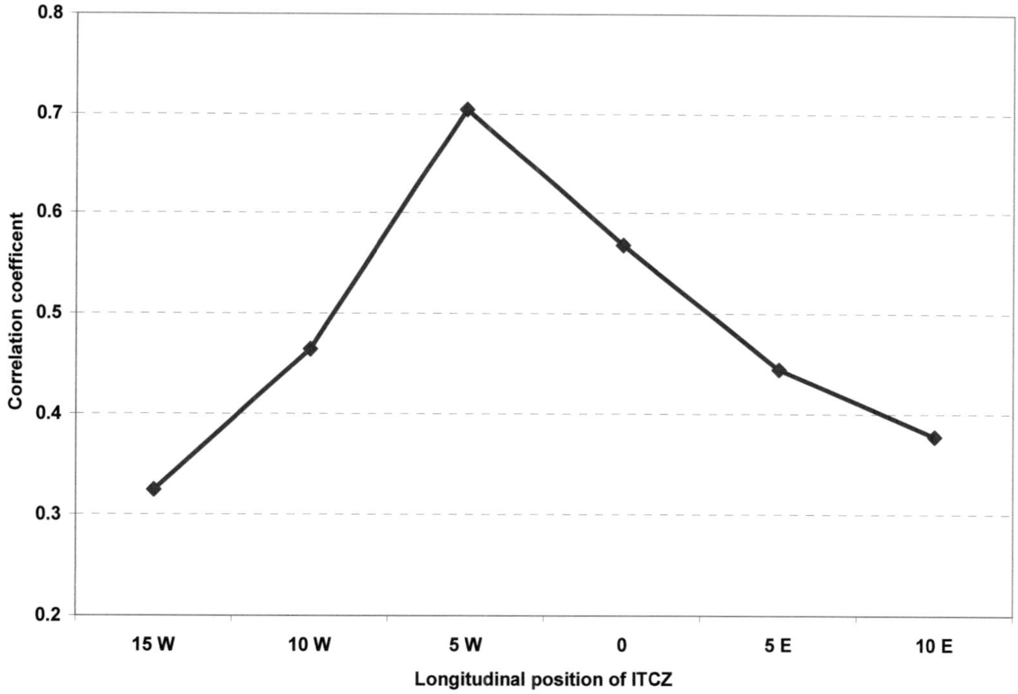

Fig. 4. The variation of correlation coefficient values between 10-day Atlantic-Western Africa ITCZ mean position anomalies during summer 2008 and 10-day mean surface temperature anomalies over UK through winter 2008/2009.

3.2 For the second case

3.2.1 Distribution of precipitation values in EM during (17-20) january 2010

Through the present work, the 6-hour NCEP/NCAR reanalysis data composites for precipitation rates over the EM region [(22° N- 40° N) latitudes and (24° E- 42° E) longitudes during the period (17 -20) January 2010 has been analyzed. Analysis of this data shows that on the first day, 17 January, the precipitation existed over the north western part of EM (Malta) and also over the south western part of EM(south west of Egypt) and Sinai and reached to its maximum value (90 mm) over South Sinai (see Table 3 and Figure 5a,b,c and d respectively). On next day, the precipitation hold eastward of EM to cover several countries include of (East Sinai, (Palestine &Israel), Jordon, North Syria and south Turkey) with maximum value of precipitation rate 90 mm/day as its clear in Table 3 and Figure 5e,f,g and h respectively. On 19 and 20 January precipitation widespread to cover all the EM region but with maximum values, 24 mm/day that less than the first two days. See Table 3 and Figure 5 (from i to p). This precipitation causing huge damages in several areas in EM and mainly lee the mountain regions. In particular, Sinai Peninsular was flash floods damage left more than 1000 homes totally destroyed, 1,076 submerged and the area suffered material losses of over US$25.3 million. Five Egyptians died in flooding in the southern Sinai desert.

Precipitation amount (mm) (6-hour) Time	Maximum amount of precipitation in (mm) over Eastern Mediterranean Maximum value	Location
17 January 2010 (0000UTC)	21 mm	Malta
17 January 2010 (0600UTC)	16	Malta and south west of Egypt
17 January 2010 (1200UTC)	27	south west of Egypt
17 January 2010 (1800UTC)	90	South Sinai
18 January 2010 (0000UTC)	90	East Sinai, (Palestine &Israel) and Jordon
18 January 2010 (0600UTC)	65	Jordon, Lebanon and south Syria
18 January 2010 (1200UTC)	55	North Syria and south Turkey
18 January 2010 (1800UTC)	50	North Syria
19 January 2010 (0000UTC)	12	Cyprus and Lebanon
19 January 2010 (0600UTC)	14	Eastern Mediterranean sea region
19 January 2010 (1200UTC)	16	Cyprus
19 January 2010 (1800UTC)	20	Eastern Mediterranean sea region
20 January 2010 (0000UTC)	22	Eastern Mediterranean region
20 January 2010 (0600UTC)	21	Eastern Mediterranean sea region
20 January 2010 (1200UTC)	24	Cyprus, west Syria and Lebanon
20 January 2010 (1800UTC)	24	Cyprus, west Syria and Lebanon

Table 3. 6-hour maximum amount of precipitation in (mm) over Eastern Mediterranean during the period of 17-20 January 2010.

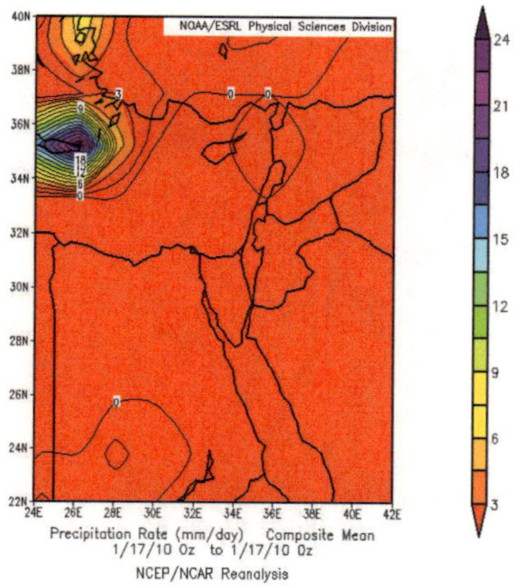
NOAA/ESRL Physical Sciences Division
Precipitation Rate (mm/day) Composite Mean
1/17/10 0z to 1/17/10 0z
NCEP/NCAR Reanalysis

(a)

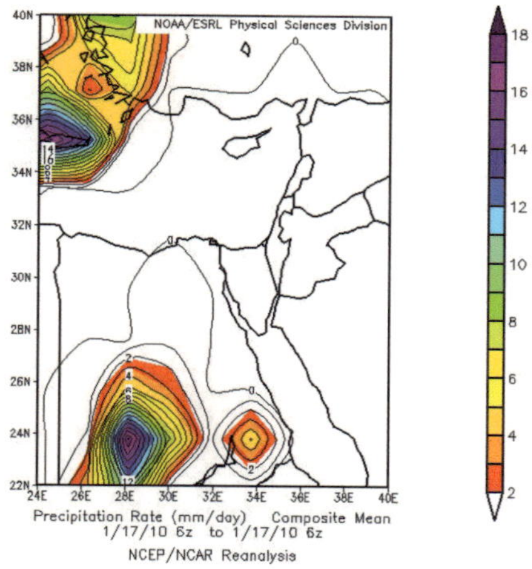
NOAA/ESRL Physical Sciences Division
Precipitation Rate (mm/day) Composite Mean
1/17/10 6z to 1/17/10 6z
NCEP/NCAR Reanalysis

(b)

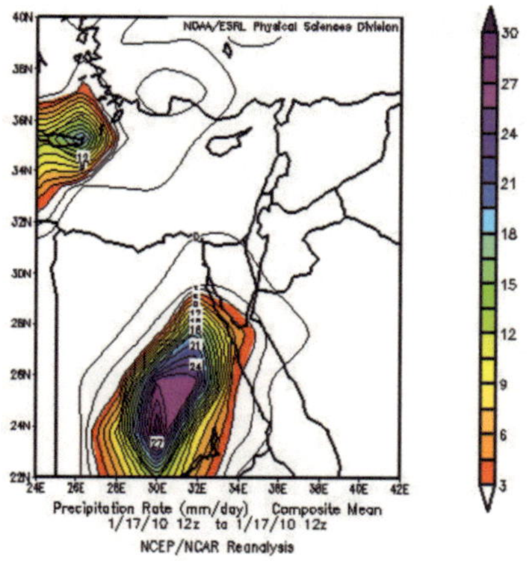
NOAA/ESRL Physical Sciences Division
Precipitation Rate (mm/day) Composite Mean
1/17/10 12z to 1/17/10 12z
NCEP/NCAR Reanalysis

(c)

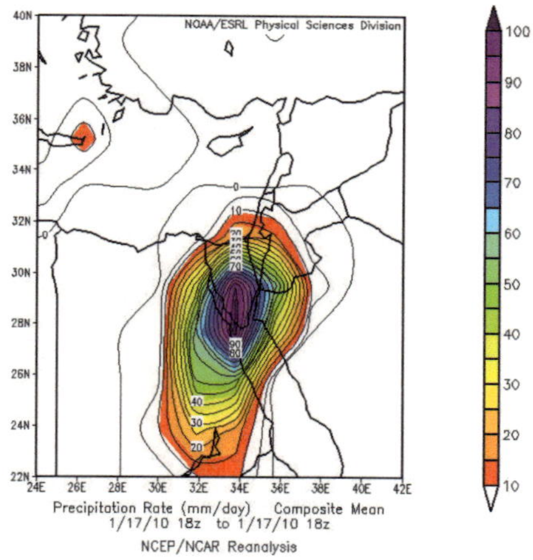
NOAA/ESRL Physical Sciences Division
Precipitation Rate (mm/day) Composite Mean
1/17/10 18z to 1/17/10 18z
NCEP/NCAR Reanalysis

(d)

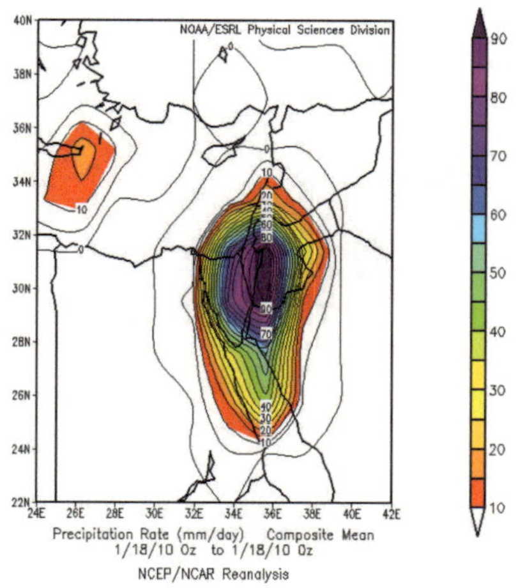
NOAA/ESRL Physical Sciences Division
Precipitation Rate (mm/day) Composite Mean
1/18/10 0z to 1/18/10 0z
NCEP/NCAR Reanalysis

(e)

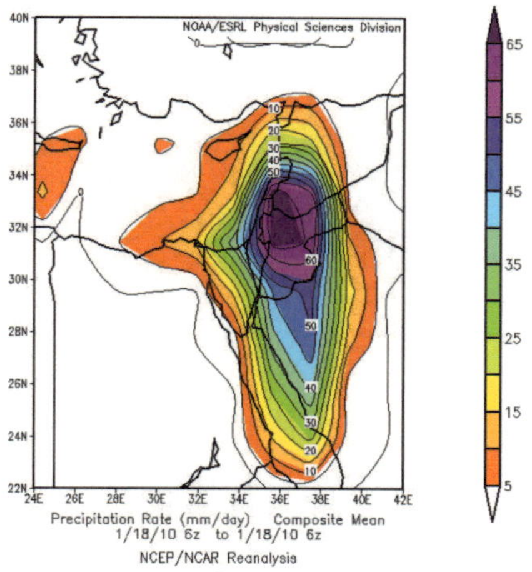
NOAA/ESRL Physical Sciences Division
Precipitation Rate (mm/day) Composite Mean
1/18/10 6z to 1/18/10 6z
NCEP/NCAR Reanalysis

(f)

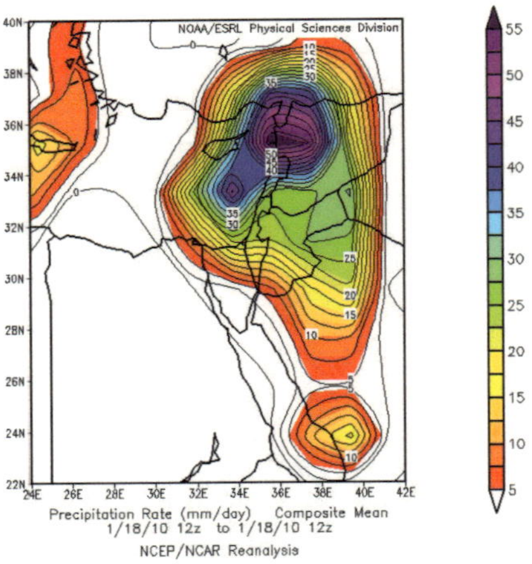
NOAA/ESRL Physical Sciences Division
Precipitation Rate (mm/day) Composite Mean
1/18/10 12z to 1/18/10 12z
NCEP/NCAR Reanalysis

(g)

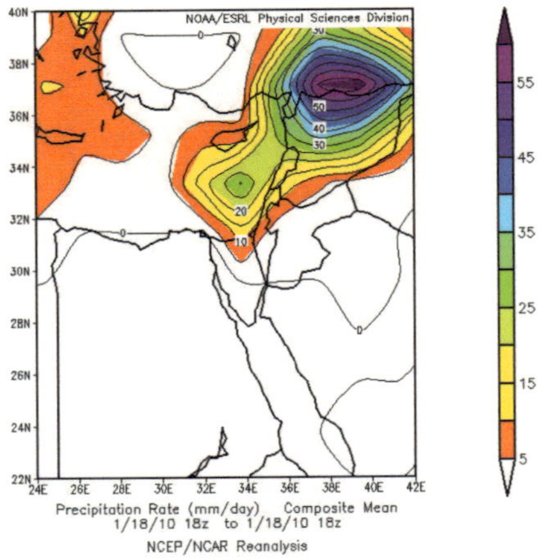
NOAA/ESRL Physical Sciences Division
Precipitation Rate (mm/day) Composite Mean
1/18/10 18z to 1/18/10 18z
NCEP/NCAR Reanalysis

(h)

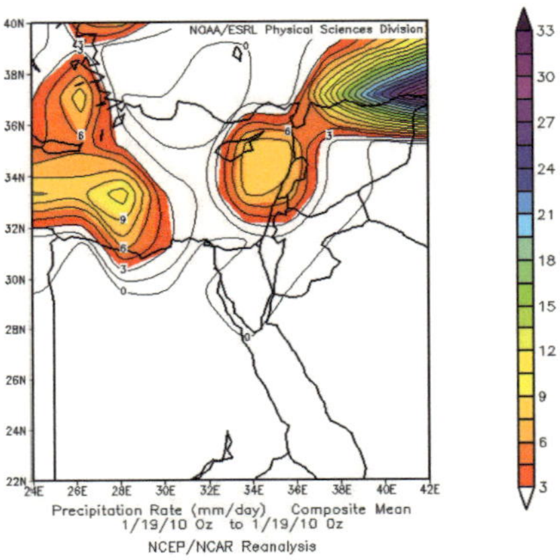
NOAA/ESRL Physical Sciences Division
Precipitation Rate (mm/day) Composite Mean
1/19/10 0z to 1/19/10 0z
NCEP/NCAR Reanalysis

(i)

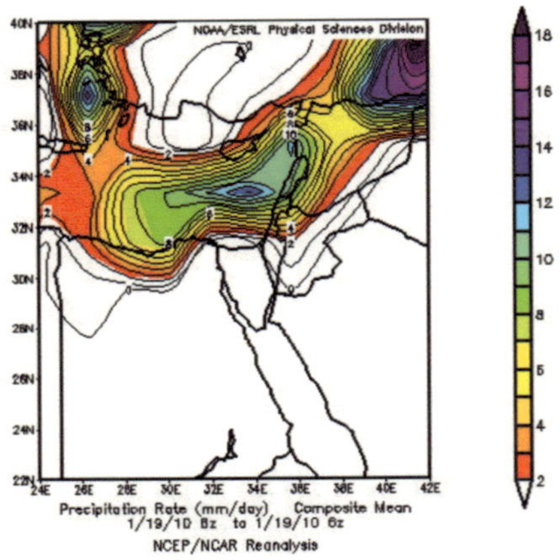
NOAA/ESRL Physical Sciences Division
Precipitation Rate (mm/day) Composite Mean
1/19/10 6z to 1/19/10 6z
NCEP/NCAR Reanalysis

(j)

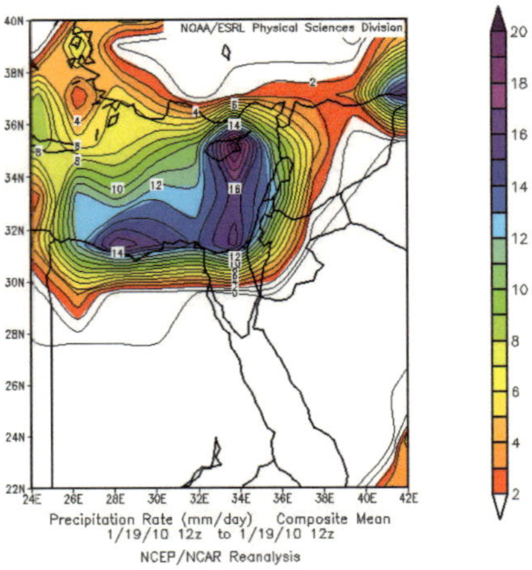
NOAA/ESRL Physical Sciences Division
Precipitation Rate (mm/day) Composite Mean
1/19/10 12z to 1/19/10 12z
NCEP/NCAR Reanalysis

(k)

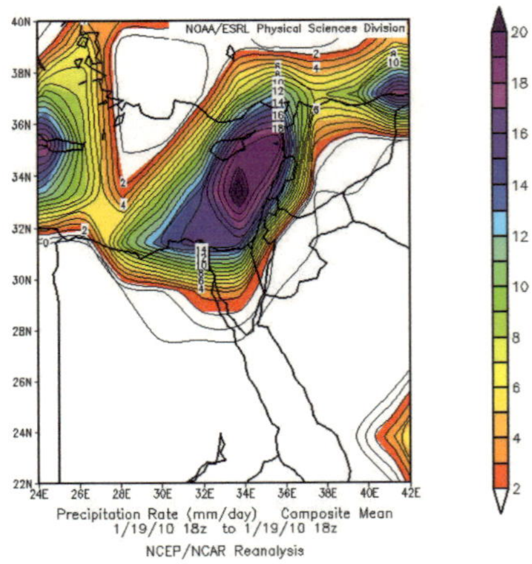
NOAA/ESRL Physical Sciences Division
Precipitation Rate (mm/day) Composite Mean
1/19/10 18z to 1/19/10 18z
NCEP/NCAR Reanalysis

(l)

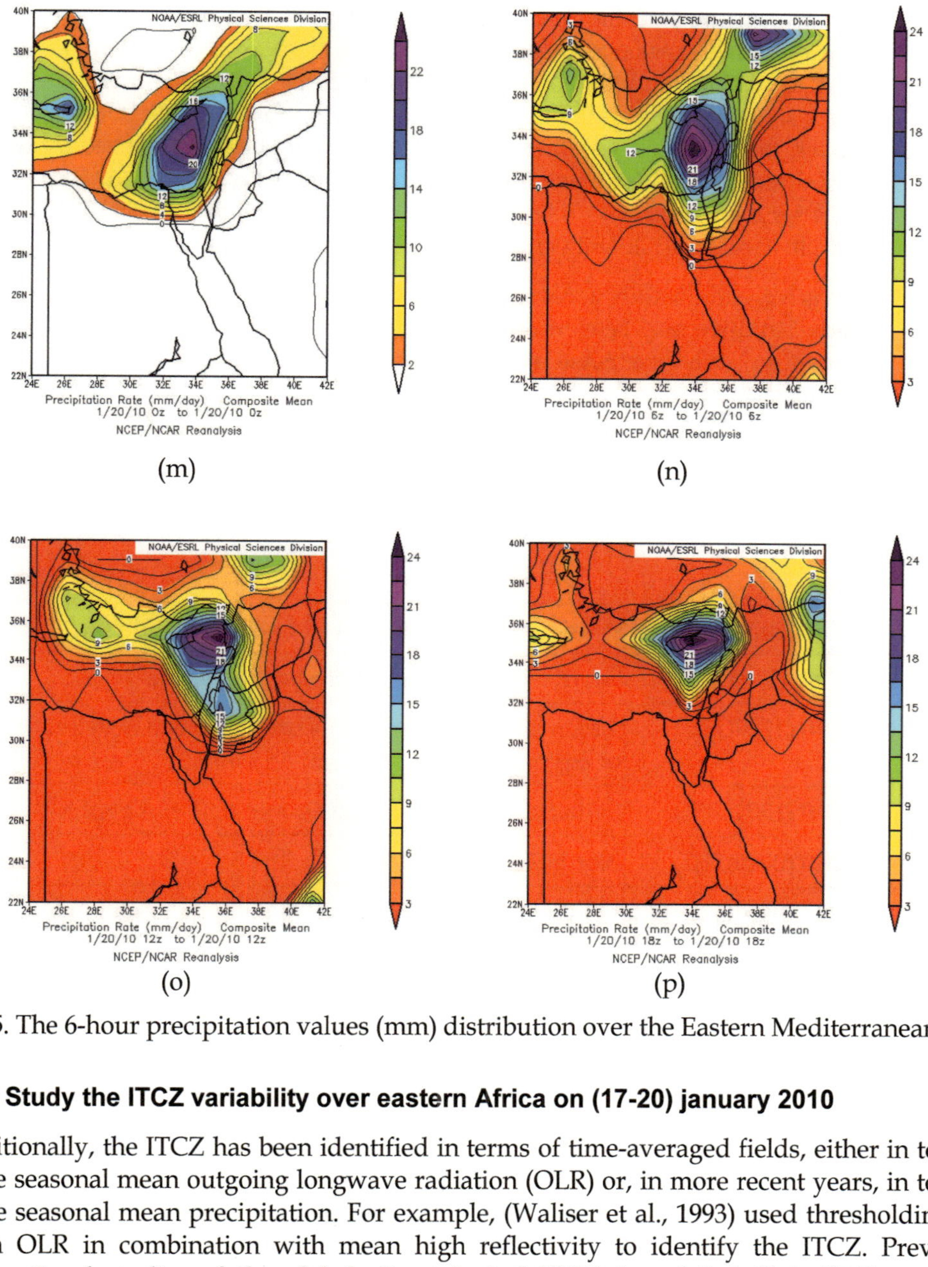

Fig. 5. The 6-hour precipitation values (mm) distribution over the Eastern Mediterranean

3.2.2 Study the ITCZ variability over eastern Africa on (17-20) january 2010

Traditionally, the ITCZ has been identified in terms of time-averaged fields, either in terms of the seasonal mean outgoing longwave radiation (OLR) or, in more recent years, in terms of the seasonal mean precipitation. For example, (Waliser et al., 1993) used thresholding of mean OLR in combination with mean high reflectivity to identify the ITCZ. Previous observational studies of the global climatological ITCZ (e.g., Mitchell & Wallace 1992; Waliser et al., 1993) focused on the annual cycle in different regions. They found very distinct longitudinal variations in the ITCZ. In the western Pacific region the summer ITCZ is broad in latitude and ill-defined due to the extensive warm pool in the ocean and monsoonal circulations. However, in the east Pacific the mean summer ITCZ is narrow and long, generally located at the southern boundary of the east Pacific warm pool, north of the strongest meridional gradient of sea surface temperature (Raymond et al., 2006). During the summer the east Pacific ITCZ is particularly visible in instantaneous satellite fields. During

Northern Hemisphere winter the ITCZ remains in the Northern Hemisphere, but its signature is considerably weaker and gets mixed in with signatures of extratropical frontal systems owing to cold air outbreaks (Wang & Magnusdottir 2006). The variability of the ITCZ presents a serious challenge to its automatic detection in instantaneous data. Here we want to focus on the ITCZ as a weather feature that has long been recognized by satellite meteorologists who analyze instantaneous fields. In the present study, 6-hour infrared (IR) satellite images for the period (17-20) January are obtained to identify the ITCZ position by using of cloud clusters. The following criteria to define the ITCZ (Bain et al.,2011).

i. The ITCZ is a predominantly zonal feature.
ii. It is cloudy but there may be cloud-free regions within the envelope of convection and the convection may be shallow (as represented by rather warm cloud-top temperatures).
iii. The ITCZ is a large-scale feature and isolated tropical disturbances, unconnected to larger cloudy regions, are not part of the ITCZ.

Variability of ITCZ (6-hour) Time	Location of Intertropical conversion zone (ITCZ)
17 January 2010 (0000UTC)	Shift to north west Sudan and north east Ethiopia
17 January 2010 (0600UTC)	South west Sudan and north east Ethiopia
17 January 2010 (1200UTC)	Sudan and extends towards north Red Sea over South east of Egypt
17 January 2010 (1800UTC)	Sudan, north Red Sea and eastern part of Egypt
18 January 2010 (0000UTC)	North Sudan and Sinai
18 January 2010 (0600UTC)	North Sudan and Sinai
18 January 2010 (1200UTC)	South Sudan
18 January 2010 (1800UTC)	South Sudan
19 January 2010 (0000UTC)	South Sudan
19 January 2010 (0600UTC)	South Sudan
19 January 2010 (1200UTC)	Extended to north Sudan
19 January 2010 (1800UTC)	Extended to Sudan, Ethiopia and Red Sea
20 January 2010 (0000UTC)	Extended eastward over south Red Sea
20 January 2010 (0600UTC)	Shift widespread eastward to Saudi Arabia
20 January 2010 (1200UTC)	Saudi Arabia
20 January 2010 (1800UTC)	Ethiopia

Table 4. The 6-hour locations of ITCZ over Eastern Africa during the period 17-20 January 2010.

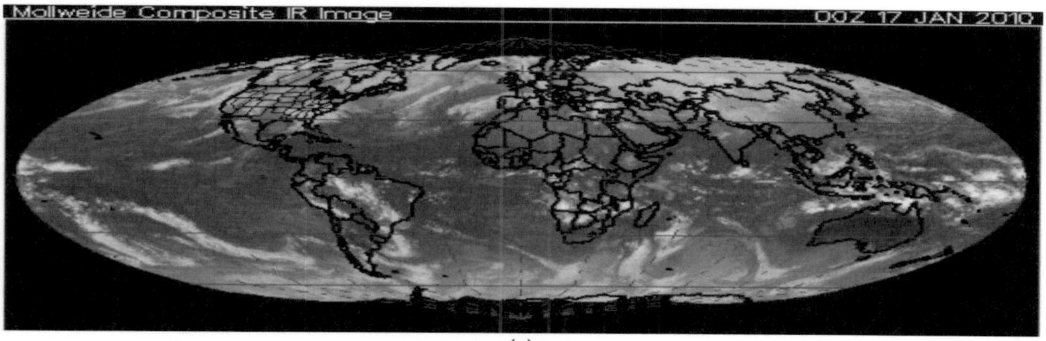
Mollweide Composite IR Image
00Z 17 JAN 2010

(a)

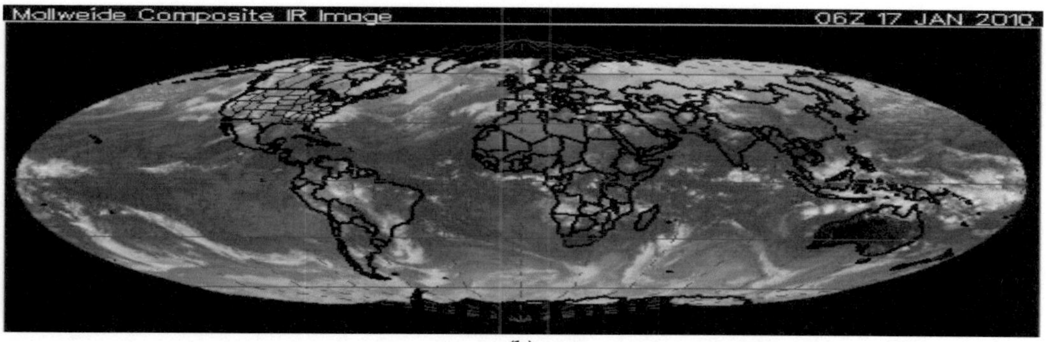
Mollweide Composite IR Image
06Z 17 JAN 2010

(b)

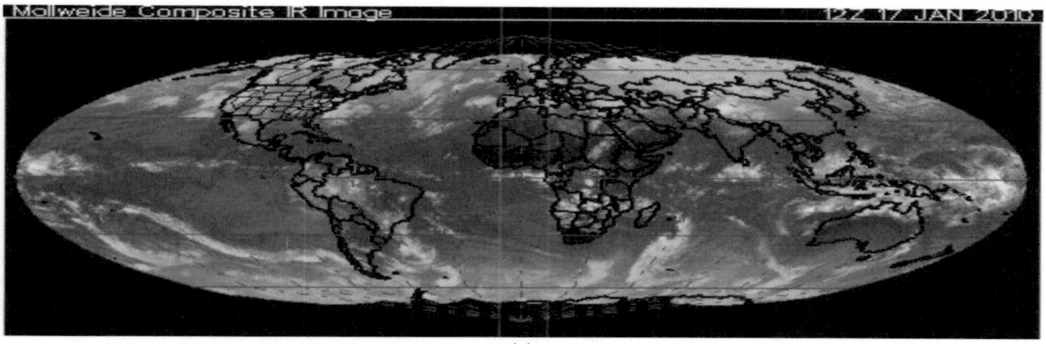
Mollweide Composite IR Image
12Z 17 JAN 2010

(c)

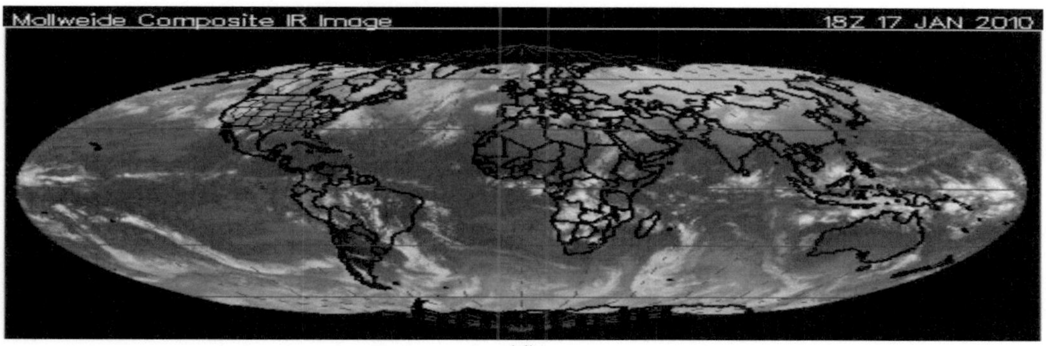
Mollweide Composite IR Image
18Z 17 JAN 2010

(d)

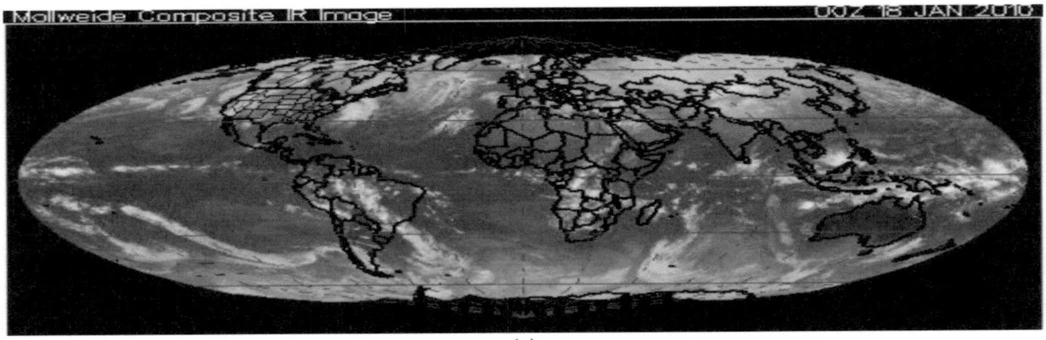
Mollweide Composite IR Image
00Z 18 JAN 2010

(e)

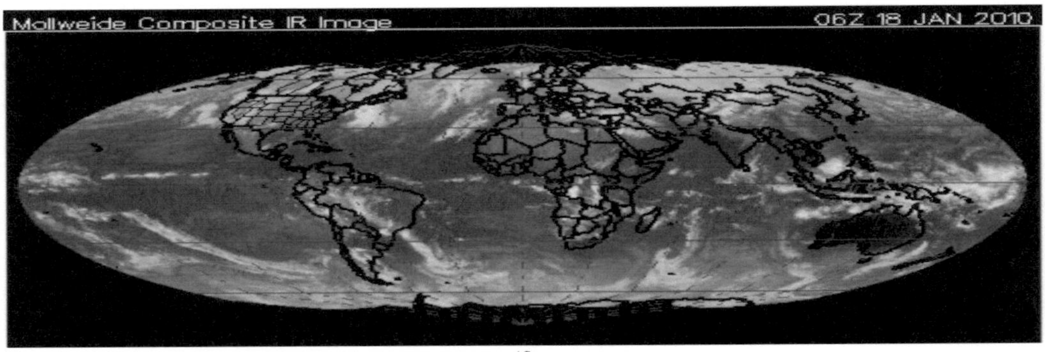
Mollweide Composite IR Image
06Z 18 JAN 2010

(f)

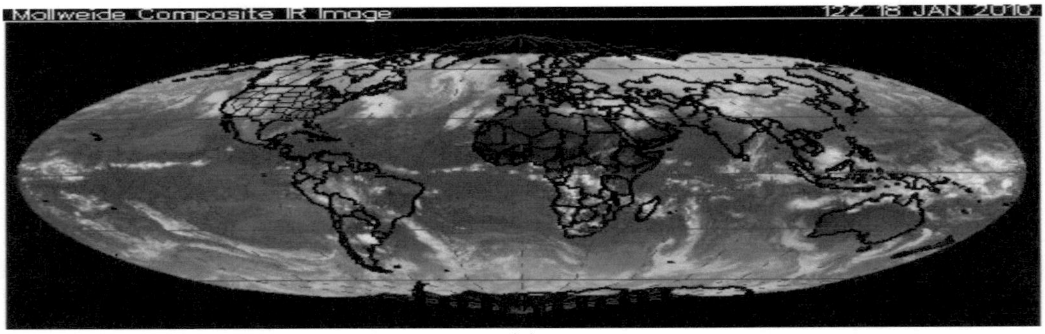
Mollweide Composite IR Image
12Z 18 JAN 2010

(g)

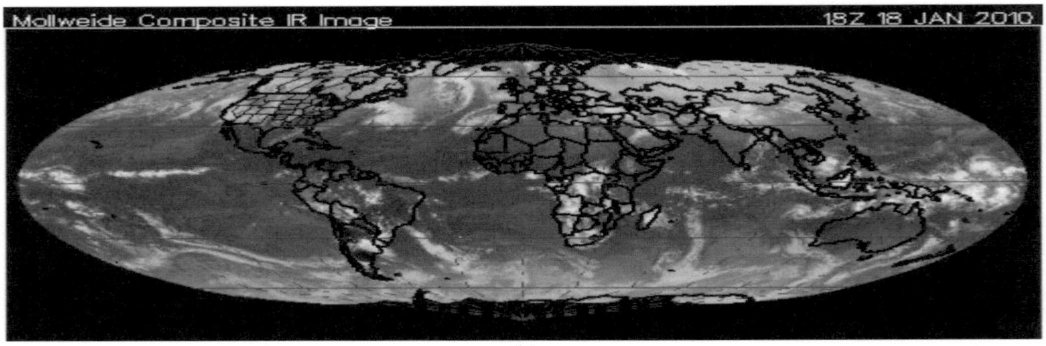
Mollweide Composite IR Image
18Z 18 JAN 2010

(h)

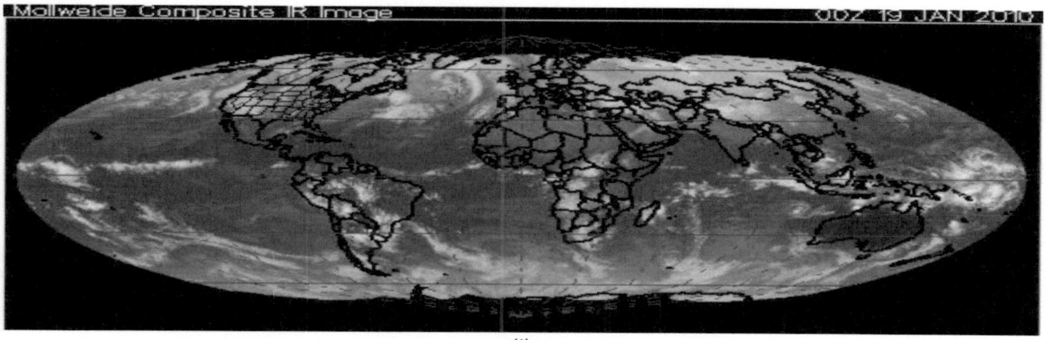
Mollweide Composite IR Image
00Z 19 JAN 2010

(i)

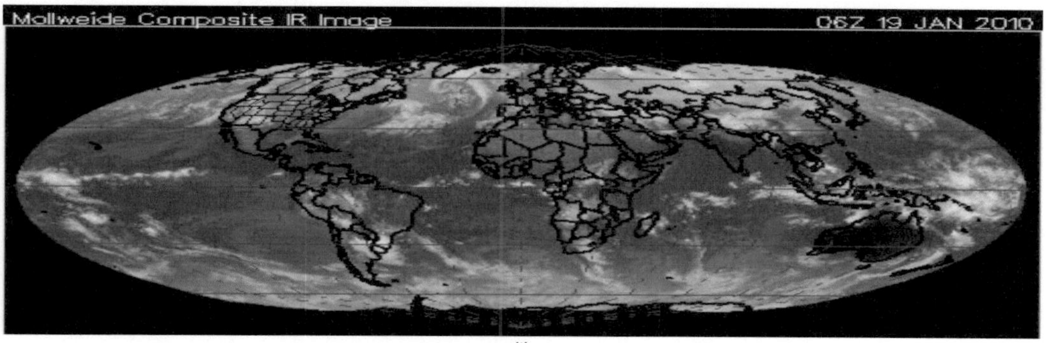
Mollweide Composite IR Image
06Z 19 JAN 2010

(j)

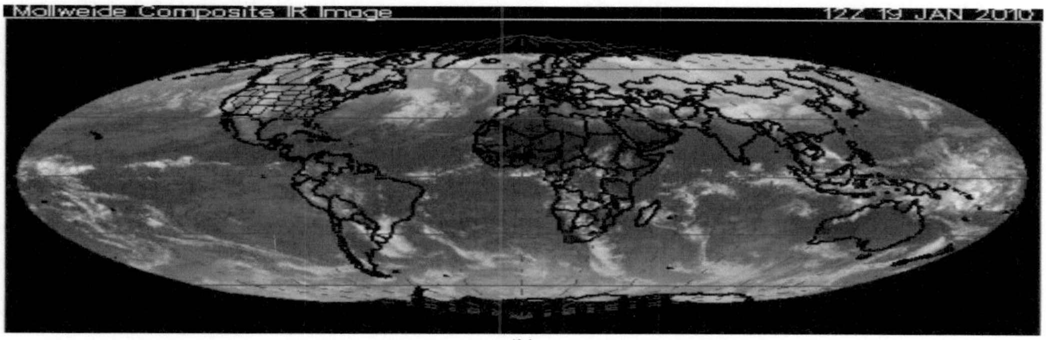
Mollweide Composite IR Image
12Z 19 JAN 2010

(k)

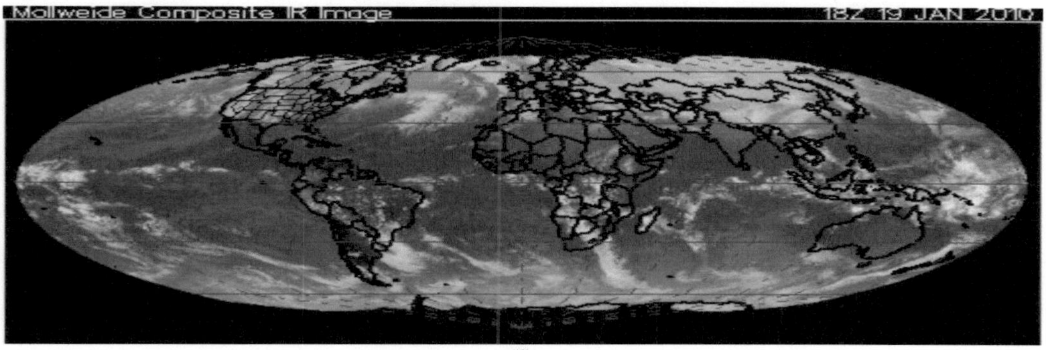
Mollweide Composite IR Image
18Z 19 JAN 2010

(l)

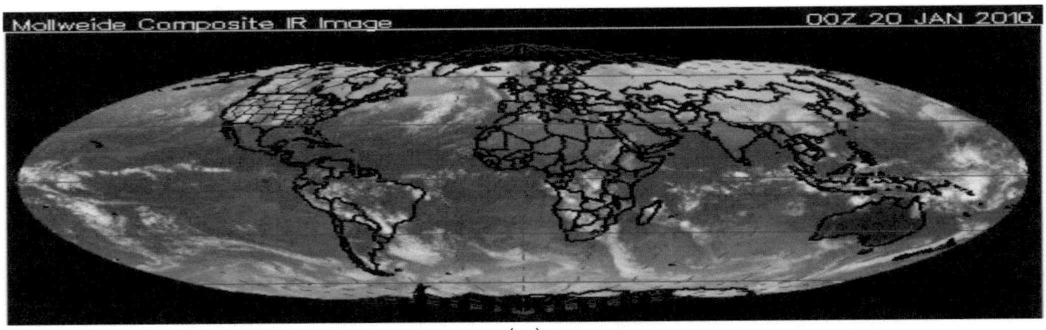

(m)

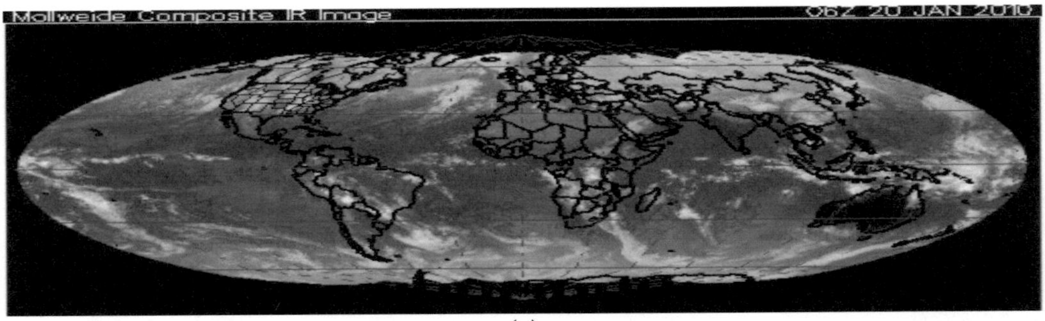

(n)

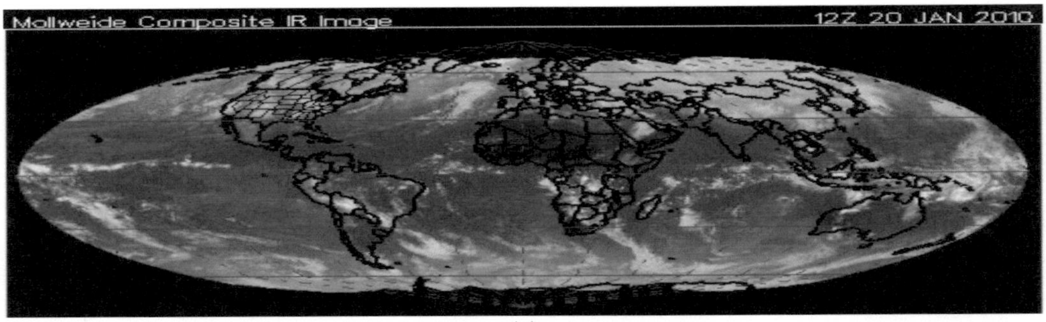

(o)

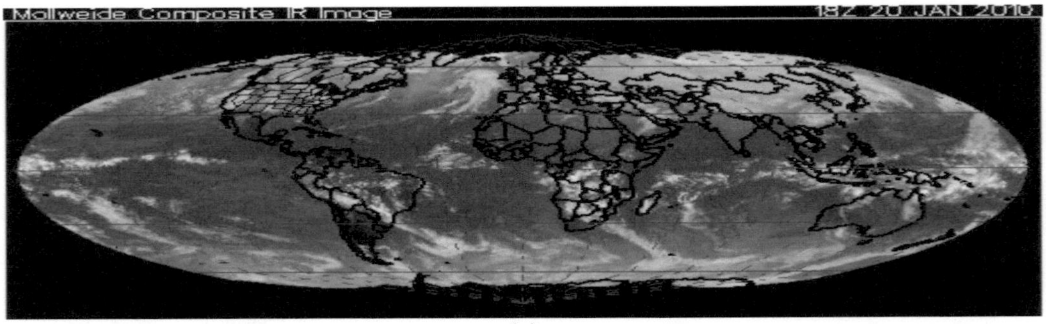

(p)

Fig. 6. The 6-hour Mollweide composite IR satellite images through the period 17-20 January 2010

We use satellite fields of infrared IR to find large-scale zonally connected regions of convection. Table 2 shows the 6 -hour location of ITCZ over eastern Africa during the period of the present case study according to the interpretation of Mollweide composite IR satellite images. From day to day it is clear that there is a shift of ITCZ over eastern Africa toward north east direction mainly over north Sudan, Ethiopia and Red Sea and reached to Sinai during the period of study. Whereas, on the first day, 17 January, ITCZ Shift to north west Sudan and north east Ethiopia and extends towards north Red Sea over South east of Egypt(see Table 4 and Figure 6a, b, c and d). During 18 January ITCZ its maximum northward extension reached to Sinai as it is clear from Table 4 and Figure 6e,f,g and h). In fact, it is abnormal that ITCZ reach to Sinai in month of January or in winter season. On 19 January, ITCZ oscillates over Sudan and extended to north Sudan, Ethiopia and Red Sea. See Table 4 and Figure 6i, j, k, and l. The eastward extension of ITCZ reach its maximum widespread eastward to reach Saudi Arabia as it is clear from Table 4 and Figure 6m, n, o, and p.

3.2.3 Relationship between variability of ITCZ over eastern Africa and occurrence of flash floods over EM on January 2010

Other studies (e.g., Serra & Houze, 2002) have referred to the ITCZ as a geographical region along which westward propagating disturbances (WPDs) tend to propagate zonally. Weak WPDs (or easterly waves) have been observed to propagate through the ITCZ cloud envelope (e.g., Scharenbroich et al., 2010) and, in general, it is impractical to attempt to separate the easterly wave signal out of the ITCZ signal. (Magnusdottir & Wang, 2008) reached the same conclusion when using spectral analysis on 40-yr European Centre for Medium-Range Weather Forecasts Re-Analysis (ERA-40) 850-hPa relative vorticity. The line of thinking that easterly waves or WPDs are inseparable from the ITCZ accommodates the idea that the ITCZ is composed of WPDs. Through the present work, in the pervious section 3.2.2, analysis of cloud clusters using of satellite images show that from day to day the location of the ITCZ is highly dynamic and changeable during the period of study. The ITCZ can form as a narrow band of convection stretched over an extensive longitudinal distance. From analysis of mean sea level pressure using of method of anomaly, it is clear there are negative anomalies (-5hpa) over EM during the period of study. As it is shown from Figure 7a, b, c and d and Table 5. In addition to that in the upper air at 500 hpa level there upper air trough of low pressure system over EM with negative anomalies less than (-75 m) of geopotential height. These anomalies persisted all the time of flash floods that existed. See Figure 8a, b, c, and d and Table 5. Also, analysis of vector wind in the tropical region revealed that there are positive anomalies more than +7 m/sec over all EM region and Red Sea through the period of study. Which mean that there was a strong westerly air current flow in EM as it is clear from Figure 9a, b, c, and d and Table 5. Meridonal wind component analysis shows that there existed a positive anomalies more than + 4 m/s over EM during the time of existing of flash floods over this region. It is clear in Figure 10a,b,c, and d and illustrated in Table 5. These results means that the south wind component is the common component in the vector wind field in EM through the period from 17-20 January 2010.The south wind component comes from the northward shift of ITCZ as unusual to exist in January month. So that, the variability of ITCZ to north and north east toward several countries in the EM region across Red Sea and Sinai carry out the convective cloud systems from tropics to EM and cause of flash floods.

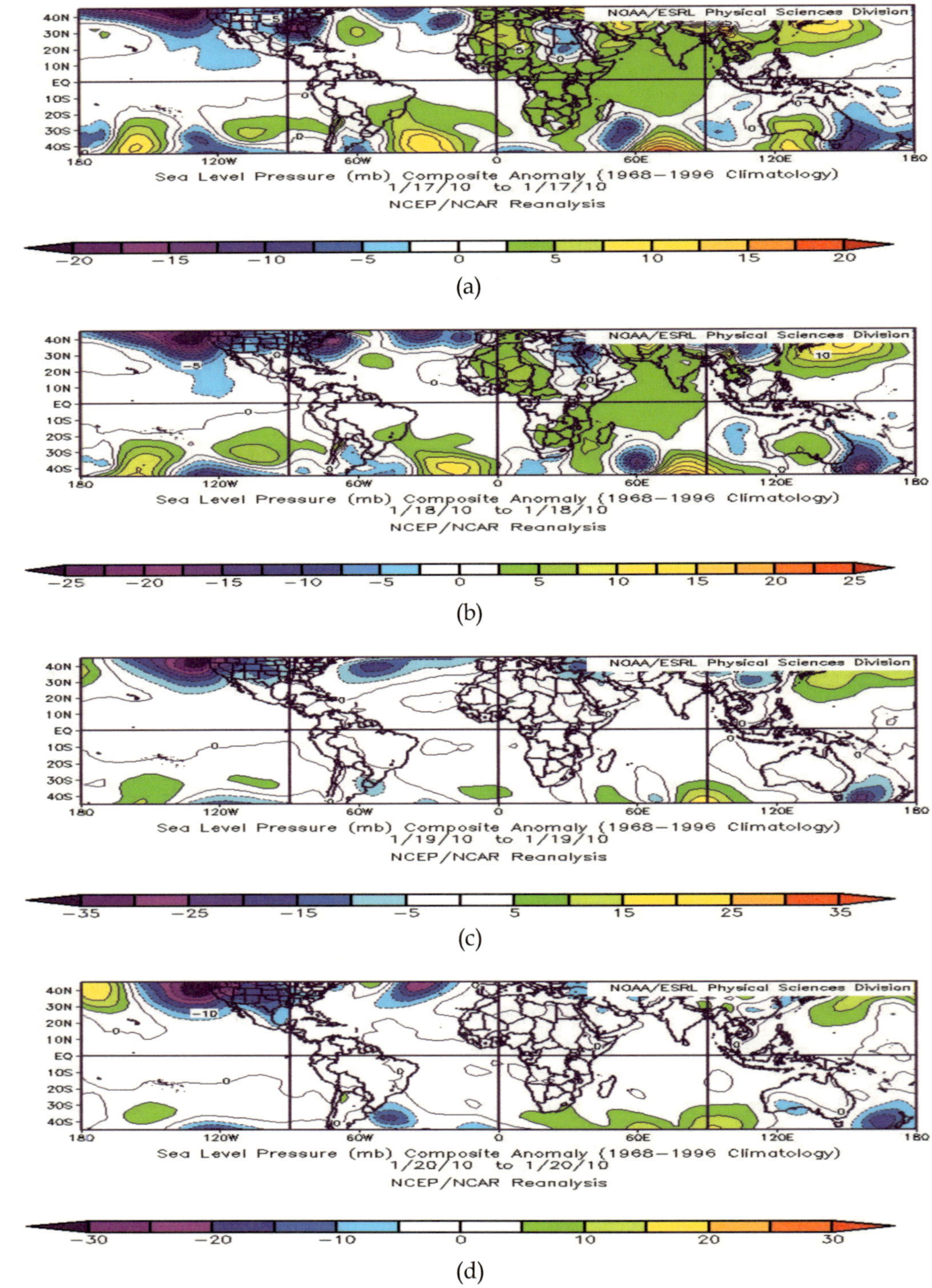

Fig. 7. The daily sea level pressure (mb) composite anomaly distribution for the period 17-20 January 2010

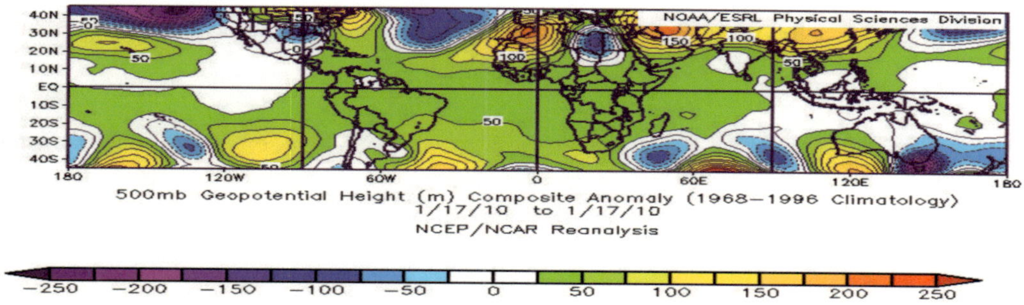

(a)

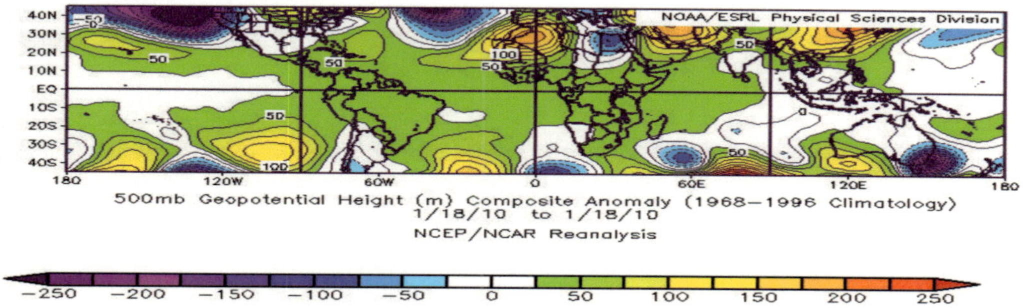

(b)

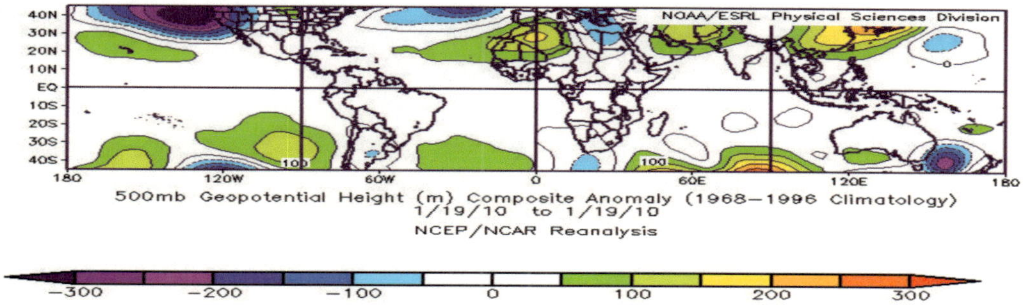

(c)

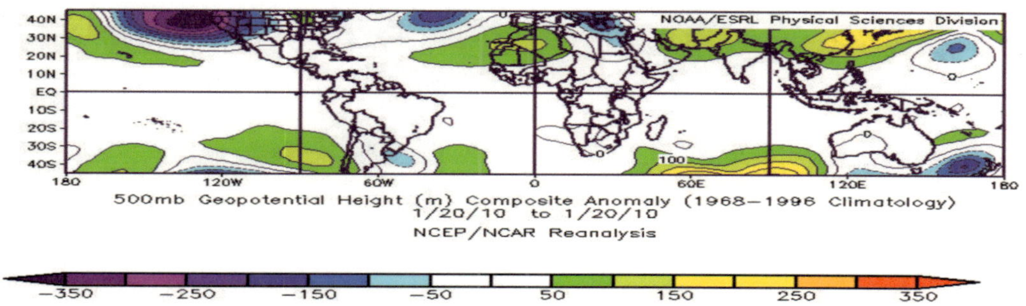

(d)

Fig. 8. The daily geopotential height (m) composite anomaly distribution for the period 17-20 January 2010

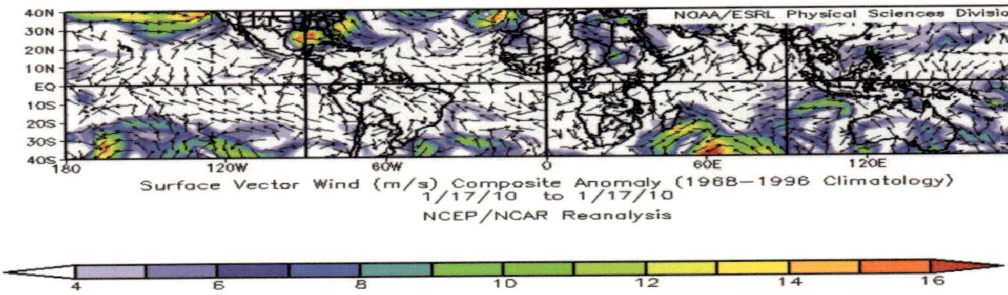

(a)

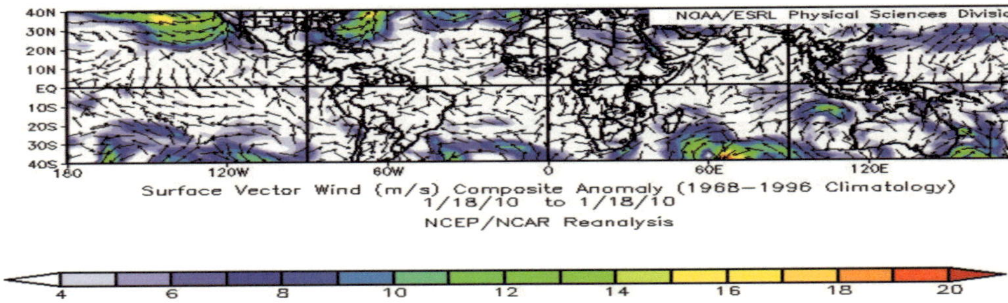

(b)

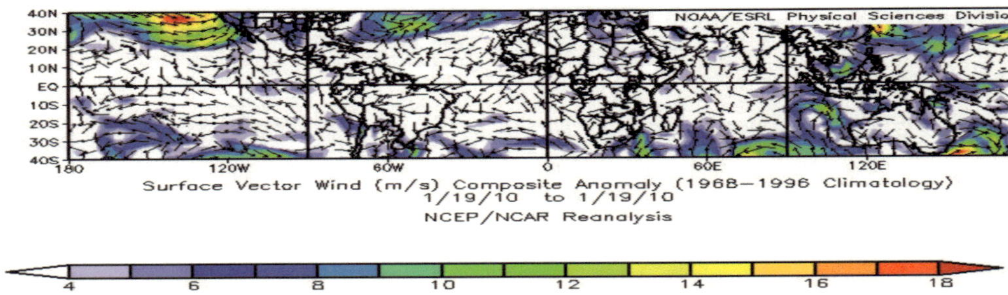

(c)

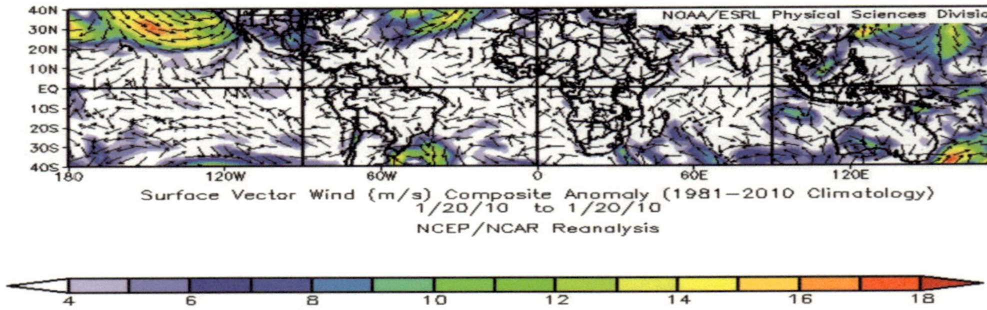

(d)

Fig. 9. The daily surface vector wind (m/s) composite anomaly distribution for the period 17-20 January 2010

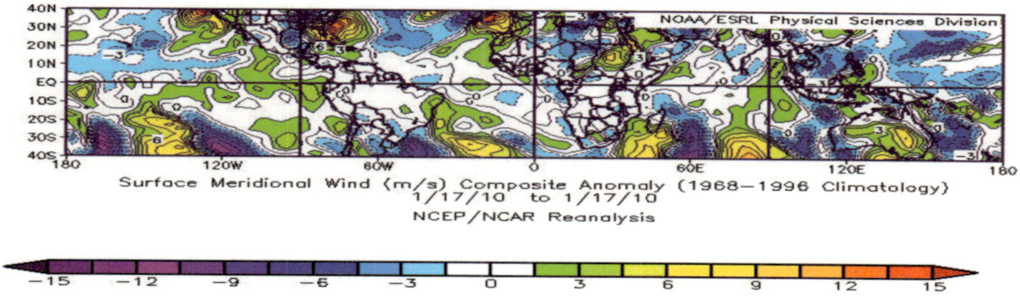

(a)

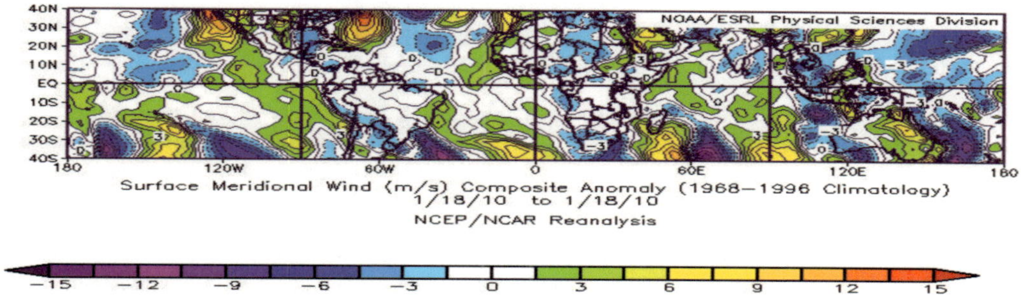

(b)

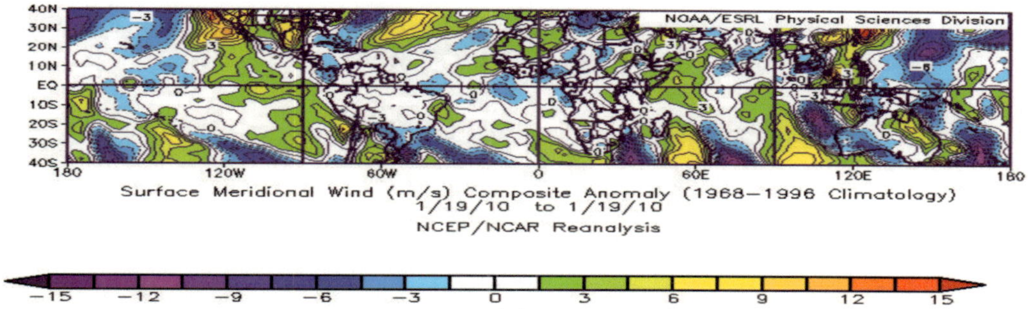

(c)

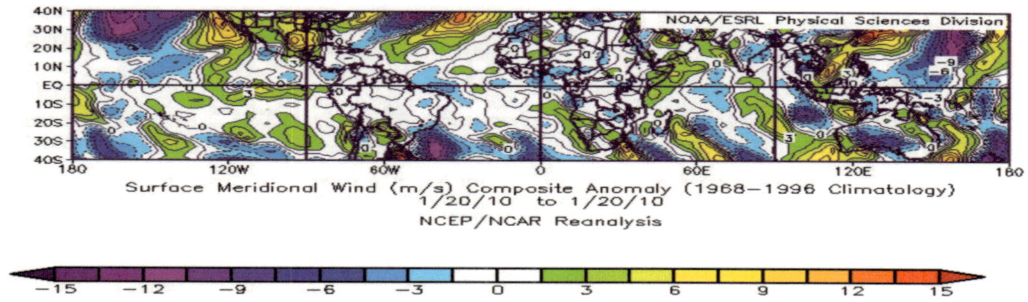

(d)

Fig. 10. The daily surface meridional wind (m/s) composite anomaly distribution for the period 17-20 January 2010

Anomalies in meteorological element Date time	Anomalies in the following meteorological elements and its location in the eastern Mediterranean region							
	Geopotential height In (m)		MSL pressure in (hpa)		Vector wind in (m/s)		Meridional wind in (m/s)	
	Value	Location	Value	Location	Value	Location	Value	Location
17 January 2010	-75	Libya and Egypt	-5	Egypt	+10	SW Sudan and medial of Red Sea	+6	Sudan, Red Sea and north of Egypt
18 January 2010	-75	Egypt	-10	Eastern Mediterranean region	+7	Libya , Sudan and central of Red Sea	+6	Red Sea and Saudi Arabia
19 January 2010	-100	North coast of Egypt	-15	Eastern Turkey	+7	Over eastern Mediterranean region	+4	Eastern Mediterranean, Red Sea and north Sudan
20 January 2010	-100	North coast of Egypt	0	Over all eastern Mediterranean region	+7	Over eastern Mediterranean region	+4	South Red Sea and eastern Mediterranean region

Table 5. Daily Anomalies amount in the meteorological elements and its location in the eastern Mediterranean region through the period 17-20 January

4. Discussion and conclusion

It is clear from the above studies that ITCZ controlling on the extreme weather events. For first case study, the UK severed from abnormal cooling through the winter season on 2009. The mean temperature for that winter was 0.5 °C below average of (1971-2000), making it the coldest winter since 1997. However, the ITCZ is an important parameter for climatic studies in the northern hemispheric circulation. Through the present work relationship between the movement of the Atlantic-Western Africa ITCZ in summer 2008 and the cooling occurred over UK in winter 2009 has been studied. The results revealed that cooling in that winter was correlated to the southward variability that existed Atlantic -Western Africa ITCZ in summer 2008. However, there is mainly a significant positive correlation coefficient value +0.7 at 5° W longitude of ITCZ. In addition to that, for second case, the results of the present study uncover that the unusual north eastward shift of ITCZ over the north Sudan, Ethiopia and Red Sea on the period of 17-20 January 2010 leads to push the tropical weather regime northward towards eastern Mediterranean. One can concluding that the extreme shift of ITCZ to north eastward over eastern Africa is causing of occurrence widespread flash floods over EM on abnormal period. So in the future works the teleconnection of interaction between tropical and midlatitude weather regimes must take in consideration to forecasting of flash floods in EM region. In fact, during winter Northern Hemisphere, winter the ITCZ lies almost in the southern Hemisphere mainly over eastern Africa and Indian ocean. Meanwhile, during Summer Northern Hemisphere summer the ITCZ lies in the

Northern Hemisphere. However, Figure 11 shows the longitudinal variations band in the ITCZ over the globe through January and July months. Satellite images show that from day to day the ITCZ is highly dynamic and changeable. This dynamic ITCZ has been the subject of dynamical modeling studies (Ferreira and Schubert 1997; Wang et al. 2010) . In a northern summer monsoon, the prevailing winds at the low levels are from the southeast. At high levels, the wind direction reverses. This configuration produces a large vertical wind shear not occur elsewhere in the tropics. In the monsoon onset process, the ITCZ shifts from near the equator to more than 10 degrees away in days. Compared with the movement of the Earth's tilt toward the Sun, this change is rapid. The shifts and preferred latitudes of the ITCZ observations and theory were investigated in several scientific studies (Bates,1970; Philander et al., 1996 and Hafez ,2003c).

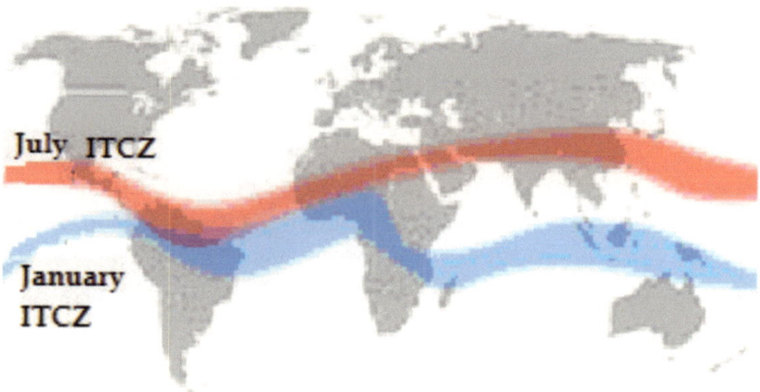

Fig. 11. The longitudinal distribution of ITCZ in winter season represents by January and in summer season represents by July.[source: Wikipedia, the free encyclopedia.mht]

5. Acknowledgments

It is a pleasure to the author to thank the Climate Diagnostics Centre for supporting the data used throughout this study. Plots and images were provided by the NOAA-CIRES Climate Diagnostics Centre, Boulder, Colorado, USA from their Web site at http://www.cdc.noaa.gov. Also, thanks to the Climate Prediction Centre for supporting the summer Atlantic - Western Africa ITCZ data which obtained through the website http://www.cpc.ncep.noaa.gov/products/monitoring_data/., and the UK Meteorological Office for its support of summary of winter 2008/2009 that obtained from the website http://www.metoffice.gov.uk/. Also, thanks to the Climate Diagnostics Centre for supporting the data used throughout this study. Plots and images were provided by the NOAA-CIRES Climate Diagnostics Centre, Boulder, Colorado, USA from their Web site at http://www.cdc.noaa.gov. Also, great thanks for dartmouth flood observatory for obtained dataset of flash floods through the website http://www.dartmouth.edu.

6. References

Bain, L. Caroline, Jorge De Paz, Jason Kramer, Gudrun Magnusdottir, Padhraic Smyth, Hal Stern, Chia-chi Wang, (2011). Detecting the ITCZ in Instantaneous Satellite Data using Spatiotemporal Statistical Modeling: ITCZ Climatology in the East Pacific. J. of Climate, 24, 216–230.

Barnolas, M. & Llasat, M. C. (2007). A flood geodatabase and its climatological applications: the case of Catalonia for the last century. Nat. Hazards Earth Syst. Sci., 7, 271–281.

Bates, J. R., (1970). Dynamics of Disturbances on the Intertropical Convergence Zone, Quart., J. R. Met. Soc., 96, 677-701.

Broccoli, A. J., Dahl, K. A., & Stouffer, R. J. (2006). Response of the ITCZ to Northern Hemisphere cooling, Geophys. Res. Let., 33, L01702, doi:10.1029/2005GL024546.

Citeau, J., Berges, J. C., Demarcq, H. & Mahe, G., (1988b). The Watch of ITCZ Migrations over Tropical Atlantic as an Indicator in Drought Forecast over Sahelian area. Ocean-Atmos. News. (45):1-3.

Cohen, J., Saito, K., & Entekhabi, D., (2001). The Role of the Siberian High in Northern Hemisphere Climate Variability. Geophys. Res. Let., 28, 2, 299-302.

Gadgil, S. & Guruprasad, A., (1990). An Objective Method for Identification of the Intertropical Convergence Zone. J. Climate, 3, 558-567.

Hafez, Y.Y., (2008). The Teleconnection Between the Global Mean Surface Air Temperature and Precipitation over Europe. I. J. Meteorology, U. K., 33, 331, 230-236.

Hafez, Y.Y., (2007). The Connection Between the 500 hpa Geopotential Height Anomalies over Europe and the Abnormal Weather in Eastern Mediterranean During Winter 2006. I. J. Meteorology, UK, 32, 324, 335-343.

Hafez, Y. Y., (2003a). Changes in Atlantic - Western Africa ITCZ Variability and Its Influence on the Precipitation Rate in Europe on Sever Rainy Summer 2002. J. Meteorology, U.K., 28, 282, 299-307.

Hafez, Y. Y. (2003b). A severe case of flash flood in Egypt. J. Enviro. Sci, Vol. 25. pp.39-58.

Hafez, Y. Y. (2003c). Changes in Atlantic - Western Africa ITCZ Variability and Its Influence on the Precipitation Rate in Europe on Sever Rainy Summer 2002. J. Meteorology, U.K., 28, 282, 299-307.

Hatzaki, M., Flocas, H. A., Oikonomou, C., & Giannakopoulos,C.(2010). Future changes in the relationship of precipitation intensity in Eastern Mediterranean with large scale circulation. Adv. Geosci.,23, 31–36, doi:10.5194/adgeo-23-31-2010.

Hess, P. G., Batissti, D. S. &Rasch, P. J., (1993). Maintenance of the Intertropical Convergence Zones and the Large-Scale Tropical Circulation on a Water Covered Earth. J. Atmos. Sci., 50, 691-713.

Ilesanmi, O. O., (1971). An Empirical Formulation of an ITD Rainfall Model for the Tropics: A Case Study of Nigeria. J. Appl. Meteor. 10, 882-891.

Kalnay, E. & Coauthors, (1996). The NCEP/NCAR Reanalysis 40-year Project. Bull. Amer. Meteor. Soc., 77, 437-471.

Kotroni, V. Katsanos, D. Michaelides, S. Yair, Y. Savvidou, K. & Nicolaides, K. (2010). High-impact floods and flash floods in Mediterranean countries: the FLASH preliminary database. Adv. Geosci., 23, 47–55.

Kraus, E. B., (1997). The Seasonal Excursions of the Intertropical Convergence Zone. Mon. Wea. Rev., 105, 1009-1018.

Llasat, M. C.: Chapter 18 (2009). Physical Geography of the Mediterranean basin, edited by: Woodward, J.,Oxford University Storms and floods, in :The Press, 504–531 pp.

Llasat, M. C. Llasat-Botija, M. Prat, M. A. Porc ́u, F., Price, C. Mugnai, A. Lagouvardos, K.

Magnusdottir & Wang, C.-C. (2008). Intertropical convergence zones during the active season in daily data. J. of Atmos. Sci., 65, 2425-2436.

Meteooffice, UK, (2009). Winter Summary, 2009, ©Crown copyright (http://www.metoffice.gov.uk/).

Michaelides, S., Yair, Y., Savvidou, K., &Nicolaides, K. (2010). High-impact floods and flash floods in Mediterranean countries:the FLASH preliminary database. Adv. Geosci., 23, 47–55.,

Nieto Ferreira, R., & Schubert, W. H. (1997). Barotropic aspects of ITCZ breakdown. J. Atmos. Sci., 54, 261–285.

Philander, S. G. H., Gu, D., Lambert, G., Lau, N.C., Li, T. & Pacanows, R. C., (1996). Why the ITCZ is Mostly North of the Equator. J. Climate, 9, 2958-2972.

Pike, A. C., (1972). The Inter-Tropical Convergence Zone Studied with an Interacting Atmosphere Ocean Model. Mon. Wea. Rev., 99, 469-477.

Raymond, D. J., Bretherton, C. S., & Molinari, J., (2006). Dynamics of the Intertropical Convergence Zone of the East Pacific. J. Atmo. Sci., 63, 2, 582-597.

Rosting, B. & Kristjansson, (2008). A Successful Resimulation of the 7-8 January 2005 Winter Storm Through Initial Potential Vorticity Modification in Sensitive Regions. Tellus A, 60, 4, 604-619.

Scharenbroich, L. Magnusdottir, G. Smyth, P. Stern, H. & Wang, C.-C. (2010). A Bayesian framework for storm tracking using a hidden-state representation. Mon. Wea. Rev., 138, 2132--2148.

Serra, Y., & Houze, Jr. R. A.(2002). Observations of variability on synoptic timescales in the east Pacific ITCZ. J. Atmos. Sci., 59, 1723-1743.

Spiegel, M. R., (1961). Theory and Problems of Statistics, Schaum, PP, 359.

Sultan, B. & Janicot, S., (2000). Abrupt Shift of the ITCZ over West Africa and Intra-Seasonal Variability. Geo. Res. Let, 27, 3353-3356.

Waliser, D. E., (1992). The Preferred Latitudes of the Intertropical Convergence Zone: Observations and Theory. Ph. D. Dissertation, Scripps Institution of Oceanography, University of California, San Diego.

Waliser, D. E., Groham, N.E. & Gautier C. (1993). Comparison of the highly reflective cloud and outgoing longwave radiation datasets for use in estimating tropical deep convection. J. Climate, 6, 331-353.

Waliser, D. E. & Somerville, C. J., (1994). Preferred Latitudes of the Intertropical Convergence Zone. J. Atmos. Sci., 51, 1619-1639.

Wang, & Magnusdottir, G. (2006). The ITCZ in the central and eastern Pacific on synoptic timescales. Mon. Wea. Rev., 134,1405-1421. Wang, C.-C., Chou C., & Lee, W.-L. (2010). The breakdown and reformation of the ITCZ in a moist atmosphere. J. of Atmos. Sci., 67,1247-1260.

7

Blocking Systems Persist over North Hemisphere and Its Role in Extreme Hot Waves over Russia During Summer 2010

Yehia Hafez
Cairo University, Faculty of Science,
Department of Astronomy, Space Science and Meteorology
Egypt

1. Introduction

The 2010 Northern Hemisphere summer included severe heat waves that impacted the European continent as a whole, along with parts of, Russia during June, July and August 2010. The 2010 summer heat wave over several parts of Russia was extraordinary, with the region experiencing the warmest July since at least 1880. During summer 2010 all of Europe lies under controlling of blocking systems that persist long time. The formation, persistence, and the role played by blocking systems in abnormal weather and climate in the northern hemisphere challenged in several scientific literatures (e.g; Rex,1951; Namias, 1964, 1978; Dickson & Namias,1976; Dole, 1983); Hafez, 1997; Lejenas, 1989; Cohen, et al., 2001; and recently Hafez, 2008b and 2011). For heat waves, (Stott et al., 2004), for the 2003 western European heat wave, they found that, human influences are estimated to have at least doubled the risk for such an extreme event. Other boundary forcing also contributed to the 2003 European heat wave, including anomalous sea surface temperatures (SSTs) (Feudale & Shukla, 2010). (Dole et al., 2011) studied the 2010 northern hemisphere summer to explore whether early warning could have been provided through knowledge of natural and human-caused climate forcings. They used Model simulations and observational data to determine the impact of observed sea surface temperatures (SSTs), sea ice conditions and greenhouse gas concentrations. They found that, analysis of forced model simulations indicates that neither human influences nor other slowly evolving ocean boundary conditions contributed substantially to the magnitude of this heat wave. Analysis of observations indicate that this heat wave was mainly due to internal atmospheric dynamical processes that produced and maintained a strong and long-lived blocking event, and that similar atmospheric patterns have occurred with prior heat waves in this region. They concluded that the intense 2010 Russian heat wave was mainly due to natural internal atmospheric variability. However, A heat wave is a prolonged period of excessively hot weather, which may be accompanied by high humidity. There is no universal definition of a heat wave, the term is relative to the usual weather in the area. Temperatures that people from a hotter climate consider normal can be termed a heat wave in a cooler area if they are outside the normal climate pattern for that area. The term is applied both to routine weather variations and to extraordinary spells of heat which may occur only once a century. Severe heat waves have caused catastrophic crop failures, thousands of deaths from hyperthermia, and widespread power outages due to increased use of air conditioning (Meehl et al., 2004). The goal of the present work is to uncover the primary causes of long persistence of blocking systems over the north atmosphere and its teleconnection with the Russian hot wave in summer 2010.

2. Data and methodology

The monthly and seasonal time series NCEP/NCAR reanalysis data composites for temperature, geopotential height at level 500 hpa over the northern hemisphere for the period (1980-2010) has been used through the present study(Kalnay et al., 1996). The daily NCEP/NCAR reanalysis data composites are used to study the persistence of blocking systems over the northern hemisphere through 91 days (1 Jun. - 30 Aug.) 2010 year. In addition to that, time series of surface temperature, geopotantial height over Russia, NAO and SOI data through the summer seasons from 1980 to 2010 had been plotted and discussed. Anomalies and correlation coefficient methods have been used to analyzed the datasets. Correlation coefficient analysis of these time series was done.

3. Results

3.1 Study of the synoptic situation and blocking high over Russia on summer of 2010

Russia is occupied a very vast area 17,075,200 sq km in Northern Asia (that part west of the Urals is included with Europe), bordering the Arctic Ocean, between Europe and the North Pacific Ocean. Climate of Russia ranges from steppes in the south through humid continental in much of European Russia; subarctic in Siberia to tundra climate in the polar north; winters vary from cool along Black Sea coast to frigid in Siberia; summers vary from warm in the steppes to cool along Arctic coast. Figure 1 is the geographical map for location of Russia. Since Russia has a vast area it is include several distinct climatic zones. Where as, the western parts of it has its common weather features rather than the eastern part. Also the climate of the north Russia is completely different of it at the southern bounders. For example, in its capital, Moscow, the climate is exposed to cold winters, warm and mild summers, and very brief spring and autumn seasons. Typical high temperatures in the warm months of July and August are around 22°C; in the winter, temperatures normally drop to approximately -12°C. Monthly rainfall totals vary minimally throughout the year, although the precipitation levels tend to be higher during the summer than during the winter. Due to the significant variation in temperature between the winter and summer months as well as the limited fluctuation in precipitation levels during the summer, Moscow is considered to be within a continental climate zone. In summer of 2010 weather regime Moscow and several parts of Russia becomes a unique severe hot weather see Figure 2. The anomalies in surface air temperature recorded its maximum value (+6° C) in that summer season through the mean average of the period (1980-2010) as it is clear in Figure 3. (Razuvaev et al., 2010) said that, the center of European Russia was well covered by meteorological observations for the past 130 years. These data, historical weather records (yearbooks or "letopisi" , which were carried on in the major Russian monasteries), and finally, dendroclimatological information, all show that this summer temperature anomaly was well above all known extremes in the past 1000 years. A 60-days-long hot anticyclonic weather system with daily temperature anomalies as high as +10° C. The extreme heat, lack

Fig. 1. The geographical map for location of Russia. [Source: climatezone.com]

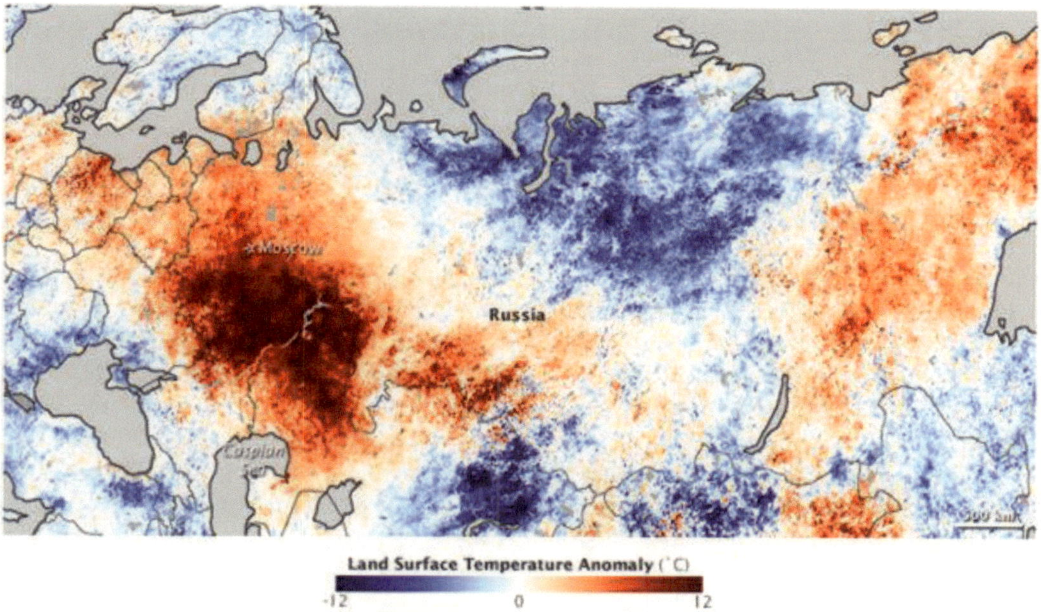

Fig. 2. Shows temperature anomalies for July 20-27 relative to the average for the same dates 2000-2008. [Source: http://earthobservatory.nasa.gov/IOTD/viewphp?id=45069]

of precipitation, and forest fires have caused hundreds of deaths and multimillion dollars in property losses. Indirect losses of lives due to this weather anomaly, with the ensuing fires and related air pollution. In the present study the synoptic situation over Russia through 2010 summer season has been studied using of temperature and geopotential height at 500 hpa level data. Figures 3 and 4 illustrated that there positive surface air and upper level temperature at 500 hpa anomalies (+6°C and + 3°C) over Russia during months of 2010

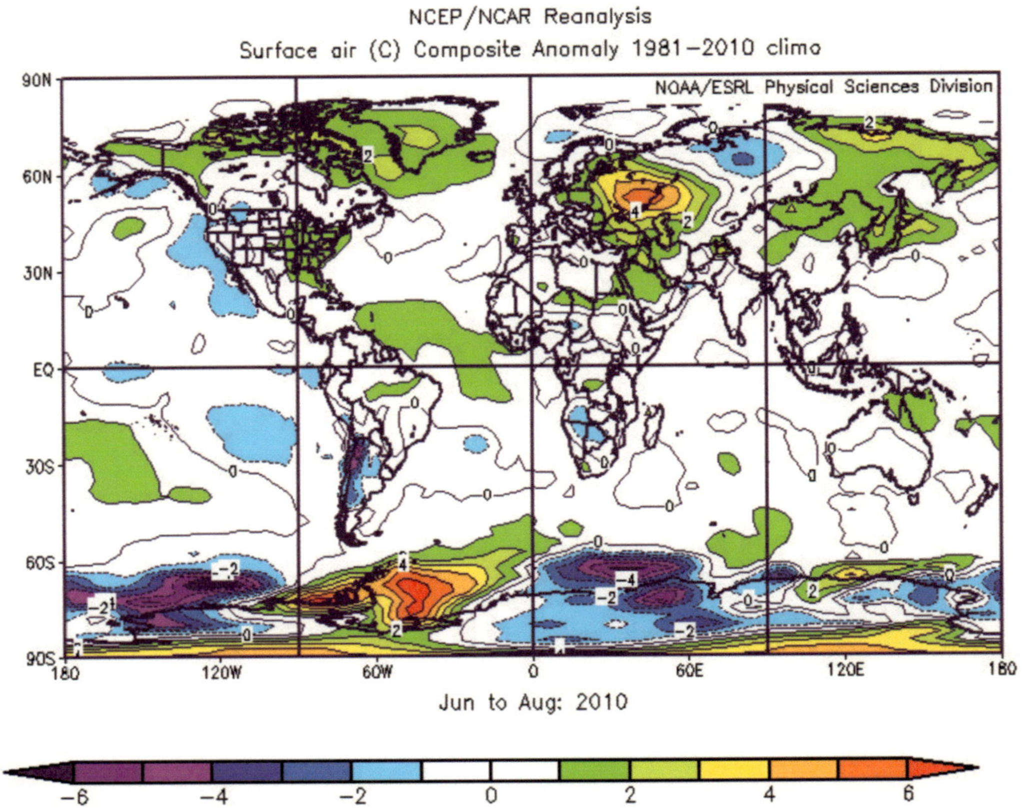

Fig. 3. The global distribution of surface air temperature (C) anomaly for summer 2010

summer season respectively. During this time land surface temperature anomalies over the western part of Russia recorded its extremely maximum value in July +12°C, see Figure 2. The synoptic and numeric criteria of the formation of blocking systems over northern hemisphere according to (Rex, 1950a, 1950b, 1951; Dole, 1978,1982, and Hafez, 1997) had been used to identify and to follow the daily persistence of the blocking systems. These criteria have basic conditions like as; the westerly air current aloft must splitting to two distinct branches, this splitting must be northern latitude of 30 N, the anomalies in geopotential height values at 500 hpa must be more than + 100 m for blocking high pressure system and these conditions must persist more than 10 days. Appling this criteria in the present study revealed that western Russia lies under the influence of blocking system about 75 days during the period of study 92 days which are the days of summer 2010. A 60 days of it through July and August months. Whereas the blocking system developed over western Russia in the next have of June 2010.This blocking episode over the northern hemisphere and mainly over Russia persisted to long time reached to the time of this season. Figure 5 shows the anomalies in geopotential height during summer 2010. It is clear that there are a notably positive anomalies over western Russia. Figures 6, 7 and 8 illustrated a 10-day anomalies of geopotential height at 500 hpa level distribution over Russia during June, July and August months of summer 2010 respectively. In fact the blocking time duration start from one week to a complete season. The historical record of blocking episodes indicated that the episodes that persist for a season are too rarely. The long

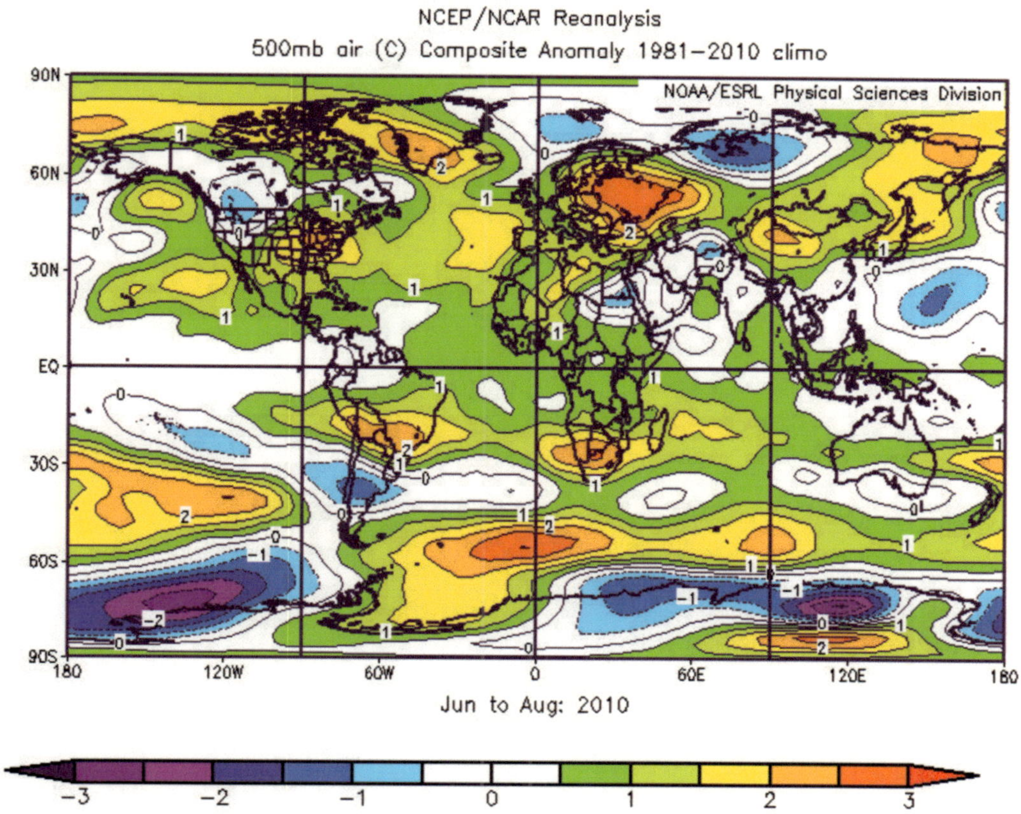

Fig. 4. The global distribution of 500 mb level air temperature (C) anomaly for summer 2010

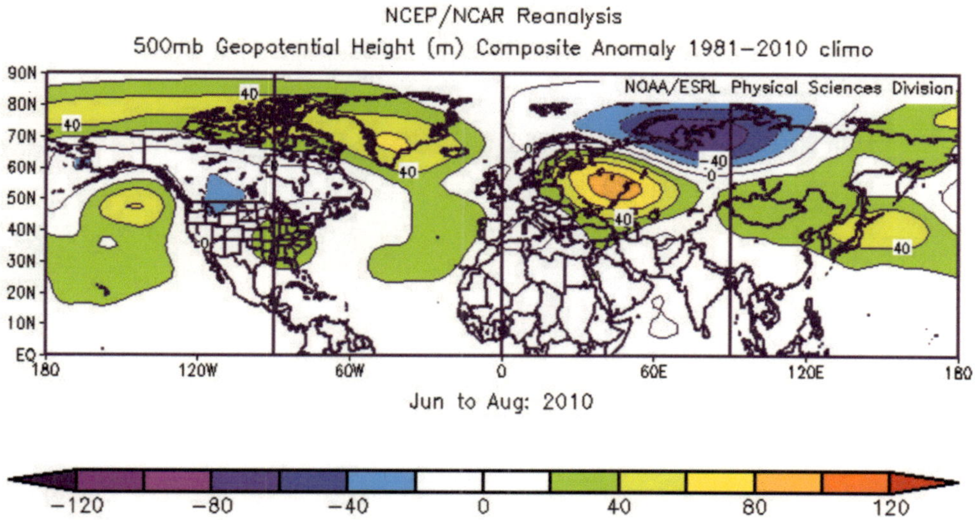

Fig. 5. The northern atmosphere distribution of geopotential height (m) anomaly for summer 2010

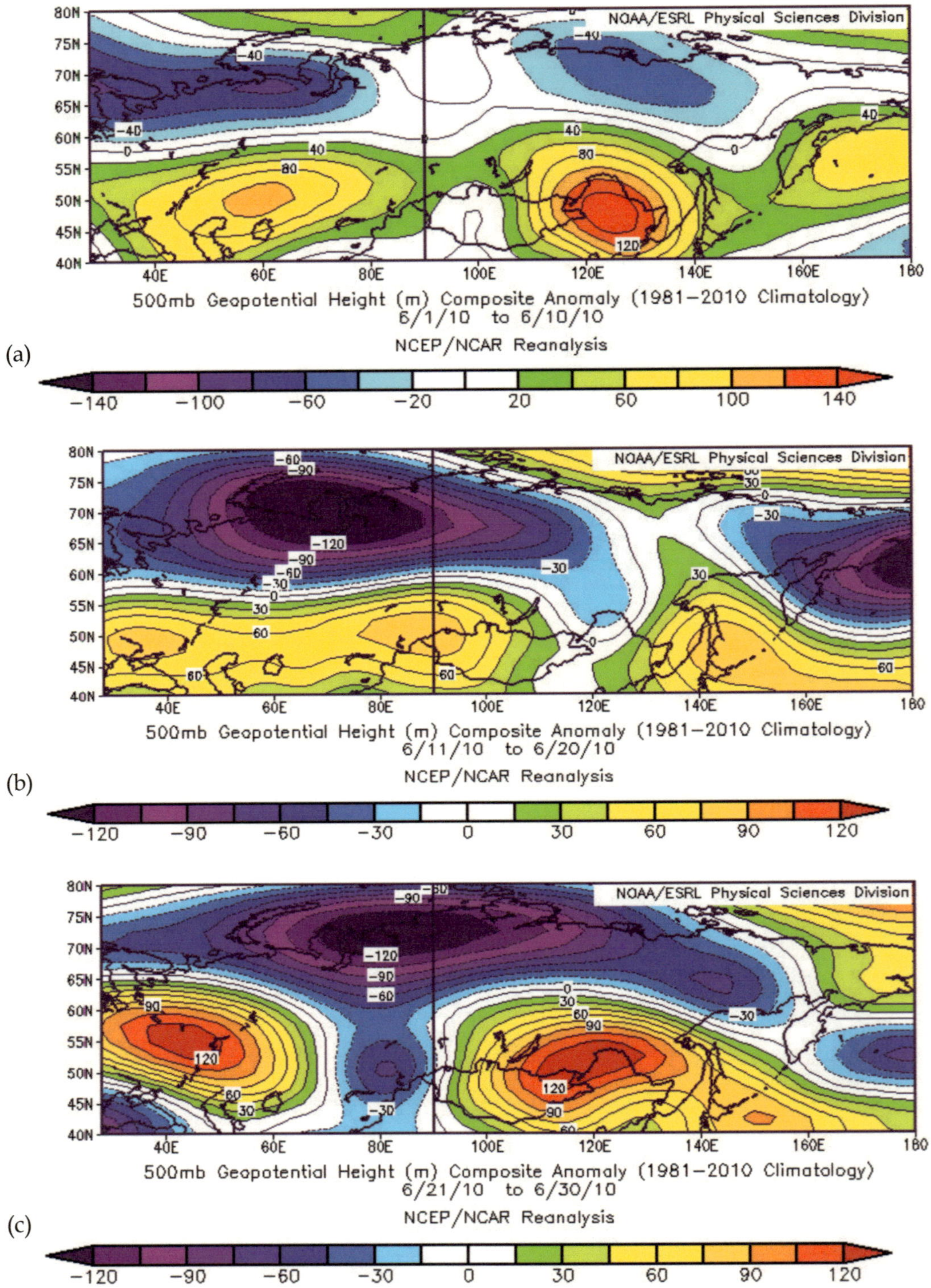

Fig. 6. A 10-day distribution of 500mb geopotential height (m) anomaly over Russia for June month of year 2010

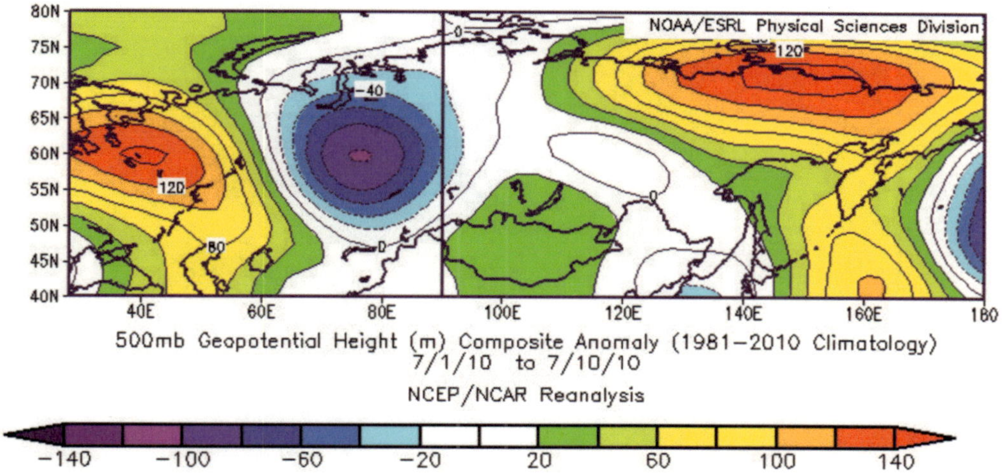

(a)

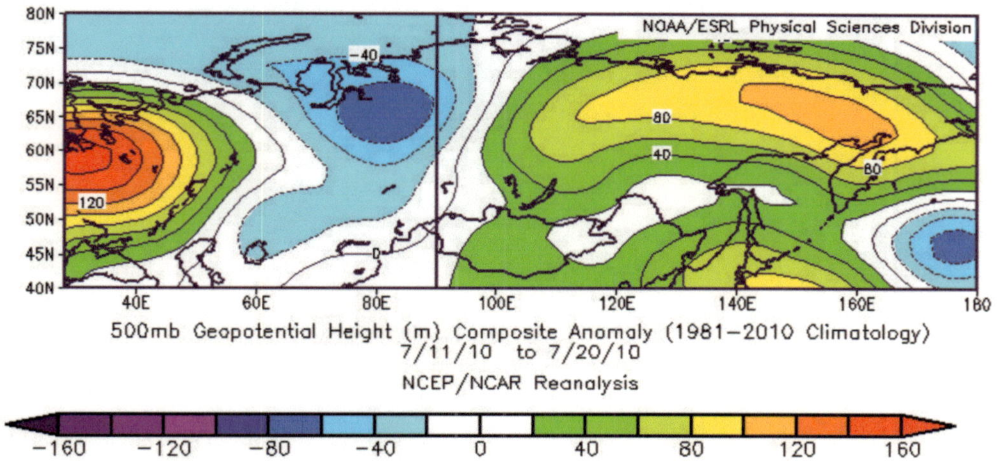

(b)

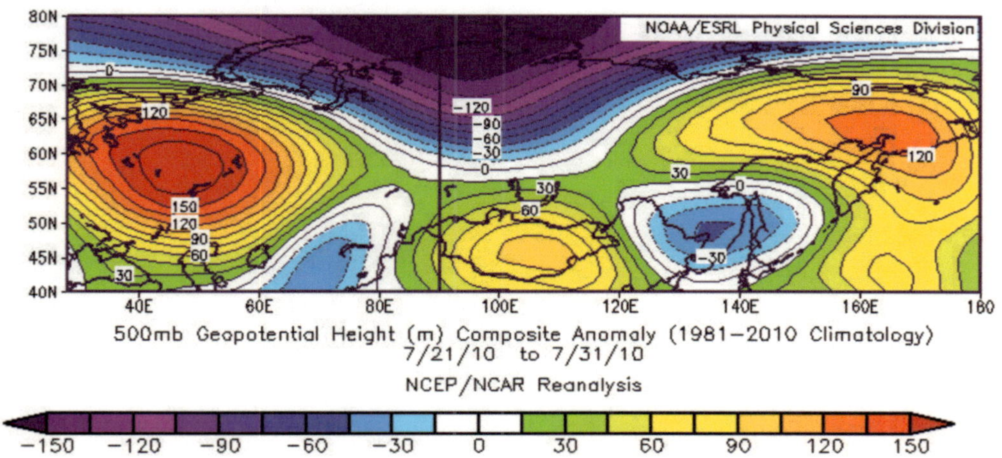

(c)

Fig. 7. A 10-day distribution of 500mb geopotential height (m) anomaly over Russia for July month of year 2010

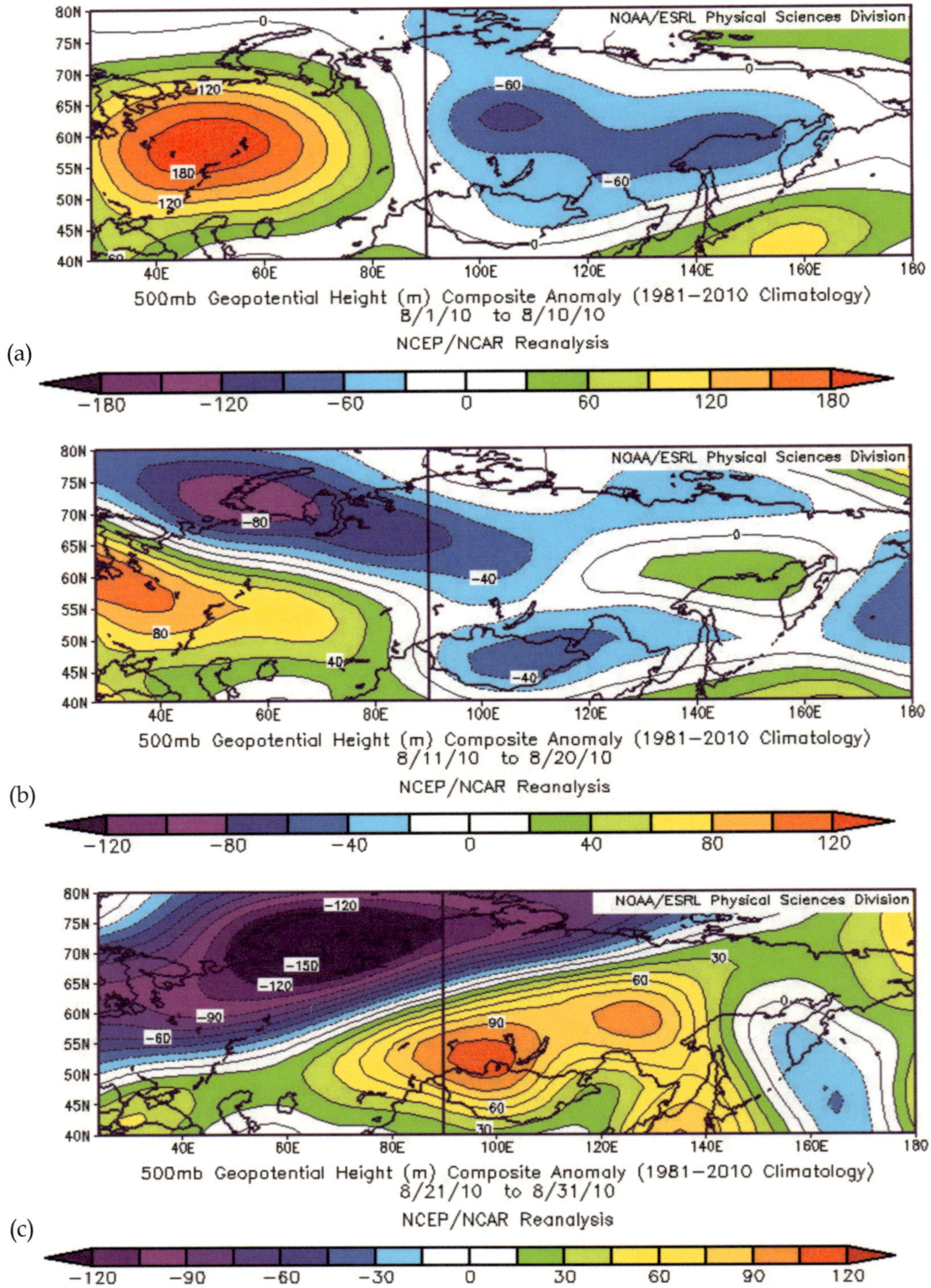

Fig. 8. A 10-day distribution of 500mb geopotential height (m) anomaly over Russia for August month of year 2010

persistence of a block means that there are an extreme disturbances in the large scale atmospheric circulation. The disturbance in the atmospheric circulation is not easy to persist for a long time like a one season in summer 2010. For the purpose of the present study we consider that, a hot wave is a heat wave with extreme anomalies in surface air temperature more than 5 °C through a ten days. A study of a 10-day anomalies in surface air temperature over Russia in the same summer has been done, see Figures 9, 10 and 11. It is noticed that the location of hot waves (hot wave in the present study indicated by positive anomalies in surface air temperature more than 5 °C through a ten days) was matched with the blocking high pressure location in the upper atmosphere.

3.3 Study the variation of NAO, SOI and geopotential height over Russia through period of (1980-2010) summer seasons

3.3.1 The variation of NAO

The North Atlantic Oscillation (NAO) is one of the major modes of variability of the Northern Hemisphere atmosphere. The NAO index is defined as the pressure gradient between Greenland and the Azores and describes the zonality of the flow in the North Atlantic region. i.e., NAO is the difference of the normalized sea level pressures between Ponta Delgada, Azores and Reykjavik, Iceland. It is particularly important in winter, when it exerts a strong control on the climate of the Northern Hemisphere. It is also the season that exhibits the strongest interdecadal variability. However, NAO is associated with changes in the system of westerly winds across the North Atlantic onto Europe. The state of this North Atlantic Oscillation (NAO) is positive when the Azores high is strong and the Icelandic low is deep and negative when reversed. Both phases are associated with changes in the intensity and location of the North Atlantic westerlies, jet stream and storm tracks, and with resulting changes in temperature and precipitation patterns The NAO can be seen as a reflection of the fluctuation of the normal winter tropospheric flow in the northern hemisphere (Rogers,1985; Hurrell, 1995; Ambaum, 2001; Rogers, & McHugh 2002). In the present work, the variability in NAO through the period (1980-2010) has been studied. The results of analysis of NAO dataset through the period (1980-2010) of summer seasons shows that NAO values in summer months (June, July and August) varies from year to year during the period of study. It is noticed that through summer of 2010 the NAO has negative values (-2.4 and -2.01) for months of June and August respectively. Meanwhile it was a very weak positive value (+0.06) for July as shown in Figure 12.The NAO value for July 2010 is the lowest positive value recorded through the period (1980-2010). So, in general one can say that, summer of 2010 has a characteristics of negative NAO phase weather regime.

In addition to that, study of daily NAO dataset represent the summer season of 2010 revealed that almost of these days had a negative NAO values. The days of positive NAO values occurred mainly in July month, as it is clear in Figure 13. The main features of this analysis is that the daily NAO has a negative phase weather conditions through that summer. On14 August 2010 the NAO recorded the lowest negative value (-2.01) recorded per a day.

3.3.2 The variation of SOI

The southern oscillation index (SOI) is defined as the difference between sea level pressure at Tahiti (145° W and 18° S) and Darwin (135° E and 16° S). SOI is coupled with El-Nino

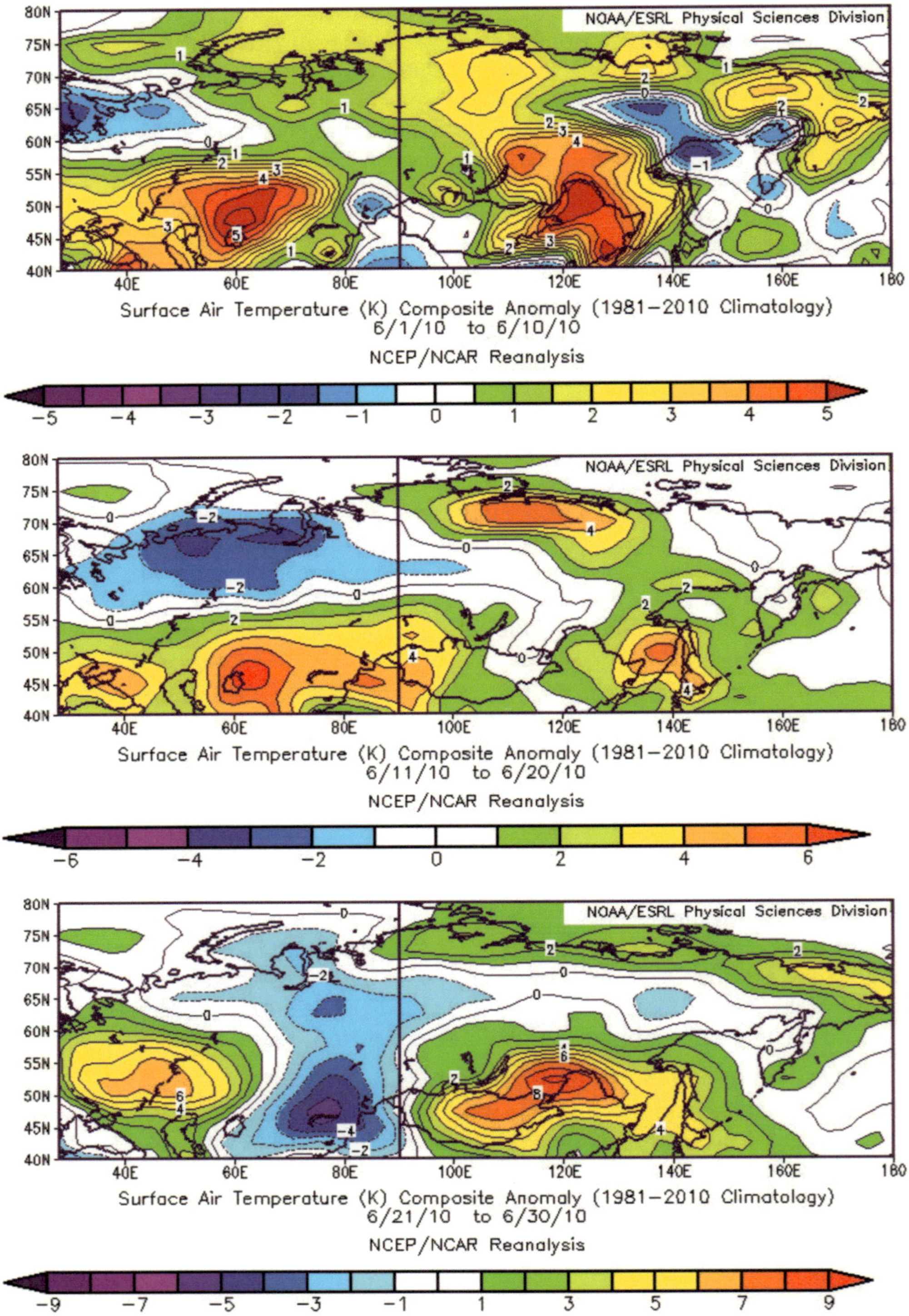

Fig. 9. A 10-day distribution of surface air temperature (°C) anomaly over Russia for June month of year 2010

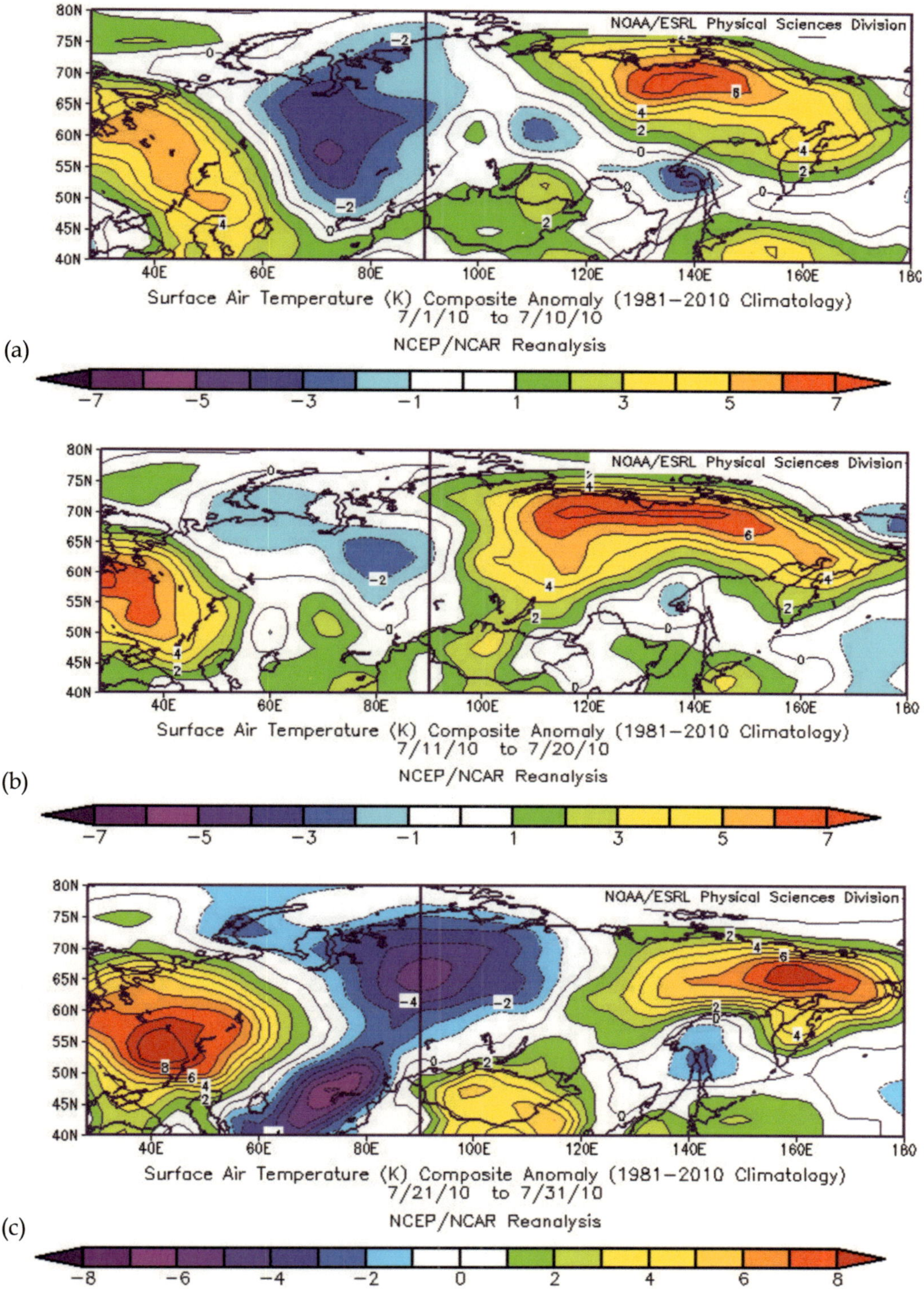

Fig. 10. A 10-day distribution of surface air temperature (°C) anomaly over Russia for July month of year 2010

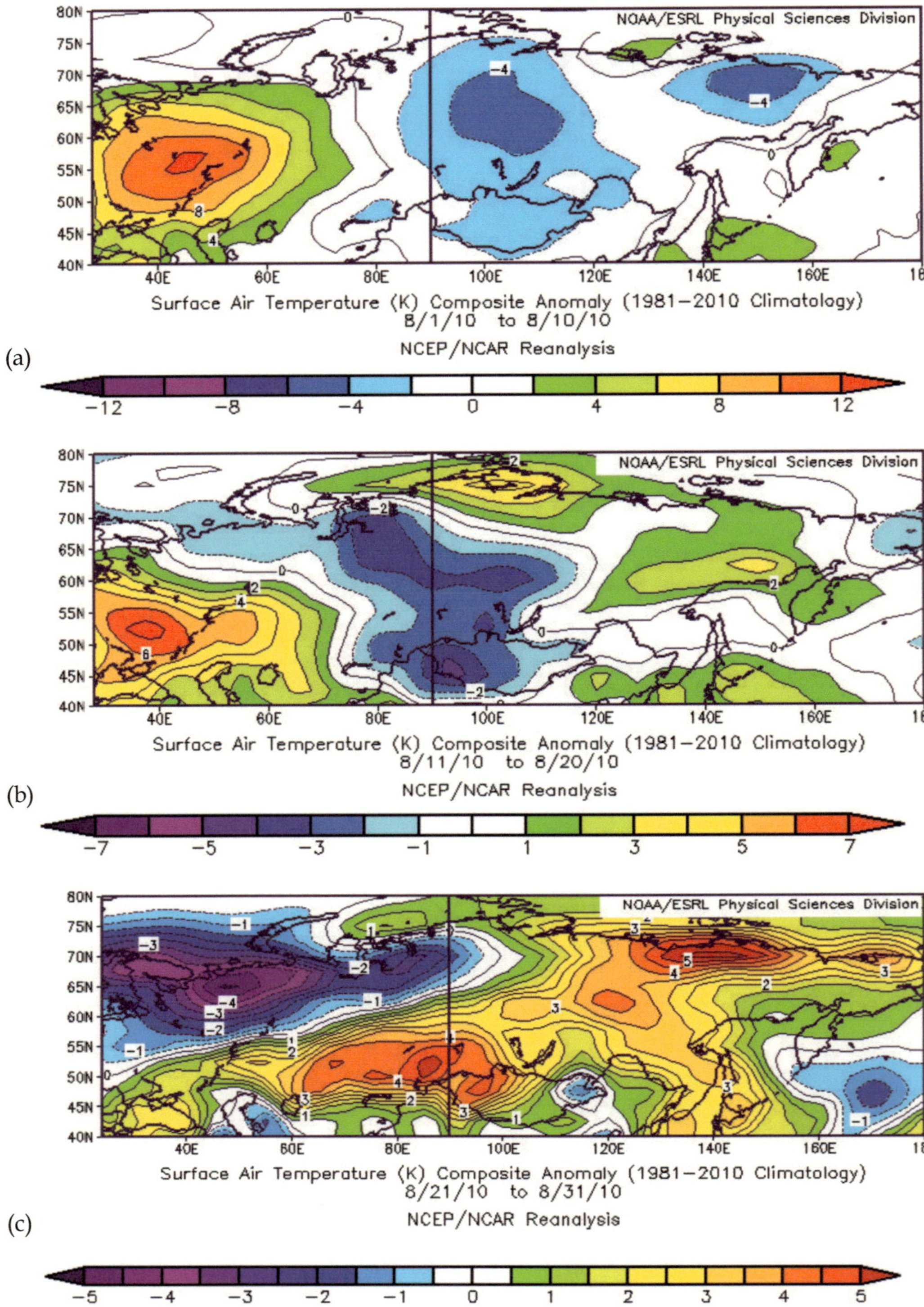

Fig. 11. A 10-day distribution of surface air temperature (°C) anomaly over Russia for August month of year 2010

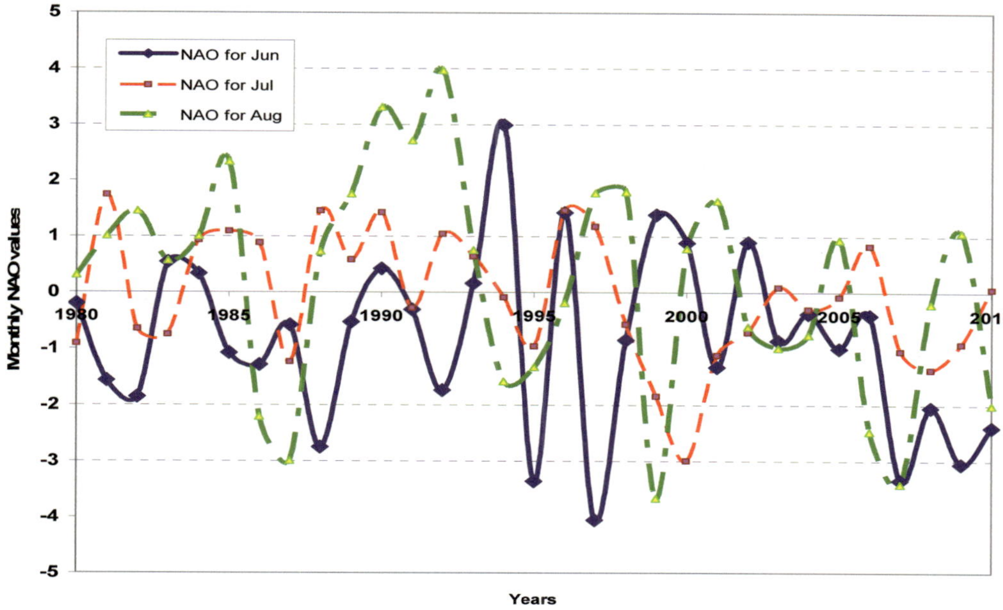

Fig. 12. The timeseries of monthly NAO values for months of June, July and August for the period (1980-2010)

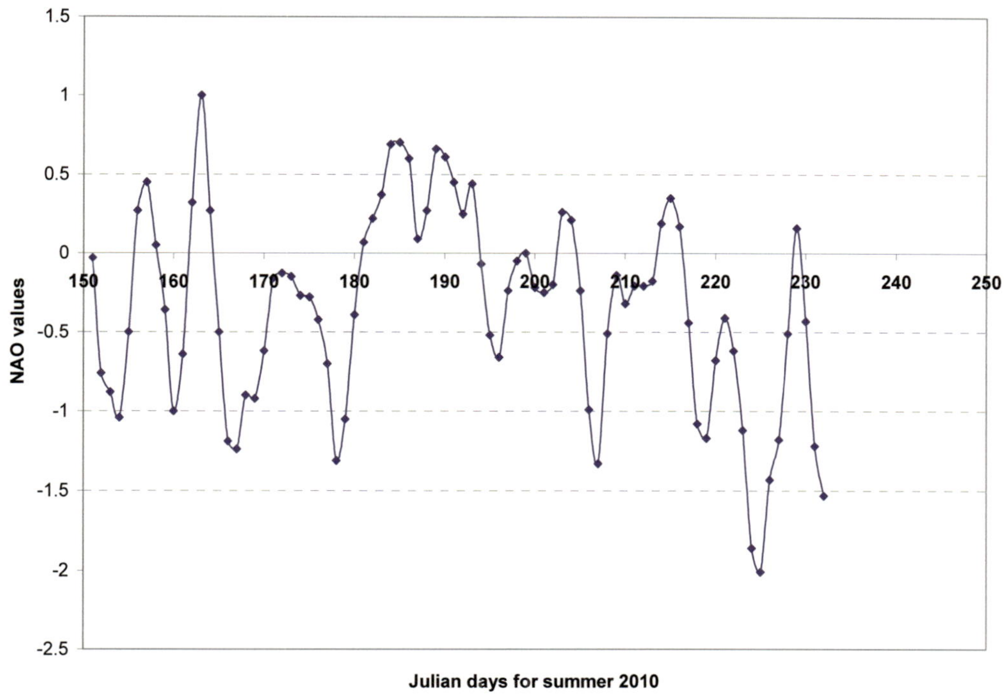

Fig. 13. The timeseries of daily NAO values for months of summer season of year 2010(NAO daily data for last week of August is missed)

which is called ENSO. Whereas, El Nino is the name given to the phenomenon, which occurs when sea-surface temperatures (SSTs) in the equatorial Pacific Ocean off the South American coast becomes warmer than normal. Nino3 is defined as the sea surface temperature at the Pacific Ocean in the region (90 °W - 150 °W, 5° S -5° N), (Trenberth,1976). Many studies have shown that the El Nino–Southern Oscillation (ENSO) has a significant influence on climate in many parts of the globe (Malmgren,1998).

In the present study, the variability in SOI through the summer seasons study period (1981-2010) has been studied. The results of analysis of monthly SOI dataset through this period of shows that sign of SOI values in summer months (June, July and August) varies from year to year. It is varied strongly mainly through the period (1980-2000). Meanwhile the SOI values has a little variation through the period (2001-2009). It is noticed that through summer of 2010 the SOI has extreme climatic positives (+20.5 and +18.8) for July and August months respectively. Also, SOI has a positive value +1.8 in June of this summer season. It is shown in Figure 14 that represents the monthly variation of SOI through the period (1980-2010). So, in general one can say that, summer of 2010 has a characteristics of extreme climatic positive SOI phase weather regime mainly on July and August. Also, study of daily SOI dataset for 92 days represent the summer season of 2010 shows that almost of these days had an extreme maximum positive SOI values. The days of extreme positive SOI values occurred mainly in July and August months, as it is illustrated in Figure 15. The main features of this analysis is that the daily SOI has an extreme positive phase weather conditions through that summer. The recorded SOI value during July 2010 never reached to this value for July month since the recording of SOI data. Whereas, 27 July 2010 observe a

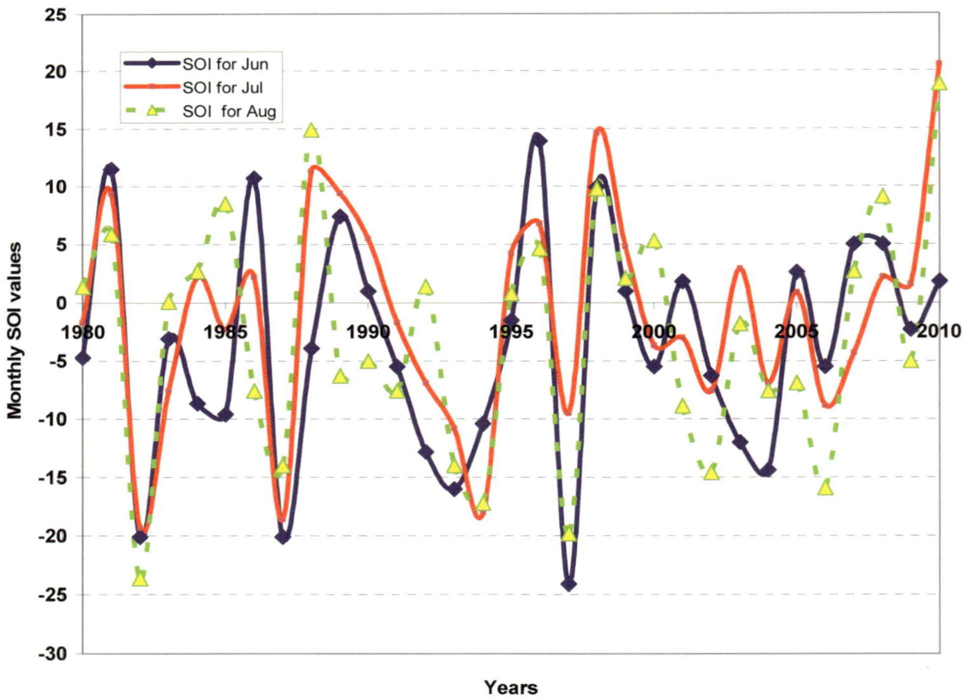

Fig. 14. The timeseries of monthly SOI values for months of June, July and August for the period (1980-2010)

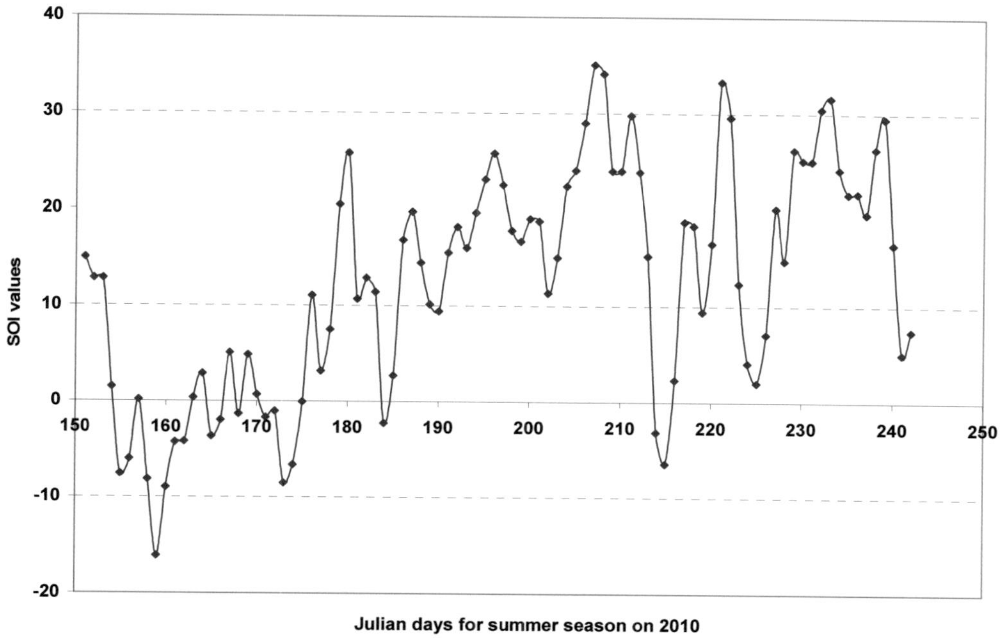

Fig. 15. The timeseries of daily SOI values for months of summer season of year 2010

daily new climatic SOI record value (+35.02). This value never reached through summer seasons of all SOI records.

3.3.3 The variation of the geopotential height

The time series of monthly variation of the geopotential height at level of 500 hpa over Russia during summer months of June, July and August through the period (1980-2010) had been plotted , analyzed and discussed. This dataset comes from the monthly mean values over the Russia area [40° N - 80° N] latitudes and [28° E -190° E] longitudes. The results revealed that the monthly geopotential height values varies from year to year for each month. The geopotential values are increase from June to July and decrease toward August. It is noticed that these values varied little variation on July months for the period (1999-2009) as shown in Figure 16. Month of July on 2010 record a first maxima of geopotential height values (5671m). Whereas, The maximum value is (5682 m) reached on 1998 July. The geopotential height values for the summer months of 2010 are greater than that on 2009.

3.3.4 The variation of the surface air temperature

The time series of monthly variation of the surface air temperature over Russia during summer months of June, July and August through the period (1980-2010) had been analyzed. This dataset comes from the monthly mean values over the Russia area. The results revealed that the variation in surface air temperature is typical in general with the variation of geopotential height values from year to year for each month. See Figures 16 and 17. Whereas, the surface air temperature values are increase from June to July and decrease toward August. It is noticed that these values varied little variation on July months for the period (1999-2009) as shown in Figure 17. Month of July on 2010 record the maximum value

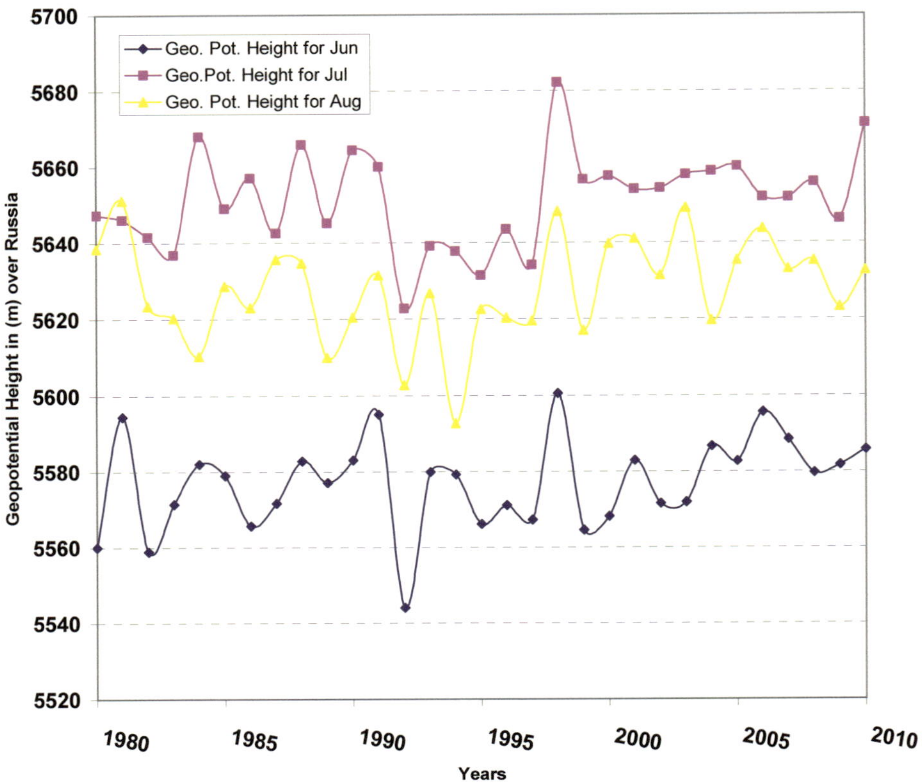

Fig. 16. Timeseries of geopotential height over Russia for summer months June, July and August through the period (1980-2010)

is (14 °C) reached. Also, Month of August 2010 record the maximum value is (14 °C) reached through the period (1980-2010). The temperature values for the summer months of 2010 are greater than that on 2009 as it is shown in Figure 17.

3.4 Teleconnection between hot waves over Russia and climatic indices NAO and SOI and blocking systems

Teleconnection patterns reflect large-scale changes in the atmospheric wave and jet stream patterns, and influence temperature, rainfall, storm tracks, and jet stream location and intensity over vast areas (e.g. Hafez, 2008b). Thus, they are often the culprit responsible for abnormal weather patterns occurring simultaneously over seemingly vast distances. Through this section, a correlation coefficient analysis according to (Spiegel, 1961) has been done between the hot waves(represents by anomalies in surface temperature) and blocking highs (represents in anomalies in geopotential height at 500 hpa), and climatic indices NAO and SOI. The datasets for all of these parameters had been taken for the period (1980-2010). The results revealed that, for June month, there exist a very strong, high significant positive correlation coefficient (+0.675) with a significant level of 0.99 % between anomalies in surface air temperature and geopotential height over Russia for month of June through the period of study. In addition to that through , anomalies in surface air temperature has a significant negative correlation coefficient (-0.205) with a significant level of 0.95 % with

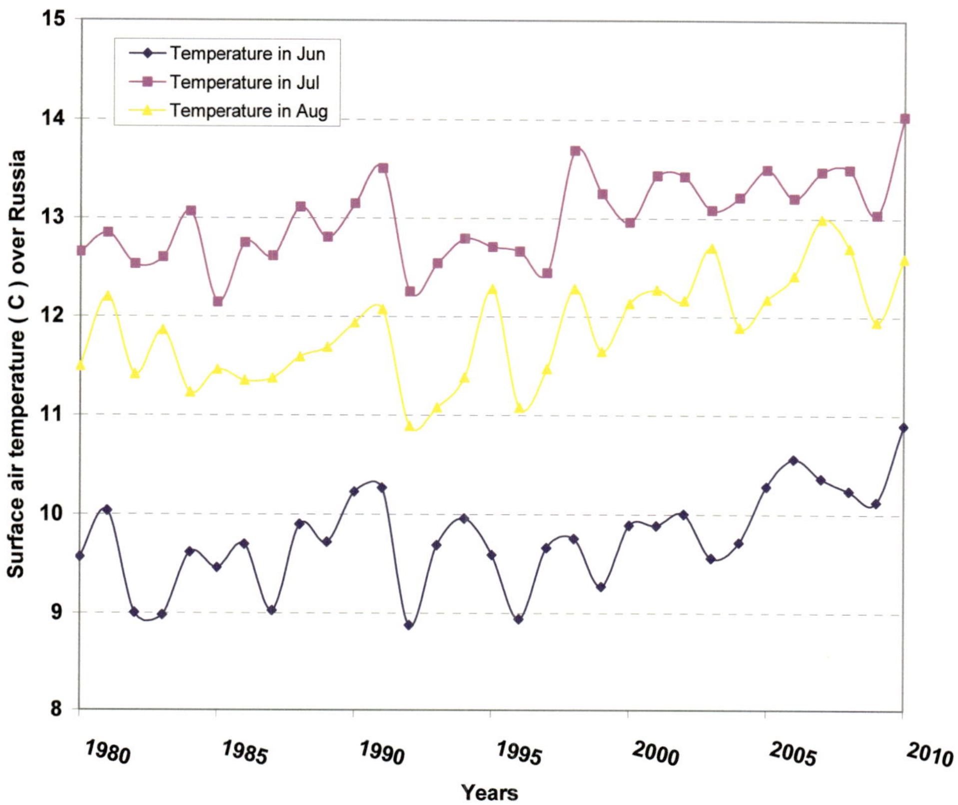

Fig. 17. Timeseries of surface air temperature over Russia for summer months June, July and August through the period (1980-2010)

NAO. Also, the results show that there is a strong significant positive correlation coefficient (0.299) with a significant level of 0.97% between the anomalies of surface air temperature over Russia and SOI as shown in Table(1).

Correlation coefficient / Weather and climatic parameters	Surface air temperature over Russia	Geopotential height at 500 hpa level over Russia	NAO	SOI
Surface air temperature over Russia	1	+0.675	-0.205	+0.299
Geopotential height at 500 hpa level over Russia	+0.675	1	-0.019	+0.343
NAO	-0.205	-0.019	1	-0.086
SOI	+0.299	+0.343	-0.086	1

Table 1. Correlation coefficient matrix for Weather and climatic parameters over Russia and climatic indices NAO and SOI for June month

The results revealed that, for July month, there exist a very strong, high significant positive correlation coefficient (+0.764) with a significant level more than 0.99 % between anomalies in surface air temperature and geopotential height over Russia through the period of study. In addition to that through , anomalies in surface air temperature has a significant negative correlation coefficient (-0.289) with a significant level of 0.95 % with NAO. Also, the results show that there is a strong significant positive correlation coefficient (+0.477) with a significant level of 0.975% between the anomalies of surface air temperature over Russia and SOI as shown in Table(2).

Correlation coefficient / Weather and climatic parameters	Surface air temperature over Russia	Geopotential height at 500 hpa level over Russia	NAO	SOI
Surface air temperature over Russia	1	+0.764	-0.289	+0.477
Geopotential height at 500 hpa level over Russia	+0.764	1	-0.06479	+0.573
NAO	-0.289	-0.064	1	+0.217
SOI	+0.477	+0.573	+0.217	1

Table 2. Correlation coefficient matrix for Weather and climatic parameters over Russia and climatic indices NAO and SOI for July month

For August month, there exist a strong, high significant positive correlation coefficient (+0.605) with a significant level more than 0.97 % between anomalies in surface air temperature and geopotential height over Russia through the period of study. In addition to that, anomalies in surface air temperature has a significant negative correlation coefficient (-0.285) with a significant level of 0.95 % with NAO. Also, the results show that there is a strong significant positive correlation coefficient (+0.264) with a significant level of 0.90 % between the anomalies of surface air temperature over Russia and SOI as shown in Table(3).

Correlation coefficient / Weather and climatic parameters	Surface air temperature over Russia	Geopotential height at 500 hpa level over Russia	NAO	SOI
Surface air temperature over Russia	1	+0.605	-0.285	+0.264
Geopotential height at 500 hpa level over Russia	+0.605	1	-0.101	+0.255
NAO	-0.285	-0.101	1	+0.018
SOI	+0.264	+0.255	+0.0184	1

Table 3. Correlation coefficient matrix for weather and climatic parameters over Russia and climatic indices NAO and SOI for August month

4. Discussion and conclusion

Abnormal hot and dry weather has hit Russian regions in the summer of 2010. Many Russians are suffering from the record-breaking heat and the worst drought in 40 years. Russia was not import grain this year despite drought wiping out a quarter of its crops. The present study the hot wave over Russia on summer 2010 are a result of extreme global atmospheric abnormal interaction between the southern hemisphere(represented by SOI) and the northern hemisphere (represented by NAO) through that season. Abnormal interaction between these two major climatic indices of pressure systems that controlling the atmospheric dynamic in the globe generates a unique blocking high over the northern hemisphere. This blocking high persisted for a long time over Russia about 80 days with anomalies in geopotential height more than +100 m in 500 hpa level. The long persistence of this episode is due to the long abnormal interaction between SOI and NAO through that summer. The blocking high over Russia associated with abnormal warming in the upper atmosphere and causing of very extreme heating in the surface. Hot waves over Russia are a results of the long persistence of the blocking high episode that existed from the abnormal atmospheric interactions in the globe. One can concluded that, extreme flow air currants in both hemispheres (represented in the extreme values of NAO and SOI)disturb the flow to be stationary and blocked the air current flow aloft over the northern hemisphere during summer 2010. The hot waves over Russia is a result from the extreme atmospheric climatic interaction between both hemispheres.

5. Acknowledgment

It is a pleasure to the author to thank the Earth System Research Laboratory, Physical Sciences Division, Climate Diagnostics Centre for supporting the data used throughout this study. Plots and images were provided by the NOAA-CIRES Climate Diagnostics Centre, Boulder, Colorado, USA from their Web site at www.esrl.noaa.gov/psd/. Also, thanks to the Climate Prediction Centre for supporting the NAO and SOI data which obtained through the website http://www.cpc.ncep.noaa.gov.

6. References

Ambaum, M. H. , Hoskins, B. & Stephenson D. B. (2001). *Arctic Oscillation or North Atlantic Oscillation?*. J. Climate., 14, 3495–3507.

Cohen J, Saito K, Entekhabi, D.(2001). *The role of the Siberian high in Northern Hemisphere climate variability.* Geophys Res Lett 2001;28(2):299-302.

Dickson, R. R., & Namis, J. (1976). *North America influences on the circulation and climate of the North Atlantic sector. Mon.* Wea. Rev., 104, 1255-1265.

Dole, R., Hoerling, M., Perlwitz, J., Eischeid, J., Pegion, P., Zhang, T., Quan, X.-W., Xu, T. & Murray, D. (2011). *Was there a basis for anticipating the 2010 Russian heat wave? Geophysical Research Letters* 38: 10.1029/2010GL046582.

Dole, R. M., and Gordon, N. D. (1983). *Persistent anomalies of the extra-tropical Northern Hemisphere wintertime circulation: Geographical distribution and regional persistence characteristics. Mon. Wea. Rev., 111,* 1567–1586.

Dole, R. M.(1982). *Persistent anomalies of the extratropical Northern Hemisphere wintertime circulation.* PH. D. THESIS. Massachusetts institute of technology.

Dole, R. M.(1978). *The objective representation of blocking patterns. In the general circulation theory, modelling and objections.* Notes from a colloquium summer. NCAR/CQ 6+ 1978- ASP, 406-426.

Feudale, L., and Shukla, J. (2010). *Influence of sea surface temperature on the European heat wave of 2003 summer.* Part II: a modeling study. *Clim. Dyn.*, doi 10.1007/s00382-010-245 0789-z.

Hafez, Y. Y. (2011). *Relationship Between Geopotential Height Anomalies Over North America and Europe and the USA Landfall Atlantic Hurricanes Activity.* The J. Amer. Sci., 7, 6, 663-671.

Hafez, Y. Y., (2008a). *The teleconnection between the global mean surface air temperature and precipitation over Europe.* J. Meteorology UK ;33(331):230-236.

Hafez, Y. Y., (2008b). *The role played by blocking over the Northern Hemisphere in hurricane Katrina.* The J. Amer. Sci., 4, 2, 10-25.

Hafez, Y. Y. (2007). *The connection between the 500 hpa geopotential height anomalies over Europe and the abnormal weather in eastern Mediterranean during winter 2006.* I. J. Meteorology, U. K., Vol., 32, No., 324, pp: 335-348.

Hafez, Y. Y. (1997). *Concerning the role played by blocking highs persisting over Europe on weather in the eastern Mediterranean and its adjacent land areas.* Ph. D. THESIS, Cairo University, Egypt.

Hurrell, J. W., (1995): *Decadal trends in the North Atlantic Oscillation: Regional temperatures and precipitation.* Science, 269, 676-679.

Kalnay, E. & Coauthors, (1996). *The NCEP/NCAR Reanalysis 40-year Project.* Bull. Amer. Meteor. Soc., 77, 437-471.

Lejenas, H., 1989: *The sever winter in Europe 1941-42: The large-scale Circulation, cut-off lows, and blocking.* Bull. Amer. Meteor. Soc., 70, 271-281.

Meehl, George A.; Tebaldi & Claudia (2004). *More Intense, More Frequent, and Longer Lasting Heat Waves in the 21st Century. Science* 305 (5686): 994. doi:10.1126/science.1098704.

Namias, J., 1964: *Seasonal persistence and recurrence of European blocking during 1958-60.* Tellus, 6, 394-407.

Razuvaev, V.; Groisman, P. Y.; Bulygina, O.; Borzenkova, I. (2010). *Extreme Heat Wave over European Russia in Summer 2010: Anomaly or a Manifestation of Climatic Trend?* American Geophysical Union, Fall Meeting 2010, abstract #GC33A-0926.

Rex, D.F. (1950a). *Blocking action in the middle troposphere and its effect upon regional climate. (I) An aero logical study of blocking action.* Tellus;2:196-211.

Rex, D.F. (1950b). *Blocking action in the middle troposphere and its effect upon regional climate. (II) The climatology of blocking action.* Tellus ;2:275-301.

Rex, D.F. (1951). *The effect of Atlantic blocking action upon European climate.* Tellus ;3:100-11.

Rogers, J. C. (1985). *Atmospheric circulation changes associated with the warming over the northern North Atlantic in the 1920s.* J. Climate Appl. Meteor., 24, 1303–1310.

Rogers, J. C. & McHugh, M. (2002). *On the separability of the North Atlantic Oscillation and Arctic Oscillation.* Clim. Dyn., 19, 599– 608.

Spiegel, M. R., (1961). *Theory and Problems of Statistics,* Schaum, PP, 359.

Stott, P. A., Stone, D. A. & Allen, M. R. (2004*). Human contribution to the European heat wave of 2003. Nature 432, 610-614.*

8

Soil-Tree-Atmosphere Water Relations

Kemachandra Ranatunga
Bureau of Meteorology, Canberra
Australia

1. Introduction

A process model of the soil-tree-atmosphere continuum, which treats the plant physiology, eco-physiology and vegetation structures in detail, needs to describe the dynamics of the water flow within this continuum. The concept of a soil-plant-atmosphere continuum (SPAC) (Philip, 1966) was first described by Huber (1924). In such a continuum, the removal of water lowers the water potential in the leaves of the plant, and water moves in the direction of decreasing potential through a continuous liquid pathway extending from the soil through the plant to the leaves.

Water uptake, transpiration, radiative transfer and sensible heat exchange are the most important processes in a soil-tree-atmosphere continuum for water relations. Biophysical exchanges of radiative energy, sensible heat and water vapor in the canopy as well as soil water dynamics and soil and root resistances are physically and physiologically interrelated processes. In order to model biophysical exchanges between canopies and the atmosphere, it is necessary to integrate these processes. The root system and soil-water dynamics are very important in below-ground water transport. Therefore, the integration of soil and root resistances into the soil-tree-atmosphere continuum is necessary to estimate the water uptake by roots. In order to model transpiration radiative transfer, sensible heat exchange and scalar variations must also be understood. The processes of radiative transfer, transpiration, sensible heat exchange, diffusion and turbulent transfer in plant canopies are intrinsically mingled together. In order to predict the exchanges of energy and mass between the tree canopy and the atmosphere in a mechanistic manner it is necessary to couple these processes directly. This means that the soil-vegetation-atmosphere transfer processes for water relations are inevitably complex.

Studies of above- and below-ground processes have been conducted by meteorologists and plant ecologists for a long time (Monteith, 1975; Gates, 1980; Grace, 1983; Landsberg and McMurtrie, 1984; Landsberg (1986). The integrated modeling process of the soil-tree-atmospheric continuum for water relations is essentially a synthesis of available physical theories describing water loss from leaves and the movement of water from soil to roots and through plants, is to examine the effects of soil and tree water status, radiative transfer in the canopy and other weather parameters on transpiration with the ultimate aim of predicting tree water uptake. A variety of different models has been developed for each of these processes. Theoretical basis and reviews of these models can be found in Ross (1981), Goel (1988), Myneni *et al.* (1989), and Myneni and Ross (1990). These developments have created a great diversity of models of soil-tree-atmosphere transfers in which the processes are partly or fully integrated. However, these models all share a common characteristic: they infer canopy functions from leaf observations. Based on how the inference is achieved, models of soil-tree-atmosphere transfers can be divided into four groups: the big-leaf models, direct scaling models, multi-layer canopy models and multi-layer soil-canopy models. Multi-layer models can be further divided into two types: incomplete multi-layer models and complete multi-layer models. This chapter describes these methods and presents the background of the soil-tree-atmospheric water transfer.

To overcome shortcomings of previous multi-layer models, accurate descriptions of basic biophysical processes are given first priority in the modeling process. Complex models, whose wide-range application is often hampered by the lack of specific data, should have their processes simplified in order to be accommodated into spatial frameworks where appropriate. Biophysical processes within simple models should consider new data sources and understanding to gain more accurate predictions (Ranatunga et al., 2008). Simplicity can be pursued only when the reality of basic biophysical processes is not compromised. The process of water movement through the soil-plant-atmosphere continuum has been widely and sucessfully used (Molz, 1981; Jarvis *et al.*, 1981; Boyer, 1985; Eckersten, 1991; Ranatunga and Murty, 1992; Nobel and Alm, 1993; Cienciala *et al.*, 1994; Friend, 1995; Williams *et al.*, 1996). The integrated modeling process is designed to cover both below- and above-ground processes, which can be combined through a methodology that links the soil-water status with the atmospheric weather conditions through canopy exchangeable water storages and estimated tree water uptake by roots and water loss by transpiration.

2. Below-ground processes

Water uptake from the soil by roots is determined by soil water content and other related soil physical properties, the root architecture system and soil and root resistances to water flow as well as demand from the canopy. As more fine-scale detailed data on the physiology of water absorption become available, the integration of this information into a quantitative framework from the root to the soil layer level can be established by coupling of the soil water dynamics and root architecture. Finally, water flow resistance from soil to root as well as for radial and axial resistances in the roots can be calculated for the entire root system.

The amount of water that a tree can remove from the soil depends on the volume of soil exploited by the root system of the tree (Landsberg and McMurtrie, 1984). Water absorption by roots from the soil is determined by three main factors:

1. soil properties such as soil water content, hydraulic conductivity and water potential,
2. root-system architecture as a network of absorbing organs, and
3. the absorption capacity of roots is dependent on water flow resistances, i.e., the soil and root resistances.

Although the water movement in soil is well represented by microscopic or macroscopic approaches, a detailed description of the root system is lacking (Molz, 1981). Water movement through plant roots has been theoretically analyzed by Landsberg and Fowkes (1978) who derived an analytical solution for a single root with a constant hydraulic resistance. Alm *et al.* (1992) extended Landsberg and Fowkes' approach to a case of variable resistance along the root by separating roots into segments, each with homogeneous resistance. With recent advances, the physiological characteristics of water uptake by roots such as radial and axial resistances can now be examined at the centimeter scale (North and Nobel, 1995). Simulation models for the below-ground tree are needed because of the difficulty in observing and quantifying the architecture of roots, the soil-water dynamics and the soil and root resistances to water flow. By coupling soil water dynamics and root architecture along with soil and root hydraulic resistances, the process of water transport from the soil to the roots and through the roots can be described. The below-ground processes and their interactions in relation to water movement are illustrated in Fig.1.

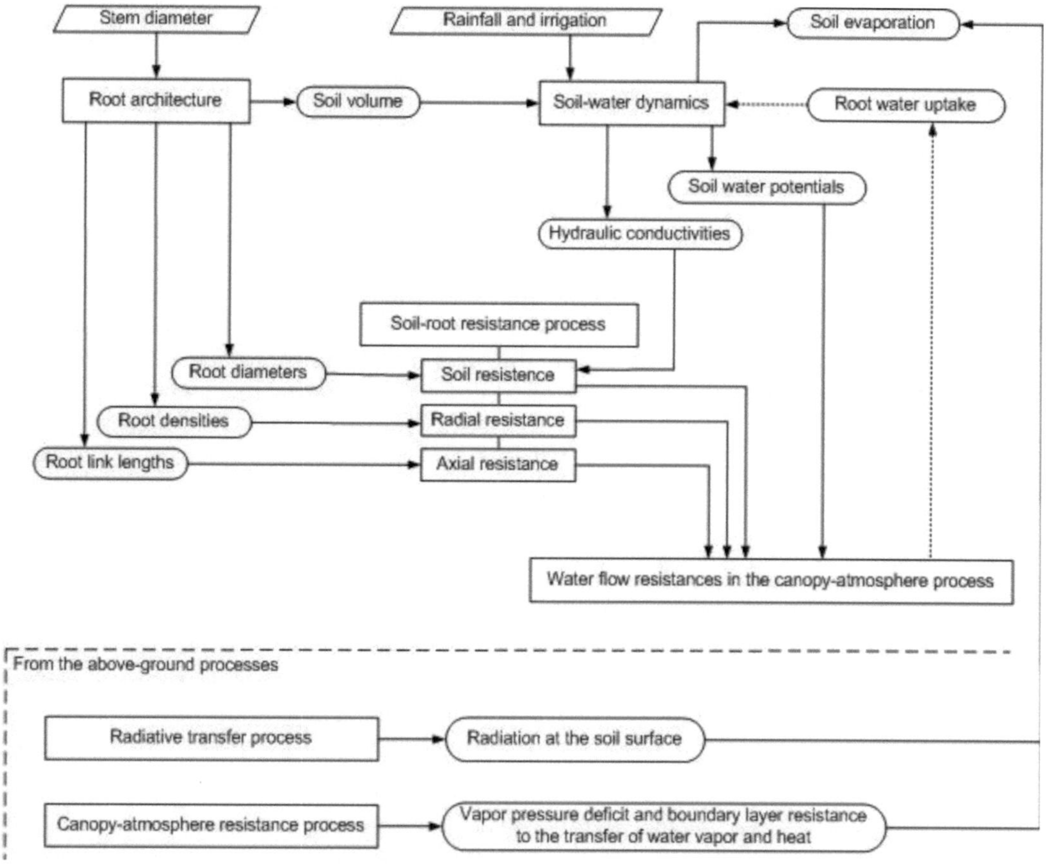

Fig. 1. A concise structural diagram of the below-ground processes and their interactions

2.1 Modelling soil water dynamics

Although soil-water fluxes are difficult to measure, the relative capabilities of existing models and the credibility of their results are still an important concern because soil water dynamics is sometimes inadequately represented in models of the soil-plant-atmosphere-water interactions and processes (Clemente *et al.*, 1994). Analytical solutions to the soil-water flow equation (Richards' equation (Richards, 1931)) are not possible for dynamic field situations (Zeng and, Decker, 2009) and most efforts have been concentrated on seeking numerical solutions (e.g. Feddes *et al.*, 1978; Broadbridge and White, 1988; Ranatunga and Murty, 1992).

A numerical solution for the unsaturated vertical soil water flow with varying water supply (irrigation and rainfall) is typically used to estimate soil water content in a vertically-structured soil profile (Ranatunga and Murty, 1992). This numerical solution also gives an implicit finite difference solution to the soil water flow equation. The assumption of only one-dimensional vertical flow is quite accurate for agricultural rooting depths (Bresler, 1991). Soil evaporation can be based on a modified version of Penman-Monteith equation (Raupach, 1991) and vapor-flow flux (within the soil profile) is neglected except at the top layer of the soil. To examine the water flow in soils, a volumetric sink term may be added to the classic Richard's flow equation for one-dimensional flow under gravity (Feddes *et al.*, 1978).

At the soil-air interface the soil can lose water to the atmosphere by evaporation of soil water where the potential rate of soil evaporation depends only on atmospheric conditions. During evaporation the requirement is that (Feddes *et al.*, 1978):

$$\psi(z,t) \geq \psi_l; \quad t > 0, z = 0$$

where ψ is the pressure head and ψ_l is the minimum pressure head allowed under air-dry conditions. Typically, ψ_l is estimated from the mean temperature and relative humidity of the surrounding air, and if it is assumed that the pressure at the soil surface is in equilibrium with the atmosphere, then, ψ_l can be derived from the following relationship (Feddes *et al.*, 1974):

$$\psi_l = \frac{RT}{Mg} In(Rh) \tag{1}$$

where R is the universal gas constant (8.314 J mol^{-1}K^{-1}), T is the absolute temperature (K), g is the acceleration of gravity (9.81 m s^{-2}), M is the molecular weight of water (0.018 kg mol^{-1}) and Rh is the relative humidity as a fraction. The simplest option and one which is often used in modeling soil-water flow, is to neglect vapor flow except at the soil surface (Campbell, 1985). Janz and Stonier (1995) incorporated the evaporation rate into the sink term of the top soil layer and were able to make reasonable predictions of soil water dynamics.

The soil water deficit is the amount of water needed to bring the soil moisture content back to field capacity, which is the amount of water the soil can hold against gravity. The soil water deficit is calculated by subtracting the total water content of the soil layer from the water content at the field capacity of the root zone. In standard models, rainfall and the amount of irrigation applied are assumed to be equally distributed among the soil layers, and adjusted by adding to the water content of the soil layer. The numerical procedure gives the soil-water content in vertically structured soil layers.

The Penman-Monteith equation provides good predictions of evaporation of forest surfaces (Saugier, 1996). It uses a modification of the form employed by Raupach (1991) as applied by Walker and Langridge (1996):

$$F_E = \left(\frac{\varepsilon Rnet_{soil} + \dfrac{\rho \lambda VPD}{Ra_{soil}}}{\varepsilon + 1 + \dfrac{R_{ss}}{Ra_{soil}}} \right) \frac{1}{\lambda} \quad (2)$$

where F_E is the evaporation rate (kg m^{-2} h^{-1}), $Rnet_{soil}$ is the net radiation at the soil surface (MJ m^{-2} h^{-1}), ρ is the air density (kg m^{-3}), ε is the dimensionless slope of the saturation specific humidity, λ is the latent heat of vaporization (MJ kg^{-1}), VPD is the vapor pressure deficit, $R_{a\,soil}$ is the boundary-layer resistance to the transfer of water vapor and heat (between the soil surface and bottom of the canopy) (s m^{-1}) and R_{ss} is soil surface resistance (s m^{-1}). R_{ss} that restricts the transfer of water from the soil surface by evaporation can be calculated as (Walker and Langridge, 1996):

$$R_{SS} = R_{SS(\min)} f\left(SWC\right)^{-1} \quad (3)$$

where $R_{ss(min)}$ is the minimum value of the soil-surface resistance under optimal conditions and SWC is the soil water content in the topmost layer.

2.2 Modelling root architecture

The architecture and space filling properties of the below-surface tree organs are necessary for a mechanistic understanding of water uptake. Pioneering work in numerical simulations of root systems was done by Lungley (1973). Based on theoretical study, Claasen and Barber (1974) assumed uniform root distributions. Rengel (1993) pointed out that the assumption of uniformity of root distribution used in models based on Claasen and Barber (1974) is an oversimplification, since the way roots fill the soil matrix is important for nutrient and water acquisition (Sattelmacher *et al.*, 1990). Fitter *et al.* (1991) also described a simulation model of root growth that simulated the development of root systems varying in several important architectural features including root link length. Root link length refers to a distance between two branching points in the root (Bernston, 1994).

Some attempts have been made to relate root biomass to stem diameter at a standardised height (Santantonio *et al.*, 1977; Brown *et al.*, 1989), but such relationships probably depend on tree species and site. Soumar *et al.* (1994) found in a study on *Sclerocaryea birrea*, that the root diameter at 1 m from the stem base is an appropriate parameter for predicting horizontal root distribution. However, relating the size of the root system to proximal root diameter may be more successful than relating it to stem diameter (Van Noordwijk *et al.*, 1994).

Fitter *et al.* (1991) have suggested that it is not possible to derive a simple analytical relationship between root system architecture and water and resource acquisition due to the complexity of the spatial arrangement of the roots within the soil. Fractal geometry is a system of geometry that is more suited to the description of many biological objects than is standard Euclidean geometry (Mandelbrot, 1983). Fractal geometry provides useful perspectives on root branching patterns (Shibusawa, 1994; Van Noordwijk *et al.*, 1994) and it is reasonable to accept that fractals may provide quantitative summaries and functional insights into root architecture (Bernston *et al.*, 1995).

The branching pattern has been identified as an important characteristic of root systems that strongly influences patterns of foraging within the soil matrix (Fitter, 1987; Hetrick, 1988) and nutrient and water acquisition (Robinson *et al.*, 1991). There have been a number of attempts to produce architectural classifications of root systems (Weaver, 1968; Krasilnikov, 1968) but none have been particularly successful, partly because of the great variability of root systems. However, following Coupland and Johnson (1965), Van Noordwijk *et al.*, 1994 identified two major classes of fractal branching patterns, i.e., herringbone and dicho-syntomous. Most monocots follow the herringbone root branching pattern comprising of a main root and laterals whereas most dicots have a dicho-syntomous branching pattern that spawns the parent branch into two daughter branches (Zamir, 2001).

The most important and easily-measurable parameter for root modelling is the proximal root diameter. The relationship between proximal diameter and the total root length of all root links obviously depends on the root branching pattern. Leonardo da Vinci (quoted by Mandelbrot, 1983 and the quoted by Van Noordwijk *et al.*, 1994) claimed that the cross sectional area of the main stem is equal to the sum of the cross sectional areas of tree branches. A relationship can be sought between the root diameter at the stem base, and functionally important root parameters (Van Noordwijk and Brouwer, 1995) such as the total root length. This approach, known as a pipestem model (Shinozaki *et al.*, 1964) states that each unit of foliage requires a unit pipe of wood to connect it to the root system, and has been used by Van Noordwijk *et al.* (1994) to understand the geometry of root branching in plants, giving the relation between proximal diameter of the root base and the diameter of the divided axes:

$$\frac{\pi Do^2}{4} = \frac{\alpha\pi}{4}\sum_{j=1}^{Nk} Di_j^2 \tag{4}$$

where D_o is proximal diameter at the root base (m), D_i is diameter of the divided axes in the link i (m), α is proportionality of cross sectional area before a branching event and the sum of the cross sectional areas after branching and Nk is the number of branch roots per branching event. In many previous studies (summarized by Van Noordwijk *et al.*, 1994) on root architecture, root link length was assumed to be a constant for a whole root system or a part of root system. A recursive programming procedure can be adopted for divisions of root branching. The number of branching steps in each lateral branch can also be in the recursive programme (Van Noordwijk *et al.*, 1994).

The length of the root axis (L), which is an important parameter for understanding how the root system fills the soil matrix can then be calculated as:

$$L = nL_m \tag{5}$$

where L_m is the mean length of the root link in the main or lateral axis and n is the number of root segments in the root system. With an estimated L_m value for each root link, it is possible to calculate the length of the root axis. Values of L decrease with increasing diameter and the longest roots occur in the small diameter classes (Kodrik, 1995).

The soil volume explored by roots in each soil layer (Vt_i) can be determined (Pregitxer et al, 1997):

$$Vt = \pi L_{RZ}^2 z \tag{6}$$

where L_{RZi} is the radius of the rooting zone in the soil layer i and z is the depth of the soil layer. The architecture of a root system can be dissected into a number of measurable variables of which the most important is root density (Dt) which is defined as the root length per unit volume and Dt is calculated as (Tardieu, 1988):

$$Dt = \frac{\left(\sum_{n=0}^{n_m} L_{m_n} + \sum_{n=0}^{n_l} L_{l_n} \right)}{Vt} \quad (7)$$

where n_m and n_l are the number of branching events in the main axis and lateral axes respectively, L_{mn} is the length of the root link in the main axis and $L_{l\,n}$ is the length of root links in lateral axes.

2.3 Water flow resistances in the soil-root system

In the soil-plant-atmosphere continuum, water flow is driven through a series of hydraulic resistances that can be identified as soil resistance, root radial resistance, and root axial resistance pertaining to the pathways of conduction through xylem vessels, and stomatal resistance with a boundary layer adjustment to the leaf surface.

Root resistance is partitioned into radial and longitudinal (axial) components and the relative importance of these has been stressed (Passioura, 1984; St Aubin *et al.*, 1986). However, reasonable quantitative data of axial resistance (R_l) and radial resistance (R_r) is rare, even though these quantities are a prerequisite for the proper modeling of water flows through roots. The procedure for resistance calculations is mechanistically more appropriate than steady-state models because it includes the effect of the soil-water influence on resistance.

Assuming that the soil hydraulic conductivity (K_s) is constant within the particular soil layer, the soil resistance associated with the root R_s ($kg^{-1}\ m^4\ s^{-1}$) was developed by Moldrup *et al.* (1992). Water uptake by roots in wet soil is generally determined by the root hydraulic resistance, which is composed of the radial resistance (R_r) from root surface to xylem, and axial resistance (R_l) along the xylem (Landsberg and Fowkes, 1978; Passioura, 1988). The sum of R_r and R_l are often called the plant resistance that is found to be independent of the water potential gradient (Abdul-Jabbar *et al.*, 1984) and transpiration rate (Neumann *et al.*, 1974). Therefore, regardless of changes in soil and plant water status with time, R_r and R_l are normally assumed to be constants (Frensch and Steudle, 1989). .

Assuming that the membranes of cells to be crossed before water reaches the xylem form concentric cylinders, it can be easily shown that radial resistances would be related to each other by (Steudle and Brinnkman, 1989):

$$R_r = R_{cell} \sum_{k=1}^{n^c} \frac{r^r}{r_k^c} \quad (8)$$

where R_r is radial resistance ($kg^{-1}\ m^2\ s^{-1}$), R_{cell} is cell resistance ($kg^{-1}\ m^2\ s^{-1}$), r^c_k is the radius of the k cell layers to be crossed by water, n^c is the average number of cells to be crossed before water reaches the xylem and r^r is the radius of the root. Thornley and Johnson (1990) explained how the resistance from root surface to the plant vary with the number of roots per unit area in a vertically-structured soil profile.

Melchior and Steudle (1993) stated that R_l can contribute substantially to the overall hydraulic resistance or even can be the limiting component. Passioura (1972), Newman (1976) and Meyer and Ritchie (1980) used a modified Poiseuille-Hagen equation to calculate R_l from root xylem dimensions:

$$R_l = \frac{4\eta L}{\rho_w^2 \pi \sum_{i=1}^{n^v} r_i^4} \qquad (9)$$

where R_l is axial resistance ($kg^{-1}\ m^4\ s^{-1}$) associated with the root link, r_i is mean effective diameter of the xylem vessels (m), η is the viscosity of water ($0.001002\ kg\ m^{-1}\ s^{-1}$)[1] , L is the length of the xylem vessel (m), n^v is average number of conducting xylem vessels in the root and ρ_w is the density of water ($kg\ m^{-3}$).

3. Above-ground processes

Canopy processes such as the leaf-energy balance and transpiration are relatively well understood. However, a sound understanding of how these processes integrate spatially and temporally within trees remains elusive. One approach to tackle this problem is the use of models that allow scaling of canopy processes at the leaf level to the whole canopy (Jarvis *et al.*, 1985; Running and Coughlan, 1988; McMurtrie, 1993; Jarvis, 1995). Both aggregated (e.g. 'big-leaf') and distributed (e.g. multi-layer or three-dimensional) approaches are commonly applied in modeling canopy processes. There are costs and benefits to both (Raupach and Finnigan, 1988): Simulations with distributed modeling approaches require assumptions about the distribution of key parameters in space (and time), but allow model parameterization using fine-scale data. Aggregation avoids the need for spatial details by building the effects of non-linearities into the model parameters. These parameters must be estimated directly from coarse-scale data (e.g. canopy rather than leaf-level data).

Tree canopy structure is a complex and dynamic outcome of the evolutionary and ecological interactions and feedbacks between vegetation and environment. Since the transpiration from a single tree is a non-linear function of absorbed solar radiation and other related environmental variables (e.g. temperature, humidity, and wind speed), it is necessary to be able to simulate the radiation regime and other related environmental variables within the canopy and before, the transpiration rate can be adequately calculated. Therefore, the above-ground modeling presented in most of the literature (Myneni and Impens, 1985; Wang and Jarvis, 1990; Whitehead *et al.*, 1990; Ryel *et al.*, 1993) is largely based on hypothesized spatial independence structures from which the canopy is made. It, therefore, precludes an aggregated approach.

Three-dimensional models of the radiative transfer and canopy architecture represent a compromise between the simplicity of the uniform canopy models and the complexity of

[1] The temperature dependence of the viscosity of water is not taken into account. The temperature of water is assumed to be 20 °C.

architectural models. A fairly straightforward method can be introduced to formulate the three-dimensional canopy in terms of the amount and spatial distribution of leaf area within the crown. Thus, each vertical canopy layer has geometrically-defined sub-canopies, assuming a vertically heterogeneous canopy consisting of horizontally homogeneous layers.

The radiation, stomatal and leaf boundary-layer resistances, micro-environmental variables and finally the energy balance vary within the canopy. Consequently, the vertical and horizontal variations within the model become apparent. Far-field variation of wind speed and air temperature can normally be simulated with depth outside the canopy, but a uniform profile of vapor pressure outside the canopy is normally assumed (these assumptions are discussed in detail later).

The prediction of the vertical distributions of micrometeorological variables is regarded as a necessary step in the prediction of sub-canopy level fluxes. The adoption of integrated stomatal resistance has received theoretical (Finnigan and Raupach, 1987; Paw U and Meyers, 1989; Kelliher *et al.*, 1994; Raupach, 1995; Leuning *et al.*, 1995) and experimental (Baldocchi *et al.*, 1987) criticism over the years. As the radiation intercepted and scalar concentrations can be calculated in each sub-canopy within the crown, it is possible to estimate the stomatal resistances individually for each sub-canopy and hence, avoid the integration of the stomatal resistance for the entire canopy. Similarly, the boundary layer resistance can be calculated for each sub-canopy and it is not required to integrate over the total leaf area of the canopy.

The above-ground canopy processes would normally include;

- the canopy architecture and leaf area - a three-dimensional analysis to estimate the beam path length of each and every volume of the sub-canopies within a paraboloid-shaped-canopy, and to estimate the vertical leaf-area distribution using a Weibull statistical pattern,
- the solar position and day length - a methodology to calculate the zenith and azimuth angles of the sun on an hourly basis and hence the day length,
- the radiative transfer through the canopy - a comprehensive methodology for radiative transfer (short-wave direct, long-wave, and diffuse) through the crown,
- the energy-balance process – an application of the energy balance equation to each volume of the sub-canopies in order to estimate leaf temperature,
- the scalar variation - to estimate wind speed, air temperature, relative humidity and solar radiation, and
- the canopy-atmosphere processes for resistance calculations - to estimate the stomatal and leaf boundary-layer resistances.

These canopy processes are discussed in detail in the following sections. The above-ground processes and their interactions in relation to water movement are given in Fig.2.

3.1 Crown architecture and leaf area estimates

The geometric form of the crown is one of the main factors that govern the productive potential of vegetation (Jahnke and Lawrence, 1965). The shape of the crown is strongly correlated with the volume and height of the stand. Biging and Dobbertin (1995) found that the geometric space occupied by the crown is highly correlated with growth. This raises the question, therefore, whether canopy form can be readily quantified.

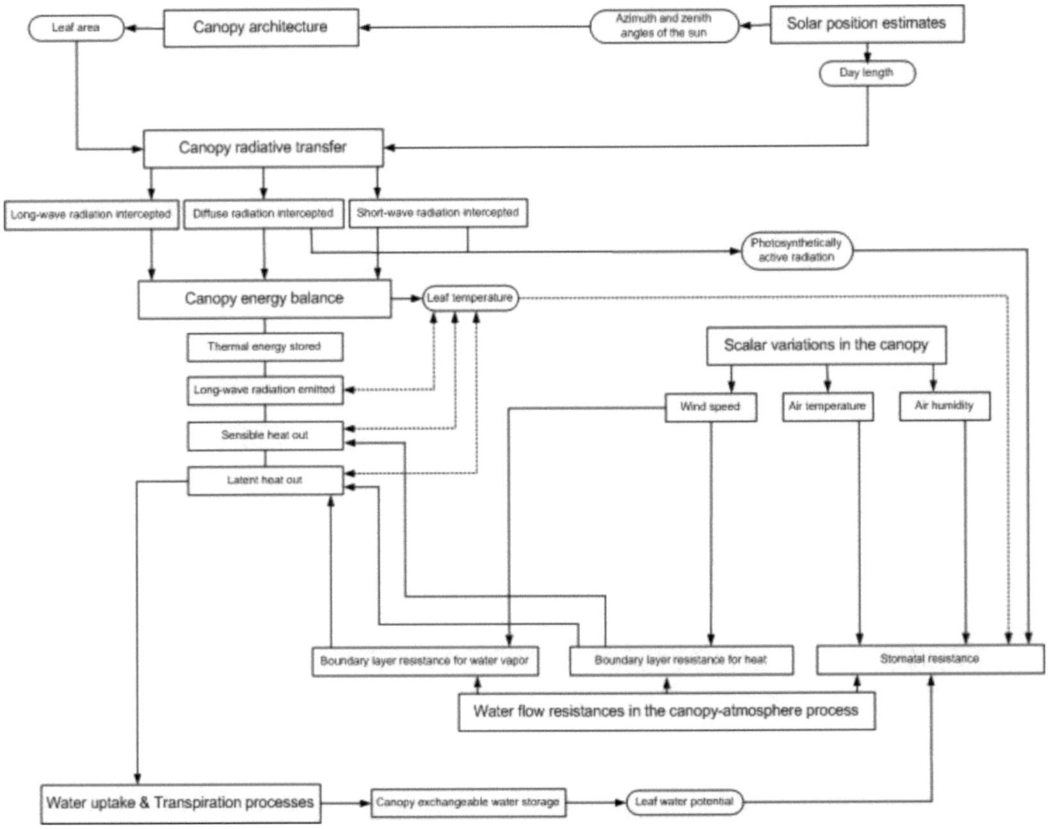

Fig. 2. A concise structural diagram of the above-ground processes and their interactions

A number of simplifying assumptions were usually made about canopy architecture to model light penetration and to assess canopy leaf area. The most elementary canopy description applied in mechanistic models is the stand-oriented approach; the canopy is assumed horizontally continuous and radiation absorption is calculated per hectare. This approach is sufficient when simulating regularly constructed, mono-species stands with closed canopies (Mohren, 1987; Nikinmaa, 1992; Bossel and Krieger, 1994). However, statistical models describing horizontally homogeneous stands are not applicable in stands where the foliage is grouped into individual tree crowns (Oker-Blom, 1986) and tree crowns display a great variety of aerial structure, which, in response to the environment, show different solutions to the problems of resource capture. As all trees depend on the radiation energy, it is evident that the vegetation structure of trees is regulated by the spatial and temporal variation of irradiance.

Models with one-dimensional canopy layers have been proposed by many authors following Monsi and Saeki (1953). Models of homogeneous continuous canopies of forests or crops have been used to provide insight into the role of specific characteristics such as leaf angle and leaf area index or vertical distribution of leaf characteristics such as leaf mass per unit area or volume (Gutschick and Weigel, 1988). Discontinuous canopies such as those formed by trees need more dimensions to analyze the vertical and horizontal variations of the radiation distribution. For individual trees, qualitative models of branching have been used to describe the development of form (Halle *et al.*, 1978), but these have had limited ecological applicability in understanding how specific architectures influence the resource capture.

Unfortunately, quantifying crown geometry is difficult because the profiles are usually irregular and simple geometric representations may not be adequate (Biging and Dobbertin, 1995). The occurrence of vertical layering or stratification in a tree canopy has been a matter of considerable attention, not least because of inconsistent definitions and methodology (Bourgeron, 1983). Richards (1983) referred to the stratification of leaf mass and of individual tree heights for which the evidence is weak. Stratification of species refers to the vertical aggregation of average mature heights of species. Extensive quantitative documentation of this phenomenon is also lacking, but there is some support for its existence (Pukkala *et al.*, 1991).

In systems with discontinuous canopies, a more detailed canopy description is necessary, since the diurnal variation of radiation availability and interception are much larger than in closed forests (Palmer, 1977). Array models have been developed that account for horizontal differences (Palmer, 1977), and these canopies were often modeled as a series of one-dimensional layers that are horizontally homogeneous. But because individual crowns are not distinguished, their application is limited to canopies which can be defined in simple geometrical terms.

Two dimensional models have been used to explore how branch angle and branch length influence leaf overlap and hence the efficiency of irradiance capture (Fisher, 1992). Two dimensional models for parallel rows have been described for rectangular (Goudriaan, 1977; Sinoquet and Bonhomme, 1992), elliptical (Charles-Edwards and Thorpe, 1976) and triangular (Jackson and Palmer, 1979) cross-sections. These models are adequate for infinite hedgerows. They consider the interception of light from only a single direction, whereas, in reality, diffuse radiation comes from many directions and the direction of the solar beam changes during the day (Bristow *et al.*, 1985). Modeling an isolated tree presents more difficulty than uniform canopies or row crops where it is possible to reduce the dimension of the problem to a single coordinate. For row crops it is still possible to develop this analysis in a fairly straightforward manner (Gijzen and Goudriaan, 1989).

Recently three-dimensional models have been used to assess light interception by individual trees. These models have generally been focused on either relatively simple canopy structures or are species-specific, or have considered the crown as a series of layers, or cells with particular foliage characteristics (Myneni and Impens, 1985). Myneni *et al.* (1990) formulated a three-dimensional leaf canopy model using a modified discrete ordinates method. Myneni (1991) used fractal models of trees to simulate leaf area distributions for modeling the radiative transfer and photosynthesis in a forest canopy. However, most of these studies have simply discussed the adaptive significance of species-specific tree crown architectures as simple allometries between crown dimensions (crown depth, width, or area) and individual sizes (mass, stem diameter at breast height), and have limited the investigations on the effects of individual crown architecture as vertical foliage distributions on the interactions between them (Mohren *et al.*, 1984; Biging and Wensel, 1990; Sinoquet and Bonhomme, 1992).

Typically, crown architecture is described in terms of the amount and spatial distribution of leaf area within the crown and defined as the set of features delineating the shape, size and geometry of the tree, and described in terms of the amount and spatial distribution of leaf area within the crown. The crown is assumed to be symmetrical around the tree stem. The radial distribution of leaf area can be approximated by a Gaussian-like pattern (Morales *et al.*, 1996); i.e. the greatest leaf area occurred half the distance between the stem axis and the edge of the crown. However, the direct measurement of leaf area in a three-dimensional canopy is almost impossible (Castro and Fetcher, 1998). Therefore, in order to simplify the modeling task, leaf area in vertically-structured canopy layers can be assumed to be homogeneous, ignoring radial distribution of leaf area around the stem (Kinerson et al., 1974). Furthermore, the vertical distribution of leaf area can be considered to fellow a Weibull statistical pattern (Yong et al., 1993).

Not only are leaves at the top of canopies (or on different sides of isolated trees) normally subject to different energy loads, but the path-lengths along which water must move to different parts of the tree are different (Landsberg, 1986). Procedures to predict the vertical profile of leaf area in the crown were developed (Kinerson and Fritchen, 1971; Schreuder and Swank, 1974; Gray, 1978; Beadle et al., 1982; Massman, 1981; Hagihara and Hozumi, 1986). The beam-path length for each sub-canopy within the canopy can be estimated for varying zenith angles of the sun. A rigorous description of the problem which involves setting up a system of mathematical equations to describe the radiation field at all points in the space within the crown should be considered in the calculation of the radiative transfer process.

It is well established that crown structure has important implications for light interception (Horn, 1971; Givnish, 1984; Kohyama, 1991). Isolated trees, in contrast to those growing in rows, often tend to form in the following shapes - cone, intermediate, cylinder or parabola. These shapes result from the typical growth habits of trees (Sinoquet and Bonhomme, 1992).

The canopy architecture is a complex and dynamic outcome of the evolutionary and ecological interactions and feedbacks between vegetation and environment. As such, canopy structure is a key feature of forest ecosystem processes (Campbell and Norman, 1989; Norman and Campbell, 1989). Thus, understanding and quantifying canopy characteristics is critical in modeling processes in the canopy and in predicting ecosystem responses (Meyers and Paw U, 1987).

3.2 Solar position and daylength estimates

Transpiration and water-use models require knowledge of the solar position (defined in terms of zenith and azimuth angles) and day length. The zenith angle of the sun has been shown to influence canopy and soil albedo (Stewart, 1971; Pinker *et al.*, 1980) and absorption (Otterman *et al.*, 1993). The absorption of solar irradiance by trees and soil under trees is significantly larger than the absorption of the bare soil surface, especially at large zenith angles. The absorption of solar irradiance by trees is larger than that by soil under trees at high zenith angles (Otterman *et al.*, 1993) so the efficiency of radiation transfer within canopies increases with increasing zenith angle. Therefore, the zenith angle dependence enters the formulation of radiation transfer to express the partitioning of absorbed solar energy between the tree and the soil components of the surface.

Kock *et al.* (1990) found that the sun zenith angle has a decisive influence on the spectral reflectance measured above forest trees, even for solar angle variations of less than 10%. Kriebel (1978) has reported the reflectance values at 0.52μm wavelength of four natural surfaces; savannah, bog, pasture land and coniferous forest. For all these surfaces, the anisotropy (ratio of the highest to the lowest reflectance value) increases with increasing zenith angle of incidence from about a factor of three to about a factor of ten or more due to the shadowing effects produced by the vertical structure of the canopies (Kriebel, 1978).

In order to estimate the light extinction coefficient, the leaf inclination function and the solar radiation, knowledge of solar position and day length at the time of interest is required. Running and Coughlan (1988) used the latitude of the modeled location and the day of the year as input to the day length calculations of transpiration and energy balance. Nikolov and Zeller (1992) employed the work done by Running and Coughlan (1988) to model solar radiation. Wang and Jarvis (1990) also incorporated the solar position into their radiative transfer model: MAESTRO, using the method developed by Barkstrom (1981). West and Wells (1992) also used the same method (Barkstrom, 1981) to estimate the solar position in a light interception model.

3.3 Radiative transfer process within the canopy

Radiation availability is one of the main driving forces behind water uptake by trees. The geometrical architecture of a tree, one of the main factors determining the radiative transfer within the canopy, defines the size, shape and geometry of the tree. The position and the size of trees and their component parts, and the orientation of leaves all play significant roles in the interaction of the tree with the incident radiation. Leaf area and the spatial arrangement of the foliage, branches, and stems determine the transmission of radiation through a forest canopy.

Radiative transfer in vegetation canopies has been studied for decades (Ross, 1981; Myneni and Ross, 1990; Law et al., 2000). A variety of models have been developed, ranging from models using the simple Beer's law to complex computer simulation models (Wang and Baldocchi, 1989; Myneni *et al.*, 1990; Andrieu *et al.*, 1995). Theoretical basics and reviews of different models can be found in Ross (1981), Goel (1988) and Myneni and Ross (1990).

The leaf-angle distribution which is a fundamental property of plant canopy structure is needed for computing distributions of leaf irradiance (Campbell, 1990) along with radiation transfer through canopies and extinction coefficients. The leaf inclination angle (θ_L) is defined as the angle between the vertical and a normal to the leaf surface, and the leaf azimuthal angle (θ_K) is the angle about the points of the compass between a normal to the leaf surface and the solar beam. Idealized leaf inclination density functions have been widely used to approximate actual leaf angle distributions (Ross, 1981). Several formulae have been given for uniform leaf inclination angles, but randomly distributed azimuthal angles (Lang, 1986; Goudriaan, 1988; Campbell, 1990). The leaf inclination described by one mean leaf angle over-evaluates the effect of the leaf orientation on radiative exchanges (Lemeur, 1973). A two-dimensional probability function $g(\theta_L\theta_K)d\theta_L$, $d\theta_K$ can be used to describe the fraction of the leaf area oriented with the inclination, θ_L and the azimuth θ_K (Lemeur, 1973).

The mathematical description of the leaf orientation distribution is a troublesome part in modeling radiative transfer through a vegetation canopy. Yet it is important because it is the only representation of the canopy geometrical structure, and directly related to the G-function (Ross, 1981) and the canopy scattering phase function - the two most important functions in the radiative transfer theory. Early practices were either to classify the leaf orientation distributions as erectophile, planophile, spherical, extremophile, or plagiophile, and give a mathematical description for each of these leaf orientation distributions (De Wit, 1965), or to represent a leaf orientation distribution by frequencies in three or nine leaf angle classes (Goudriaan, 1977, 1988).

3.3.1 Radiation transfer through the canopy

Several studies have been done to examine the influence of canopy architecture on penetration of radiation in theoretical models (Kira *et al.*, 1969; Oker-Blom and Kellomaki, 1983; Campbell and Norman, 1989). Many of these studies used the Beer-Lambert model, which defines canopy light penetration as proportional to the cumulative leaf area index (e.g. Monsi and Saeki, 1953). Small within-species variations in light extinction have often been attributed to changes in solar altitude during sampling of under-canopy light flux density (Campbell and Norman, 1989) rather than to intra-specific variability. Recent results, however, suggested substantial variation in light extinction for several tree species (Gholz *et al.*, 1991). Models have often been used to estimate light penetration in forest stands of different *LAI* (Dalla-Tea and Jokela, 1991; Gholz *et al.*, 1991).

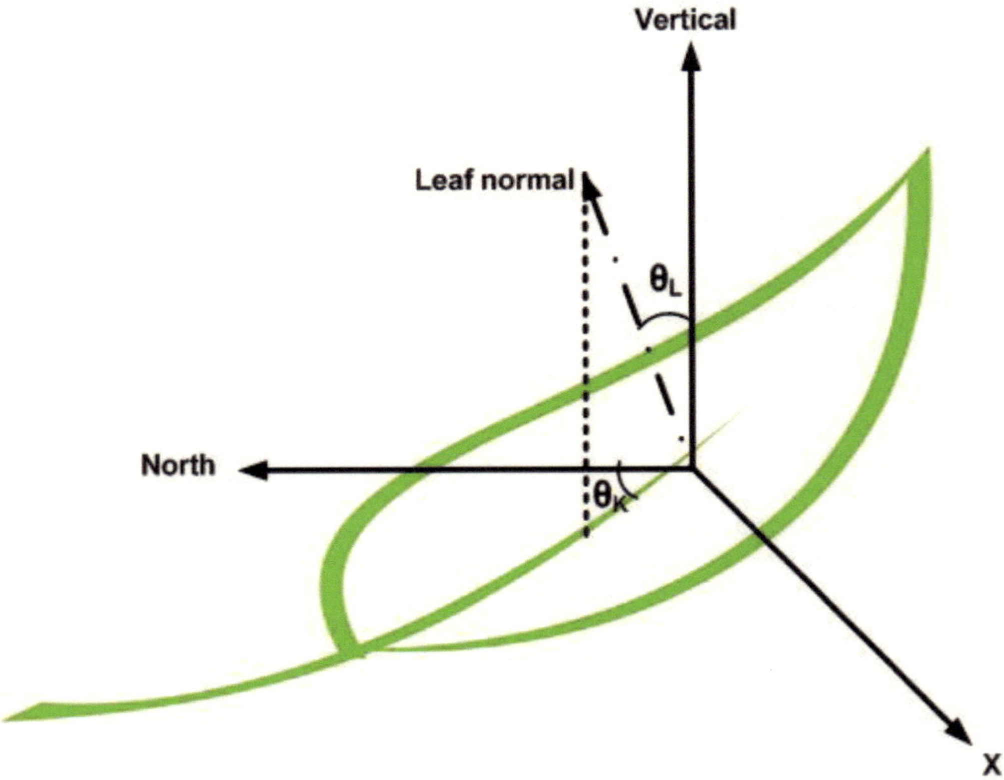

Fig. 3. Diagrammatic representation of the orientation of the normal to a leaf surface

The leaf angle distribution, which is a mathematical description of the angular orientation of the leaves in the canopy, is needed for computing distributions of leaf irradiance along with radiation transfer through canopies and extinction coefficients (Campbell, 1990). The leaf angle distribution consists of two angles i.e., the leaf inclination angle (θ_L) and the leaf azimuthal angle (θ_K). The θ_L is defined as the angle between the vertical and a normal to the leaf surface, and the θ_K is the angle about the points of the compass between a normal to the leaf surface and the solar beam (Campbell, 1990) (Fig. 3). Idealized leaf inclination density functions have been widely used to approximate actual leaf angle distributions. If all leaves are inclined at a constant angle (mean leaf angle), θ_0, then, the inclination angle distribution, $g(\theta_L)$ is given by (Ross, 1981):

$$g(\theta_L) = \delta(\theta_L - \theta_0)\sin\theta_L \tag{10}$$

where $\delta(\theta_L - \theta_0)$ is the Dirac delta function. A horizontal distribution results when $\theta_0 = 0$, a vertical or cylindrical distribution when $\theta_0 = \pi/2$, and a conical distribution when θ_0 is between these values. In most plants the upper (adaxial) surfaces of the leaves face the upper hemisphere so that θ_L varies between 0° and 90° whereas θ_K varies between 0° and 360°. The leaf inclination described by one mean leaf angle over-evaluates the effect of the leaf orientation on radiative exchanges (Lemeur, 1973). A two-dimensional probability function $g(\theta_L\theta_K)d\theta_L$, $d\theta_K$ can be used to describe the fraction of the leaf area oriented with the inclination, θ_L and the azimuth θ_K (Lemeur, 1973). This function expresses the probability that a leaf has an inclination within θ_L and $\theta_L+d\theta_L$ and an azimuth within θ_K and $\theta_K+d\theta_K$. Therefore, integration over θ_L and θ_K leads to (Ross, 1975):

$$\int_{\theta_K=0}^{\theta_K=2\pi}\int_{\theta_L=0}^{\theta_L=\pi/2} g(\theta_K,\theta_L)d\theta_K d\theta_L = 1 \tag{11}$$

It is possible to estimate the value of light extinction coefficient for different sets of orientation of the leaves and the direction of incoming radiation by considering the projected area of solids having the angle distributions for the given distribution function (Monteith, 1975). The value of the light extinction coefficient for direct beam radiation is related to its mean cosine of incidence on the leaf surfaces. For a single flat leaf, the cosine of the angle of irradiance is described by θ_L, θ_K, and φ_L, and the cosine $t(\varphi_L,\theta_L,\theta_K)$ is determined as (Ross, 1975):

$$t(\varphi_L,\theta_L,\theta_K) = \cos\varphi_L\cos\theta_L + \sin\varphi_L\sin\theta_L\cos\theta_K \tag{12}$$

The distribution of t is uniform with θ_K if no preferred azimuthal angle is assumed. The cumulative distribution function of t can be found by increasing θ_K from $-\pi$ to 0. With symmetrical geometry , the other half of the azimuthal circle (from 0 to π) can be omitted. θ_K, at the range from $-\pi$ to 0, is equivalent to the range of 0 - 1 for the cumulative distribution probability S. The average value of t can be found as the integral of t with respect to S. This is the same quantity as the average projection of the leaves into the direction of the solar beam called the O function by De Wit (1965) and the G function by Ross (1975) and Goudriaan (1989).

3.3.2 Radiation scattering

As pointed out by Myneni *et al.* (1988), the canopy scattering phase function determines how realistically the physics of the radiation transport process in vegetation is represented (Ross,

1981). Therefore, the scattering phase function should be one of the most important factors determining the level of accuracy of the radiative transfer process.

With the assumption of isotropic scattering for leaf elements, Sellers (1985) found a simple expression for the scattering phase function, and solved the equations analytically. However, the isotropic scattering assumption is a rather crude approximation for real leaf elements. As a consequence of this assumption, scattering does not change with the relative magnitudes of the reflection and transmission coefficients for a given leaf scattering coefficient. The unrealistic implications of this shortcoming can be seen easily in the expression for the single scattering albedo for a semi-infinite horizontal leaf canopy. It is a quarter of the leaf scattering coefficient (the leaf scattering coefficient equals the sum of the leaf reflection and transmission coefficients). Since only single scattering is considered, radiation which is transmitted through a leaf is never able to escape from the canopy and cannot contribute to the canopy reflection. Thus, the transmission coefficient is unrelated to the canopy single scattering albedo, and the only relevant leaf optical parameter is the leaf reflection coefficient. For most situations, canopy hemispheric reflectance as given by the isotropic two-stream model, proposed by Dickinson (1983) and further developed by Sellers (1985), hardly changes with the relative magnitudes of the reflection and transmission coefficients. Although the isotropic two-stream approximation model has widely been applied (Sellers, 1985), comprehensive testing of the model has been limited.

Although Hassika *et al.* (1997) stated that the radiation rescattered by the crowns can be neglected, estimates using the Kubelka-Munk equations that indicate this term is small, up to secondary scattering may be good enough for the canopy radiative process.

3.3.3 Radiative transfer process with the canopy

Radiative transfer models are based on more or less abstract representations of reality, usually in relation to the aim and scale of the simulation; models based on a uniform-canopy hypothesis and on the architectural description of tree crowns represent the opposite extremes of this range. Radiative models of the first type can be used to simulate light interception by agricultural crops and uniform forests (Wang and Baldocchi, 1989) and to provide insight into the role of specific characteristics such as leaf angle and leaf area index or vertical distribution of leaf characteristics such as leaf mass per unit area or volume (Gutschick and Weigel, 1988). Models in the second group work at the same scale as leaves or leaf groups and are used to investigate the relationship between canopy architecture and radiative regime in complex heterogenous canopies (Myneni *et al.*, 1990). For individual trees, qualitative models of branching have been used to describe the development of form (Halle *et al.*, 1978), but these model applications have had limited ecological applicability in understanding how specific architectures influence light capture. Two dimensional models have been used to explore how branch angle and length influence leaf-overlap and, hence, the efficiency of light capture (Fisher *et al.*, 1981). These approaches were limited in that they considered the interception of light from only a single direction, whereas, in reality, diffuse radiation comes from many directions and the direction of the solar beam changes during the day. Neither of these model types can address some of the research tasks typical for forest ecology (Pukkala *et al.*, 1993).

Myneni *et al.* (1990) formulated a modified discrete ordinates method for a three-dimensional leaf canopy. Myneni (1991) used fractal models of trees to simulate the radiative transfer and photosynthesis in a forest canopy. Recently, three-dimensional models such as Myneni *et al.*'s (1990) have been used to assess light interception by individual trees. These models have generally focused on either relatively simple canopy structures or are species-specific, or have considered the crown as a series of layers, or cells with particular foliage characteristics (Myneni and Impens, 1985; Ryel *et al.*, 1993). The foliage intercepts most of the radiation captured by the canopy, and an increased foliage area decreases the penetration of radiation. It is normally assumed that only the crown intercepts radiation; stems, branches and fruits are 'invisible' for the light beams (Jarvis and Leverenz, 1983; Cannel *et al.*, 1987).

Short-wave direct radiation (*Rbs*) and diffuse radiation (*Rds*) are the main spectral characteristic for many radiative transfer models Myneni (1991). However, Kjelgren and Montague (1998) stated that trees over asphalt had consistently higher leaf temperatures than those over turf due to the interception of the greater upwards long-wave radiation fluxes due to the higher surface temperatures of the asphalt over turf. When bare soil between trees exits, upwards long-wave radiation may contribute substantially to the energy balance in the canopy. Therefore, the radiation above, within and below the canopy can be separated in terms of short-wave radiation (*Rb*) and long-wave radiation (*Rlw*), and *Rb* is separated into *Rbs* and *Rds* radiation.

The characterization of the magnitude and direction of the diffuse and direct components of incident radiation is required as the light distribution within the canopy depends on the spectral optical properties of the leaves and the soil. Therefore, the optical properties of the vegetation elements and the soil are important factors to determine radiative transfer within the canopy.

3.4 Canopy-energy balance

Radiant energy received from the sun and the atmosphere is exchanged for latent and sensible heat by plants and the soil surface. Sensible heat is transferred because of the temperature difference between trees and surrounding air, and it moves by convection, advection and diffusion. The relative amounts of latent and sensible heat exchanged by plants and soil surfaces are understood to be mediated by resistance to water movement in aqueous and vapour phases between the plant or the soil and the atmosphere.

Gates (1962) pioneered leaf energy budget investigations in plant ecology. Although the interaction between a canopy and its radiant energy environment can be extremely complex, a simplified approach that serves to introduce basic concepts is now well-established. According to energy conservation, the heat gain by radiation should be equal to the heat losses in the canopy when leaf temperature is constant.

The leaf temperature of trees is higher at the upper part than at the lower in daytime (He *et al.*, 1996; Zermeno and Hipps, 1997) and the difference between leaf temperatures at a depth of 20 m and 11 m is about 12 ^{0}C around noon (Miyashita and Maitani, 1998). Therefore, it is clear that energy balance cannot be applied uniformly throughout the canopy. Describing the long-wave radiative transfer is further complicated in plant canopies because leaf elements are also long-wave radiation sources (Rotenberg et al., 1999). This fact complicates the description of long-wave radiative transfer in plant canopies. To make the situation even worse, in order to solve the leaf temperature from the energy balance equation, the absorbed long-wave radiation, which in turn depends on the temperature of the leaf, must be known in advance. Solving the long-wave radiative transfer problem relies on iterative search methods which can be computationally intensive for a canopy process model. Complicating this, the energy balance equation has (potentially up to) four roots because long wave radiation is proportional to the fourth power of temperature. At least one of the roots should be bio-physically sensible. Unfortunately there is no guarantee that iterative search methods will converge on the correct root and sometimes yield bifuricated or chaotic solutions (Baldocchi, 1994). To overcome this difficulty, a polynomial expression can be introduced in order to solve the energy balance equation analytically for leaf temperature. Doing so avoids the involvement of a complex algorithm with numerous partial derivatives and iterations in energy balance calculations such as that used by Cienciala *et al.* (1994). Applying the energy balance calculation in a horizontally- or/and vertically-structured canopy allows us to investigate spatial variations of leaf temperature.

The sensible heat flux can be considered to be proportional to the difference between the radiometric temperature which is identical to canopy surface temperature and the air temperature at a reference height, and inversely proportional to aerodynamic resistance.

The large amount of standing biomass in a forest causes a substantial amount of net radiation to be transformed into canopy heat storage, further affecting evapotranspiration (Jarvis, 1976). The heat storage, which is mostly in biochemical reactions represents a fraction of radiation over a very short period (Montheith, 1975) is about 8% of the *Rnet* (Jones, 1992).

3.5 Scalar variations within the canopy

Many canopy models have been developed to describe the exchange of sensible and latent heat between plant canopies and the atmosphere (Baldocchi, 1993). An important function of these models is to predict the mean profile of humidity and temperature of the air in the canopy, because transpiration at each canopy level depends on the air temperature and humidity at that level.

To calculate these profiles, some assumptions are made about the turbulent transport processes within the canopy. The most common assumption has been that turbulent convection conveys scalars (heat and vapour pressure), down local concentration gradients by a turbulent diffusion process. As such, these models have been based on *K*-theory (Waggoner and Reifsnyder, 1968). The *K*-theory has been challenged by observations of fluxes of scalars moving in directions opposed to their local concentration gradients within plant canopies (Denmead and Bradley, 1985). New theories have been developed to explain counter-gradient transport, but are yet to produce an accurate method to demonstrate scalar transport processes within the canopy. Raupach (1989), for instance, stated that because of the strong influence of source distribution and the relatively weaker influence of the fine details of the inhomogeneous wind field, a rather simple model for the wind field in the canopy is probably all that is necessary to calculate scalar concentrations in the vertical dimension quite accurately. Simple, half-order closure modelling which assumes a uniform scalar profile does not yield large errors in the computation of flux densities because the source-sink formulation of fluxes is relatively insensitive to changes in scalar concentrations in the profile and the scalar gradients are small (Baldocchi, 1992).

Sensible and latent heat transfer processes consider the environmental gradients that occur between some reference level outside the canopy and sub-canopies inside the canopy in a three-dimensional modeling framework. Gradients of these quantities exist between the bulk air and sub-canopies within the canopy and drive the fluxes from (or to) the canopy. To evaluate these gradients, for example in a tree canopy, linkages between the strengths of the respective sources and sinks and scalar concentration in the sub-canopy must be considered. These linkages arise because the rate at which material is released (or taken up) affects the local scalar concentration in sub-canopies within the canopy, and the rate of leaf emission, (or uptake), depends on the local scalar concentration. Implicitly, it can be assumed that the atmosphere within the sub-canopy is well mixed and no inter- or intra-layer mixing within the canopy. It can be assumed further that water vapor outside any sub-canopy (but inside the canopy) is removed without affecting the micro-environment along its pathway to the bulk air outside the canopy. However, it is recognized that these assumptions are not accurate, but a reasonable answer to a difficult, if not unquantifiable, situation. Despite these assumptions, the humidity deficit between the various sub-canopies and the bulk air is not equal as long as the modelling process simulates;

1. surface temperature with respect to the sub-canopy location within the canopy, and
2. vertical profiles of air temperature and wind speed outside the canopy.

The diurnal variations of wind speed and air temperature and humidity of air can be calculated using empirical equations given in the literature (Landsberg and James, 1971; Dogniaux 1977; Wu, 1990; Eckersten, 1991; Paw U *et al.*, 1995). The shape of the diurnal scalar curves is modeled with a variety of methods with varying degrees of complexity. These methods include linear models and curve-fitting models based typically on sine or Fourier analysis (De Wit *et al.*, 1978; Worner, 1988; Fernandez, 1992).

The vertical wind profile affects the boundary layer resistance for heat and water vapor transfer (Landsberg, 1986). Although tree edges can modify the wind speed, and wind direction both inside and outside of the canopy, it is normally assumed that wind speed does not change while passing through the canopy.

Air temperature affects latent and sensible heat flux. A simple method is applied to determine the air temperature inside the canopy from the air temperature outside the canopy and sensible heat flux. Diurnal variation of air temperature can also be estimated using a sinusoidal progression. There are several studies that have been carried out to determine the vertical profile of the air temperature (Ta) above the canopy (Wu, 1990), but such knowledge does not exist in the same simple manner for profiles within the canopy. Paw U *et al.* (1995) used a form of surface renewal analysis by assuming that under unstable conditions (canopy warmer than the air), any rise in the temperature profile represents air being heated by the canopy, and under stable conditions (canopy cooler than air) any temperature drop represents air being cooled by the canopy. The air temperature normally changes in a sinusoidal progression during the day and a decreasing exponential curve at night (van Engelen and Guerts, 1983).

To estimate relative humidity (Rh) rather than interpolating it over time using synoptic values, it is more appropriate to use linear interpolation of the absolute air humidity (Ha) (Eckersten, 1991).

A knowledge of solar radiation is of interest in studies relating to crop evapotranspiration, forest transpiration and for solar energy applications. The incident global radiation (350-3000 nm) on the earth's surface (*Rnet*) is a product of incident radiation outside the atmosphere (*Ro*) and the atmospheric transmissivity, which is dependent upon the degree of cloudiness.

If uniform cloudiness is assumed over the day, the mean irradiance (Rn^{M}) is computed from the ratio of the daily sum of actual global radiation to the daily sum of global radiation from a clear sky. The effects of cloudiness on radiation within the day can be approximated by introducing a factor that varies according to cloudiness.

3.5.1 Separation of solar radiation into direct and diffuse components

For the radiation-mediated processes of a canopy energy study, it is not sufficient to know the total incoming radiation, but estimates of the direct short wave, diffuse short wave and long wave components are required. Bristow *et al.* (1985) established a relationship to estimate hourly diffuse transmittance (*Tdd*) from hourly total transmittance (*Ttd*). Diffuse radiation can then be obtained by multiplying *Tdd* by potential solar radiation. The difference between total incident radiation and diffuse radiation is the direct beam radiation.

Apart from Bristow *et al.* (1986), there are several references for the partition of global radiation available (Weiss and Norman, 1985), but no general agreement on the method to be used to estimate the proportions. There are several methods to calculate the proportions of *Rb* and *Rds* in global radiation, both empirical and theoretical (Liu and Jordan, 1960; Weiss and Norman, 1985; Bristow et al., 1986). Although empirical methods give better results, their validity is limited to a particular place and time. Theoretical methods are preferable because they are more general, but no method is completely satisfactory for all latitudes and seasons (Castro and Fetcher, 1998). Many factors including clouds, aerosols, etc., affect the scattering of radiation in the atmosphere and therefore the proportion of *Rds*. Consequently, a theoretical method such as that described in this section for the partition of *Rb* and *Rds* in global radiation can be used.

3.5.2 Long-wave radiation

Long-wave radiation comes from objects with extended radiating surfaces such as clouds, sky, rocks, soil, water, and vegetation or animals. Arbitrary limits of 3 and 100 μm are usually taken to define the long-wave spectrum (Monteith and Unsworth, 1990). Downward long-wave radiation ($Rlw\downarrow$) can be given as a function of air temperature (*Ta* in ^{0}C) and the vapour pressure ($e_a(Ta)$ in hPa) (Brutsaert, 1982).

3.5.3 Photosynthetically active radiation

In micrometeorology, special attention is paid to photosynthetically active radiation (R_{PAR}), which is defined as radiation in the spectral region between 400 and 700 nm (Monteith, 1975). R_{PAR} is needed when stomatal resistance is calculated. The calculation of R_{PAR} is performed by means of a conversion factor C_{PAR} as a function of solar zenith angle (φ_L) is given by Perelyot (1970).

3.6 Water flow resistance process in the canopy-atmosphere process

The turbulent transport between canopies and the bulk of the atmosphere depends on the turbulent nature of the planetary boundary layer (Meyers and Paw U, 1986 and 1987). In the transfer of water vapor to and from trees, some exchanges occur by molecular diffusion such as the passage of water through stomata. The flux of diffusing gas (kg m^{-2} s^{-1}) can then be equated to the concentration difference (kg m^{-3}) over a diffusion resistance (s m^{-1}) as given by Fick's law. In the process of the diffusion of water vapor away from the leaves, the stomatal resistance (R_t) accounts for diffusion from the evaporation sites within the leaf to the leaf surface, while the leaf boundary layer resistance (R_a) accounts for diffusion from the surface to the well-mixed surrounding air. Both the stomatal resistance and the leaf boundary layer resistance are highly dependent on the size, shape and surface properties of the leaves, and wind velocity.

3.6.1 Stomatal resistance

Empirically established stomatal resistance, which is the most important factor determining transpiration from high vegetation especially forests, has marked variation within the canopy as well as over the day and the season. Stomatal resistance (R_t) depends both on soil and atmospheric factors. These factors are short-term changes in leaf water potential, vapour pressure deficit, solar flux, leaf temperature, ambient carbon dioxide concentration and significant drying of the soil (very negative soil water potential). Therefore, there is clearly a need to describe the response of R_t to atmospheric factors as well as soil water status. Estimation of stomatal resistance has generally involved two approaches;

a. serially integrating the stomatal resistance of individual layers, weighted by leaf area, or
b. using measured values of latent heat flux and other relevant variables in a stand-level equation.

Baldocchi *et al.* (1991) presented an excellent overview of the strengths and weaknesses of different approaches for estimating canopy stomatal resistance. As discussed in their paper, the above mentioned approaches may not yield the same results. The former is primarily a physiological parameter whereas the latter involves additional eco-physiological factors within the canopy. The latter also includes the contribution of soil evaporation. Baldocchi *et al.* (1991) developed a multi-layer canopy stomatal conductance model in which the spatial variation of canopy structure and the radiative transfer within the canopy were taken into account.

Jarvis (1976) has modelled the stomatal resistance as a function of solar radiation, temperature, specific humidity deficit, leaf water potential, and ambient CO_2, using a non-linear least squares technique. Based on the Jarvis work, Stewart (1988) developed a model in which the stomatal resistance was related to solar radiation, temperature, specific humidity deficit, soil moisture deficit and leaf area index. A semi-empirical stomatal resistance model was proposed by Ball *et al.* (1987). After analyzing Ball's model, Aphalo and Jarvis (1993) have proposed a new model, which views the model as a description of the relationship between CO_2 flux rate and stomatal conductance, rather than as a model of stomatal conductance alone. Ball's empirical model was later modified by Leuning (1995).

Lacking independent estimates of canopy surface resistance for independent assessments of the Penman-Monteith equation, many researchers assumed that canopy surface resistance is equivalent to the integrated stomatal resistance. The adaptation of this assumption has received theoretical (Finnigan and Raupach, 1987; Kelliher *et al.*, 1994; Raupach, 1995) and experimental (Baldocchi *et al.*, 1987) criticism over the years. Under the optimal environmental conditions required to achieve 'minimum' resistance (ample water supply, non-extreme temperature, and fully developed non-senescent leaves), stomatal resistance (R_t) varies through the canopy only in response to variation in photosynthetically active radiation (Kelliher *et al.*, 1994). However, R_t decreases with radiation and increases with the vapor pressure deficit of the atmosphere (Lohammar *et al.*, 1980). Soil water content varies with uptake by roots as well as with rainfall and irrigation, and soil water potential directly affects R_t (Jones, 1982). Air temperature changes in a sinusoidal fashion during the day (Goudriaan and van Laar, 1994) and affects R_t (Aphalo and Jarvis, 1991). White *et al.* (1999) modified R_t by describing the response of light, air temperature and vapor pressure deficit, but soil water potential directly affects R_t in some circumstances (Jones, 1992), but in many cases, soil water potential is considered to affect leaf water potential, which in turn controls R_t (Lynn and Carlson, 1990). Therefore, R_t as a function of leaf water potential may be more accurate.

A procedure to describe the behavior of R_t can be given as a function of photosynthetically active radiation, vapor pressure difference, air temperature and soil moisture deficit. Although R_t is considered to be influenced by changes of CO_2 concentration (Hall, 1982), it is typically not included, because in most cases it was found to be almost constant (about 4 ppm variation) (Yang *et al.*, 1998). Baker (1996) also supported the case for insignificant variation of the CO_2 profile in forests.

As the radiation interception and scalar profiles are formulated for each sub-canopy within the crown, it is possible to estimate R_t individually for each sub-canopy, and thus avoid the integration of R_t for the whole canopy. Similarly, R_a calculated for each sub-canopy in the canopy is not required to be integrated over the total leaf area of the canopy. One can observe that stem resistance also contributes to the overall resistance of the canopy-atmosphere system. However, in resistance studies by Melchior and Steudle (1993), it was found that resistance to water flow was usually negligible where the xylem had already matured.

3.6.2 Leaf boundary layer resistance

The average thickness of the boundary layer is related to the leaf size. Thus small leaves have thin boundary layers which give small boundary layer resistances whereas large leaves have thick boundary layers with large boundary layer resistances and temperatures which may differ substantially from that of the surrounding air (Grace, 1983). At high wind speeds, the boundary layer is thinner than at low speeds and the resistance correspondingly smaller. The canopy slows down the air flow and creates a turbulent boundary layer. Transport of heat or water vapor through this layer occurs by turbulent diffusion, at a rate determined by the turbulent structure of the air which, in turn, is determined by the wind speed and the aerodynamic roughness of the canopy. The main determinants of boundary layer resistance are therefore leaf size and wind speed, with leaf form exerting a secondary effect through its effect on turbulence (Nikolov *et al.* 1995).

Values of boundary layer resistance (R_a) for individual leaf components can be estimated using engineering equations or empirical relationships. A comprehensive analysis of how to quantify the leaf boundary layer can be found in work done by Campbell (1977), Grace *et al.* (1987), Gates (1980), Monteith and Unsworth (1990) and Nikolov *et al.* (1995).

4. Combining below- and above-ground processes

A tree is an organism with leaves and has a capacity to store water in its boles. The transport of water through the water storage in the tree causes hysteresis between rates of soil water uptake and transpiration (Jones, 1982). Landsberg (1986) pointed out that it becomes necessary to include the fluxes in and out of storage in models that predicts the time course of leaf and other tissue water potentials. Trees use stored water to keep stomata open and maintain transpiration in the face of limiting soil moisture or excessive atmospheric demand. Therefore, water movement can be modelled in terms of water potentials and resistances via exchangeable water storages.

Having identified the major soil and environmental variables affecting the movement of water from soil to the atmosphere through trees and the subsequent changes in flow associated with resistances in the previous sections, it is possible to combine all this understanding of below- and above-tree water movement into a predictive model. Soil and root resistances, and soil water potential come from the below-ground models whereas the above-ground models contribute stomatal (R_{tp}) and boundary layer ($R_a{}^{W}{}_p$) resistances, vapor densities at the surface and the air temperature. Initial rates of water uptake (F_{Up}) and transpiration (F_{TP}) are used to estimate the exchangeable water storage (V_p) and leaf water potential (ψ_{Lp}). Since ψ_{Lp} is a function of R_{tp}, an iterative procedure is required to estimate the final value of F_{TP}.

It is possible to eliminate the intermediate water potentials mathematically by neglecting any storage of water at the surface or in the tree (Thornley and Johnson, 1990). Campbell (1991) has employed a multi-layered root zone to include variable rooting density conditions into the water uptake estimations.

Monteith (1980) stated that latent energy for evaporation must be supplied from an external source (according to the law of conservation of energy), and the saturated water vapor in contact with the wet surface must be swept away and replaced by dry air which becomes saturated in turn. To sustain vaporization, however; there must not only be a continual supply of energy, but also an inward flow of liquid water from the soil or the plant (McIlroy, 1984). Accepting this, it is reasonable to assume that there is exchangeable water stored in the plant (Weatherley, 1970 and Jarvis, 1975). Until recently, tree water storage has been largely ignored in soil-tree-atmosphere models. The exchangeable water in the plant allows transpiration to exceed, equal or be less than water uptake by the roots at any given time. This concept of exchangeable water in the plant was used by Kowalik and Ekersten (1984) to formulate a continuous simulation model for transpiration and was solved numerically.

4.1 Modelling water uptake and transpiration

The flow of water through the soil-tree-atmospheric continuum can be divided into three components: (a) water uptake by roots; (b) exchangeable water in storages; and (c) water loss by leaves. All necessary components related to (a) are given in Section 2 whereas for (c), they are given in Section 3.

The essential concept is that the water storage in the canopy for a given period is governed by water lost by transpiration and water supplied by roots. The volumetric change of the water storage in the sub-canopy P during one time step is the difference between water uptake and transpiration:

$$\delta V_p = \int_{t-\delta t}^{t} \left(F_{U_P} - F_{T_P}\right)dt \tag{13}$$

where F_{Up} is root water uptake to computational sub-canopy P (kg h^{-1}), δV_P is change of the amount of exchangeable water stored for a given time (kg) and F_{TP} is transpiration from the sub-canopy P (kg h^{-1}). The purpose of introducing the exchangeable water stored in the tree is to show the effects of stored water in a coupled soil-tree-atmospheric model on the transpiration flux and leaf water potential. V_P is the state variable. The water uptake by trees, F_U,and transpiration from the leaves, F_T, are the main driving variables.

It is possible to eliminate intermediate water potentials mathematically by neglecting any storage of water at the surface or in the tree. Thus, the water flow from the soil to the plant (Thornley and Johnson, 1990) was written as:

$$F = \frac{\psi_S - \psi_L}{R_s + R_r} \tag{14}$$

where R_s is flow resistance from bulk soil to root surface (kg^{-1} m^4 s^{-1}), R_r is flow resistance from root surface to xylem (kg^{-1} m^4 s^{-1}), ψ_s is soil water potential (J kg^{-1}) and ψ_L is leaf water potential (J kg^{-1}).

Equation 14 is not particularly useful because it assumes a constant rooting density with respect to depth in the soil. To extend the equation to include variable rooting density conditions, (the root volume is assumed to be made up of zones with constant root densities), Campbell (1991) has employed a multi-layered root zone as follows:

$$F_U = \int_{i=1}^{n} \left(\frac{\psi_{S_i} - \psi_L}{R_{S_i} - R_{r_i}}\right) \tag{15}$$

The amount of water uptake by roots for the sub-canopy P (kg h^{-1}) can be computed by expanding equation 8 as follows (after Thornley and Johnson, 1990; Campbell, 1991; Eckersten, 1991):

$$F_{U_P} = \int_{i=1}^{l} \left(\frac{\psi_{S_i} - \psi_{L_P} - Z_P}{R_{S_i} + R_{r_i} + R_{l_i} + R_{st}}\right) LA_P 3600 \tag{16}$$

where ψ_{si} is soil water potential in the soil layer i (J kg^{-1}), ψ_{Lp} is leaf water potential in the sub-canopy P (J kg^{-1}), l is number of soil layers, R_{si} is soil resistance for the water flow from soil to the root surface in the soil layer i (kg^{-1} m^4 s^{-1}), R_{ri} is root radial resistance for the water flow in roots in the soil layer i ($kg^{-1}\ m^4\ s^{-1}$), R_{li} is axial resistance for the water flow in roots in the soil layer i ($kg^{-1}\ m^4\ s^{-1}$), R_{st} is stem resistance (assumed negligible), z_p is gravitational potential ($J\ kg^{-1}$) and LA_P is leaf area of the sub-canopy P (m^2).

The transpiration rate is driven by the difference in vapour pressure between that inside the stomatal cavities and that of the air outside. When the air in stomatal cavities is assumed to be saturated, the transpiration rate in sub-canopy P ($kg\ h^{-1}$) can therefore be calculated as (Eckersten, 1991):

$$F_{T_P} = \frac{\rho c_P\left(e_S(Ts)_P\right) - e_a(Ta)_P}{\lambda\gamma(Rt_P + Ra_P^W\ )}3600 \quad (17)$$

where ρ is the specific density of moist air ($kg\ m^{-3}$), γ is the psychrometric constant ($kPa\ K^{-1}$), λ is the latent heat of vaporization of water ($MJ\ kg^{-1}$), c_P is specific heat per unit mass of air ($MJ\ kg^{-1}\ K^{-1}$), $e_a(Ta)_P$ is vapour density (kPa) at Ta_P in the sub-canopy P, $e_s(Ts)_P$ is vapour density (kPa) at Ts_P in the sub-canopy P, $R_a{}^W{}_p$ is leaf boundary layer resistance for water vapor transport in the sub-canopy P ($s\ m^{-1}$) and R_{tp} is stomatal resistance in the sub-canopy P ($s\ m^{-1}$).

A linear relationship between exchangeable water storage and leaf water potential is employed after Federer, (1979); Kowalik and Eckersten, (1984); Eckersten, (1991) and Cienciala *et al.* (1994), as follows:

$$V_P = \left(\psi_{L_P} - \psi_{L\min}\right)\left(\frac{V_{\max_P}}{\psi_{L_{\max}} - \psi_{L_{\min}}}\right) \quad (18)$$

where $\Psi_{L\min}$ is minimum leaf water potential ($J\ kg^{-1}$), $\Psi_{L\max}$ is maximum leaf water potential ($J\ kg^{-1}$) and $V_{\max\ p}$ is maximum easily-exchangeable water. A linear relationship between Ψ_L and exchangeable water storage in the canopy was also suggested by Tyree (1988) from experimental data taken from Brough *et al.* (1986). The idea of a minimum leaf water potential originated from Cowan (1965), who called it the 'supply function'. Jarvis (1975) suggested that the threshold leaf water potential equates to minimum stomatal resistance until the onset of leaf water stress. The leaf water potential is not uniform throughout the tree (Landsberg and McMurtrie, 1984). Ψ_{Lp} can be calculated from equation 16, provided that V_p is calculated in an iteration. As stated earlier, it is considered that water in the tree is assumed to be in storages in each sub-canopy P, and thus, the variation of leaf water potential within the sub-canopy can be predicted.

5. Further research

The soil-tree-atmosphere water relations consists of important physiological and physical processes which control the soil water dynamics, water uptake by roots, energy and water transfer from the canopy. The following issues have identified for further advancement in the soil-tree-atmospheric water relations.

1. The resulting increase in humidity from soil evaporation can be added to the canopy processes in at least the lower part of the canopy.
2. The trade-off between small and large sub-canopies is that large sub-canopies can affect the accuracy while small sub-canopies can increase the model computational time

considerably. The optimum size of the sub-canopies can only be found by trial and error and requires further investigation.

3. The assumption of a horizontally-homogeneous, but vertically-heterogeneous canopy has not escaped severe criticism. To have an accurate radiation environment in the crown, the radial distribution of leaf area may be introduced in the canopy.
4. In air temperature calculations, the turbulent parcel of air is trapped in the canopy for some period of time, and then is ejected after being modified by canopy-atmosphere transfer may be shown to be unrealistic. Even though this may give some plausible results, a realistic mechanism needs to be developed.
5. A uniform profile of relative humidity is assumed which may not be realistic. In order to be compatible with other canopy processes in a vigorous three-dimensional canopy, spatial variation of humidity should be incorporated.
6. Precipitation could also be a strong factor for latent heat flux. However, it should be noted that the incorporation of the canopy convective heat flux from free water is essential for acomplete energy balance calculation.
7. Even if a close correlation exists between stomatal aperture and soil moisture (through leaf water potential), the existence of a root signaling process based solely on increased abscisic acid (ABA) in the xylem sap can be postulated.
8. An important question to be answered by future research is how a model can successfully predict transpiration rates over time periods where the physical environment has been subjected to frequent perturbations.

6. References

Abdul-Jabbar, A.S., Lugg, D.G., Sammis, T.W., and Gay, L.W. (1984). A field study of plant resistance to water flow in alfalfa. Agronomy J. 76: 765-769.

Alm, D.M., Cavelier, J., and Nobel, P.S. (1992). A finite element model of radial and axial conductivities for individual roots: Development and validation for two desert succulents. Ann. Bot. 69: 87-92.

Andrieu, B., Ivanov, N., and Boissard, P. (1995). Simulation of light interception from a maize canopy model constructed by stereo plotting. Agric. For. Meteorol. 75: 103-119.

Aphalo, P. J., and Jarvis, P. G. (1993). An analysis of Ball's empirical model of stomatal conductance. Ann. Bot (London) 72: 321-327.

Baker, J.M. (1996). Use and abuse of crop simulation models. Agron. J. 88: 689-690.

Baldocchi, D.D. (1992). A Langrangian walk model for simulating water vapour, carbon dioxide, and sensible heat flux densities and scalar profiles over and within a soybean canopy. Boundary Layer Meteorol. 61: 113-144.

Baldocchi, D.D. (1993). Scaling water vapour and carbon dioxide exchange from leaves to a canopy: Rules and tools. In: J.R. Ehleringer, and C.B. Field, (Eds.) Scaling Physiological Processes-Leaf to Globe. Academic Press, London, pp. 77-108.

Baldocchi, D.D. (1994). An analytical solution for coupled leaf photosynthesis and stomatal conductance models. Tree Physiol. 14: 1069-1079.

Baldocchi, D.D., Hicks, B.B., and Camara, P. (1987). A canopy stomatal resistance model for gaseous deposition to vegetated surfaces. Atmos. Environ. 21: 91-101.

Baldocchi, D.D., Luxmoore, R.J., and Hatfield, J.L. (1991). Discerning the forest from the trees: an essay on scaling canopy stomatal conductance. Agric. For. Meteorol. 54: 197-226.

Ball, M. C., Woodrow, I.E., and Berry, J. A.(1987). A model predicting stomatal conductance and its contribution to the control of photosynthesis under different environmental conditions. In: I. Biggins, (Ed), Progress in Photosynthesis Research. Martinus Nijhoff Publishers, Netherlands, pp. 221-224.

Barkstrom, B. (1981). What time does the sun rise and set? Byte 6: 94-114.

Beadle, C.L., Talbot, H., Jarvis, P.G. (1982). Canopy structure and leaf area index in a mature Scots pine forest. Forestry 55:105-123.

Bernston, G.M. (1994). Modelling root architecture: Are there tradeoffs between efficiency and potential of resource acquisition? New Phytol. 127: 483-493.

Bernston, G.M., Lynch, J.P., and Snapp, S. (1995). Fractal geometry and plant root systems: Current perspectives and future applications. In: P. Baveye, J.Y. Parlange, and B.A. Stewart, (Eds.), Fractals of Soil Science. Lewis Publishers, New York, pp.123-154.

Biging, G.S., and Wensel, L.C. (1990). Estimation of crown form for six conifer species of northern California. Can. J. For. Res. 20: 1137-1142.

Biging, G.S., Dobbertin, M. (1995). Evaluation of competition indices in individual-tree growth models. For. Sci. 41: 360-377.

Bossel, H., and Krieger, H. (1994). Simulation of multi-species tropical forest dynamics using a vertically and horizontally structured model. For. Ecol. Manage. 69: 123-144.

Bourgeron, P.S. (1983). Spatial aspects of vegetation structure. In: F.B. Golley (Ed.), Tropical Rain Forest Ecosystems: Structure and Function. Elsevier, Amsterdam, pp. 29-47.

Boyer, J.S. (1985). Water transport. Annual Review of Plant Physiol. 36: 473-516.

Bresler, E. (1991). Soil spatial variability. In: Hanks, J., and Ritchie, J.T. (Eds.), Modelling Plant and Soil Systems. Agronomy Monograph 31, pp. 145-179.

Bristow, K.L., Campbell, G.S., and Calissendorff, C. (1986). The effects of texture on the resistance to water movement within the rhizosphere. Soil Sci. Soc. Am. J. 48: 226-270.

Bristow, K.L., Campbell, G.S., and Saxton, K.E. (1985). An equation for separating daily solar irradiation into direct and diffuse components. Agric. For. Meteorol. 35: 123-131.

Broadbridge, P., and White, I. (1988). Constant rate rainfall infiltration: A versatile nonlinear model. I. Analytical solution. Water Resource Res. 24: 145-154.

Brough, D.W., Jones, H.G., Grace, J. (1986). Diurnal changes in water content of the stems of apple trees, as influenced by irrigation. Plant Cell Environ. 9: 1-7.

Brown, S., Gillespie, A.J.R., and Lugo, A.E. (1989). Biomass estimation procedures for tropical forests with applications to forest inventory data. For. Sci. 35: 889-902.

Brutsaert, W. (1982). Evaporation into the Atmosphere: Theory, History, and Applications. Reidel, Dordrecht, The Netherlands, 299 pp.

Campbell, G.S. (1977). An Introduction to Environmental Biophysics. Springer, New York, 245 pp.

Campbell, G.S. (1985). Soil Physics with BASIC: Transport Models for Soil-plant Systems. Elsevier, Amsterdam, 213 pp.

Campbell, G.S. (1990). Derivation of an angle density function for canopies with ellipsoidal leaf angle distributions. Agric. For. Meteorol. 49: 173-176.

Campbell, G.S. (1991). Simulation of water uptake by plant roots. In: J. Hanks and J.T Ritchie (Eds.), Modelling Plant and Soil Systems. No. 31, Agronomy series. ASA, CSSA, SSSA pp.273-286.

Campbell, G.S., and Norman, J.M. (1989). The description and measurement of plant canopy structure. In: G. Russell, B. Marshall, and P.G Jarvis (Eds.), Plant Canopies: Their Growth, Form, and Function. Cambridge University Press, Cambridge, pp. 1-19.

Cannel, M.G.R., Milne, R., Sheppard, L.J., and Unsworth, M.H. (1987). Radiation interception and productivity of willow. J. Appl. Ecol. 24: 261-278.

Castro, F.D., and Fetcher, N. (1998). Three dimensional model of the interception of light by a canopy. Agric. For. Meteorol. 90: 215-233.

Charles-Edwards, D.A., and Thorpe, M.R. (1976). Interception of diffuse and direct beam radiation by a hedgerow apple orchard. Ann. Bot. 40: 603-613.

Cienciala, E., Eckersten, H., Lindroth, A., and Hallgren, J. (1994). Simulated and measured water uptake by *Picea abies* under non-limiting soil water conditions. Agric. For. Meteorol. 71: 147-164.

Claasen, N., and Barber, S.A. (1974). A method for characterizing the relation between nutrient concentration and flux into roots of intact plants. Plant Physiol. 54: 564-568.

Clemente, R.S., De Jong, R., Hayhoe, H. N., Reynolds, W. D., and Hares, M. (1994). Testing and comparison of three unsaturated soil water flow models. Agric. Water Manage. 25: 135-152.

Coupland, R. T., Johnson, R. E. 1965 Rooting characteristics of native grassland species in Saskatchewan. Journal of Ecology 53, 475-507.

Dalla-Tea, F. and Jokela, E. J. (1991). Needlefall, canopy light interception, and productivity of young intensively managed slash pine stands. For. Sci. 37: 1298-1313.

De Wit, C.T (1965). Photosynthesis of Leaf Canopies. Agric. Res. Rep. 663. Institute of Biological and Chemical Research on Field Crops and Herbage, Wageningen, pp. 1-57.

De Wit, C.T., Goudriaan, J., and van Laar, H.H. (1978). Simulation, Respiration and Transpiration of Crops. Pudoc, Wageningen, The Netherlands, 342 pp.

Denmead, O.T., and Bradley, E.F. (1985). Flux-gradient relationships in a forest canopy. In: B.A. Hutchison, and B.B. Hicks (Eds.), Forest-atmosphere Interactions. Reidel, Dordrecht, The Netherlands, pp. 421-442.

Dickinson, R.E. (1984). Modelling evapotranspiration for three-dimensional global climate models. Geophys. Monogr. 29: 58-72.

Dogniaux, R. (1977). Computer procedure for accurate calculation for radiation data related to solar energy utilization. Proceedings of the UNESCO/WMO Symposium on Solar Energy, 30 Aug.-3 Sept. 1976, Geneva (WMO No. 477), pp. 191-197.

Eckersten, H. (1991). Simulation model for transpiration, evaporation and growth of plant communities: SPAC-GROWTH Model Description. Division of Agricultural Hydro-technics, Report 164, Dept. of Soil Sci., Swed. Univ. of Agric. Sci., Uppsala, 24pp.

Feddes R.A., Kowalik, P.J., and Zaradry, H., (1978). Simulation of Field Water Use and Crop Yield. Centre for Agricultural Publishing and Documentation, Wageningen, 298 pp.

Feddes, R.A., Bresler, E., Neuman, S.P. (1974). Field test of a modified numerical model for water uptake by root systems. Water Resour. Res. 10: 1199-1206.

Fernandez, C.J. (1992). Simulation of normal annual and diurnal temperature oscillations in non-mountainous mainland United States. Agron. J. 84: 244-251.

Finnigan, J.J., and Raupach, M.R. (1987). Modern theory of transfer in plant canopies in relation to stomatal characteristics. In: Zeiger, E., Farquhar, G., and Cowan, I. (Eds.), Stomatal Function. Stanford Uni. Press, pp. 385-429.

Fisher, J.B. (1992). How predictive are computer simulations of tree architecture?. Int. J. Plant Sci. 153: 137-146.

Fisher, M.J., Charles-Edwards, D.A., and Ludlow, M.M. (1981). An analysis of the effects of repeated short-term soil water deficits on stomatal conductance to carbon dioxide and leaf photosynthesis by the legume *Macroptilium atropurpureum* cv, Siratro. Aust. J. Plant Physiol. 8: 347-357.

Fitter, A.H. (1987). An architectural approach to the comparative ecology of plant root systems. New Phytol. 106: 61-77.

Frensch, J., and Steudle, E. (1989). Axial and radial hydraulic resistance to roots of maize (*Zea mays L.*). Plant Physiol. 91: 719-726.

Friend, A.D. (1995). PGEN: an integrated model of leaf photosynthesis, transpiration, and conductance. Ecol. Model. 77: 233-255.

Gates, D.M. (1962). Energy Exchange in the Biosphere. Harper and Row, New York, 445 pp.

Gates, D.M. (1980). Biophysical Ecology. Springer-Verlag, New York, 611 pp.

Gholz, H.L., Vogel, S.A., Cropper, W.P. Jr., Mckelvey, K., Ewel, K.C., Teskey, R.O., and Curran, P.J. (1991). Dynamics of canopy structure and light interception in *Pinus elliottii* stands, north Florida. Ecol. Monogr. 61: 33-51.

Gijzen, H., and Goudriaan, J. (1989). A flexible and explanatory model of light distribution and photosynthesis in row crops. Agric. For. Meteorol. 48: 1-20.

Givnish, T.J. (1984). Leaf and canopy adaptations in tropical forests. In: E. Medina, H.A. Mooney, and C. Vazquez-Yanes (Eds.), Physiological Ecology of Plants of the Wet Tropics. DrW. Junk Publishers, The Hague, pp. 51-84.

Goel, N.S., 1988. Models of vegetation canopy reflectance and their use in estimation of biophysical parameters from reflectance data. Remote Sensing Reviews 4: 1-212.

Goudriaan, J. (1977). Crop micro-meteorology: a simulation study. Simulation monographs, Pudoc, Wageningen, 249 pp.

Goudriaan, J. 1988. The bare bones of leaf angle distribution in radiation models for canopy photosynthesis and energy exchange. Agric. Forest. Meteorol. 43: 155-169.

Goudriaan, J., and van Laar, H.H. (1994). Modelling Potential Crop Growth Processes: Textbook with Exercises. Kluwer Academic Press. The Netherlands, 239 pp.

Grace, J. (1983). Outline Studies in Ecology: Plant-atmospheric Relations. Chapman and Hall, London, 92 pp.

Grace, J.C., Jarvis, P.G., and Norman, J.M. (1987). Modelling the interception of solar radiant energy in intensively managed stands. N. Z. J. For. Sci. 17: 193-209.

Gray, H.L. (1978). The vertical distribution of needles and branch-wood in thinned and unthinned 80-year-old lodgepole pine. Northwest Sci. 52: 303-309.

Gutschick, V.P., and Weigel, F.W. (1988). Optimizing the canopy photosynthetic rate by patterns of investment in specific leaf mass. Am. Nat. 132: 67-86.

Hagihara, A., and Hozumi, K. (1986). An estimate of the photosynthetic production of individual trees in a *Chamaecyparis obtusa* plantation. Tree Physiol. 1: 9-20.

Hall, A.E. (1982). Mathematical models of plant water loss and plant water relations. In: Lang, O.L., Nobel, P.S., Osmand, C.B., and Ziegler, H.(Eds.), Physiological Plant Ecology II, Encyclopaedia of Plant Physiology. Vol.12B. Springer-Verlag, Berlin, pp. 231-261.

Halle, F., Oldemann, R.A.A., Tomlinson, P.B. (1978). Tropical Trees and Forests: An Architectural Analysis. Springer, The Netherlands, 345 pp.

Hassika, P., Berbigier, P., and Bonnefond, J.M. (1997). Measurement and modelling of the photosynthetically active radiation transmitted in a canopy of maritime pine. Annales des Sciences Forestieres Paris 54: 715-730.

He, J., Chee, C.W., and Goh, C.J. (1996). Photoinhibition of Heliconia under natural tropical conditions: The importance of leaf orientation for light interception and leaf temperature. Plant Cell Environ. 19: 1238-1248.

Hetrick, B.A. (1988). Physiological and topological assessment of effects of a vesicular-arbuscular mycorrhizal fungus on root architecture of big bluestem. New Phytol. 110: 85-96.

Horn, H.S. (1971). The Adaptive Geometry of Trees. Princeton University Press. New Jersey, 342 pp.

Huber, B. (1924). Die Beurteilung des Wasserhaushalts der Pflanze. Ein Beitrag zur vergleichenden Physiologie. Jahrb. Wiss. Bot. 24: 1-120.

Jackson, J.E., and Palmer, J.W. (1972). Interception of light by model hedgerow orchards in relation to latitude, time of year and hedgerow configuration and orientation. J. Appl. Ecol. 9: 341-357.

Jahnke, L.S., and Lawrence, D.B. (1965). Influence of photosynthetic crown structure on potential productivity of vegetation, based on primarily on mathematical models. Ecol. 46: 319-326.

Janz, T.C., Stonier, R.J. (1995). Modeling water flow in cropped soils: Water uptake by plant roots. Environment International 21: 704-709.

Jarvis, P.G. (1975). Water transfer in plants. In: D.A. de Vries and N.H. Afgan (Eds.), Heat and Mass Transfer in the Biosphere. 1. Transfer Processes in Plant Communities. Halsted, Washington, D.C., pp. 369-394.

Jarvis, P.G. (1976). The interpretation of the variations in leaf water potential and stomatal conductance found in canopies in the field. Philos. Trans. R. Soc. London. Ser. B. 273: 593-610.

Jarvis, P.G. (1995). Scaling processes and problems. Plant Cell Environ. 18: 1079-1089.

Jarvis, P.G., and Leverenz, J.W. (1983). Productivity of temperature, deciduous, and evergreen forests. In: O. Lange, P. Nobel, C. Osmand and H. Ziegler (Eds.), Physiological Plant Ecology IV. Ecosystem Processes: Mineral Cycling, Productivity and Man's Influence. Springer, New York, pp. 233-280.

Jarvis, P.G., Edwards, W.R.N., and Talbot, H. (1981). Models of plant and crop water use. In: D.A. Rose and D.A. Charles-Edwards (Eds.), Mathematics of Plant Physiology. Academic Press, London, pp. 151-194.

Jarvis, P.G., Miranda, H.S., and Muetzelfeldt, R.I. (1985). Modelling canopy exchanges of water vapour and carbon dioxide in coniferous forest plantations. In: B.A. Hutchison, and B.B. Hicks (Eds.), Forest Atmosphere Interactions. Reidel, Dordrecht, The Netherlands, pp. 521-542.

Jones, H. G. (1992). Plant Microclimate: A Quantitative Approach to Environmental Plant Physiology (second edition) Cambridge University Press, New York, 428 pp.

Jones, H.G. (1982). Plants and Microclimate. Cambridge University Press, Cambridge, UK, 345 pp.

Kelliher, F.M., Leuning, R., Raupach, M.R. and Schulze, E.D. (1994). Maximum conductances for evaporation from global vegetation types. Agric. For. Meteorol. 73: 1-16.

Kinerson, R.J., Fritchen, L.J. (1971). Modelling a coniferous forest canopy. Agric. Meteorol. 8: 439-445.

Kinerson, R.J., Higginbotham, K.O., and Chapman, R.C. (1974). Dynamics of foliage distribution within a forest canopy. J. Appl. Ecol. 11: 347-353.

Kira, T., Shinozaki, K., and Hozumi, K. (1969). Structure of forest canopies as related to their primary productivity. Plant Cell Physiol. 10: 129-142.

Kira, T., Shinozaki, K., and Hozumi, K. (1969). Structure of forest canopies as related to their primary productivity. Plant Cell Physiol. 10: 129-142.

Kjelgren, R., and Montague, T. (1998). Urban tree transpiration over turf and asphalt surfaces. Atmospheric Environ. 32: 35-41.

Kock, B., Ammer, U., Schneider, T., and Wittmeir, H. (1990). Spectroradiometer measurements in the laboratory and in the field to analyse the influence of different damage symptoms on the reflection spectra of forest trees. Int. J. Remote Sens. 11: 1145-1163.

Kodrik, M. (1995). Distribution and fractional composition of below ground spruce biomass and length in Biosphere Reserve Pol'ana Ekologia (Bratislava) 14: 413-417.

Kohyama, T. (1991). A functional model describing sapling growth under a tropical forest canopy. Functional Ecol. 5: 83-90.

Kowalik, P.J., and Eckersten, H. (1984). Water transfer from soil through plants to the atmosphere in Willow energy forest. Ecol. Model.. 26: 251-284.

Krasilnikov, P.K. (1968). On the classification of the root systems of trees and shrubs. In: Methods of Productivity Studies in Root Systems and Rhizophore Organisms. USSR Academy of Sciences, NAUKA, Leningrad, pp. 101-114.

Kriebel, K.T. (1978). Measured spectral bidirectional reflection properties of four vegetated surfaces. Appl. Optics 17: 253-259.

Landsberg, J.J. (1986). Physiological Ecology of Forest Production. Academic Press, London, 191 pp.

Landsberg, J.J. and Fowkes, N.D. (1978). Water movements through plant roots. Ann. Bot. 42: 493-508.

Landsberg, J.J., and James, G.B. (1971). Wind profiles in plant canopies: studies on an analytical model. J. Appl. Ecol. 8: 729-741.

Landsberg, J.J., and McMurtrie, R. (1984). Water use by isolated trees. Agric. Water Manage. 8: 223-242.

Lang, A.R.G. (1986). Leaf area and average leaf angle from transmission of direct sunlight. Aust. J. Bot. 34: 349-355.

Law, B.E., Cescatti, A., and Baldocchi, D.D. (2000). Leaf area distribution and radiative transfer in open-canopy forests: implications for mass and energy exchange. Tree Physiology, 21 (12-13): 777-787.

Lemeur, R. (1973). A method for simulating the direct solar radiation regime in Sunflower, Jerusalem artichoke, Corn, and Soyabean canopies using actual stand structure data. Agric. Meteorol. 12: 229-247.

Leuning, R. (1995). A critical appraisal of a combined stomatal-photosynthesis model for C_3 species. Plant Cell and Environ. 18: 339-355.

Leuning, R., Kelliher, F.M., dePury, D. and Schulze, E.D. (1995). Leaf nitrogen, photosysthesis, conductance and transpiration: scaling from leaves to canopies. Plant Cell and Environ. 18: 1129-1146.

Liu, B.Y., and Jordan, R.C. (1960). The interrelationship and characteristic distribution of direct, diffuse and total solar radiation. Solar Energy 4: 1-19.

Lohammar, T., Larrsson, S., Linder, S., and Falk, S.O., (1980). FAST simulation models of gaseous exchange in Scots pine. In: T. Persson (ed), Structure and Function of Northern Coniferous Forets - An Ecosystem Study. Ecol. Bull. 32: 505-523.

Lungley, D.R. (1973). The growth of root systems: A numerical computer simulation model. Plant Soil 38: 145-159.

Lynn, B., and Carlson, T.N. (1990). A model illustrating plant versus external control of transpiration. Agric. For. Meteorol. 52: 5-43.

Mandelbrot, B.B. (1983). The Fractal Geometry of Nature. Freeman, New York, 468 pp.

Massman, W.J. (1981) Foliage distribution in old-growth coniferous tree canopies. Can. J. For. Res.12:10-17.

McIlroy, I.C. 1984. Terminology and concepts of natural evaporation. Agric. Water Manage. 8: 77-98.

McMurtrie, R.E. (1993). Modelling of canopy carbon and water balance. In: D.O. Hall, J.M.O. Scurlock, H.R. Bolhar-Nordenkamf, R.C. Leegood, and S.P.Long. Photosynthesis and Production in a Changing Environment: A Field and Laboratory Manual. Chapman and Hall, London, 231 pp.

Melchior, W., and Steudle, E. (1993). Water transport in onion roots; Change of axial and radial hydraulic conductivities during root development. Plant Physiol. 101: 1305-1315.

Meyer, W.S., and Ritchie, J.T. (1980). Resistance to water flow in the sorghum plant. Plant Physiol. 65: 33-39.

Meyers, T.P., and Paw U, K.T. (1986). Testing of a higher-order closure model for modelling airflow within and above plant canopies. Boundary Layer Meteorol. 37: 297-311.

Meyers, Y.P., and Paw U, K.T. (1987). Modelling plant canopy micro-meteorology with higher-order closure principles. Agric. For. Meteorol. 41: 143-163.

Miyashita, K., and Maitani, T. (1998). Vertical structure of temperatures of tree's leaves and gate wall at Rashomon doline. Bulletin of the Research Institute for Bioresources, Okayama University 5: 169-181.

Mohren, G.M.J. (1987). Simulation of Forest Growth, Applied to Douglas fir Stands in The Netherlands. Pudoc, Wageningen, 213 pp.

Mohren, G.M.J., Van Gerwen, and Spitters, C.J.T. (1984). Simulation of primary production in even-aged stands of Douglas fir. For. Ecol. Manage. 9:27-49.

Moldrup, P., Rolston, D.E., Hansen, J.A.A. and Yamaguchi, T. (1992). A simple mechanistic model for soil resistance to plant water uptake. Soil Sci. 150: 87-93.

Molz, F.J., (1981). Models of water transpiration soil-plant system: A review. Water Resour. Res. 17: 1245-60

Monsi, M., and Saeki, T. (1953). Uber den lichtfaktor in den pflanzengesellschaften und seine bedeutung fur die stoffproduktion. Jpn. J. Bot. 14: 22-52.

Monteith, J.L. (1965). Light distribution and photosynthesis in field crops. Ann. Bot. 29: 17-37.

Monteith, J.L. (1975). Principles of Environmental Physics. Edward Arnold, London, 241 pp.

Monteith, J.L. (1980). The development and extension of Penman's evaporation formula. In: D. Hillel (Ed.), Application of Soil Physics. Academic Press, New York, pp. 247-253.

Monteith, J.L., and Unsworth, M.H. (1990). Principles of Environmental Physics. Edward Arnold, 291 pp.

Morales, D., Jiménez, M.S., González-Rodriguez, A.M., and _ermák, J. (1996). Laurel forests in Tenerife, canary Islands: II. Leaf distribution patterns in individual trees. Trees 11: 41-46.

Myneni, R. B., Gutschick, V. P., Asrar, G., and Kanemasu, E.T. 1988. Photon transport in vegetation canopies with anisotropic scattering - part I. Scattering phase functions in one angle. Agric. For. Meteorol. 42: 1-16.

Myneni, R.B. (1991). Modelling radiative transfer and photosynthesis in three-dimensional vegetation canopies. Agric. For. Meteorol. 55: 323-344.

Myneni, R.B., and Impens, I. (1985). A procedural approach for studying the radiation regime of infinite and truncated foliage spaces. 1. Theoretical considerations. Agric. For. Meteorol. 34: 3-16.

Myneni, R.B., and Ross, J. (1990). Photon-vegetation Interactions. Applications in Optical Remote Sensing and Plant Ecology. Springer-Verlag, Berlin, 565 pp.

Myneni, R.B., Asrar, G., and Gerstl, A.W. (1990). Radiative transfer in three dimensional leaf canopies. Transport Theory Stat. Phys. 19: 205-250.

Myneni, R.B., Ross, J., and Asrar, G. 1989. A review on the theory of photon transport in leaf canopies. Agric. For. Meteorol. 45: 1-153.

Neumann, E.L., Thurtell, G.W., and Stevenson, K.R. (1974). *In situ* measurements of leaf water potential and resistance to water flow in corn, soybean and sunflower at several transpiration rates. Can. J. Plant Sci. 54: 175-184.

Newman, E.I. (1976). Water movement through root systems. Phil. Trans. R. Soc. B. 273: 467-478.

Nikinmaa, E. (1992). Analysis of the growth of Scots pine: Matching structure with function. Acta For. Fenn. 235: 68-88.

Nikolov, N.T., and Zeller, K.F. (1992). A solar radiation algorithm for ecosystem dynamic models. Ecol. Model. 61: 149-168.

Nikolov, N.T., Massman, W.J., and Schoettle, A.W. (1995). Coupling biochemical and biophysical pressures at the leaf level: an equilibrium photosynthesis model for leaves of C_3 plants. Ecol. Model. 80: 205-235.

Nobel, P.S. and Alm, D.M. (1993). Root orientation vs. water uptake simulated for monocotyledonous and dicotyledonous desert succulents by a root-segment model. Functional Ecology 7: 600-609.

North, G.B., and Nobel, P.S. (1995). Hydraulic conductivity of concentric root tissues of *Agave deserti* Engelm. under wet and drying conditions. New Phytologist 130: 47-57.

Oker-Blom, P. (1986). Photosynthetic radiation regime and canopy structure in modelled forest stands. Acta For. Fenn. 197: 44-58.

Oker-Blom, P., and Kellomaki, S. (1983). Effect of grouping of foliage on the within-stand and within-crown light regime: comparison of random and grouping canopy models. Agric. For. Meteorol. 28: 143-155.

Otterman, J., Novak, M.D., and Starr, D.O.C. (1993). Turbulent heat transfer from a sparsely vegetated surface: two component representation. Boundary Layer Meteorol. 64: 409-420.

Palmer, J.W. (1977). Diurnal light interception and computer model of light interception by hedgerow apple orchards. J. Appl. Ecol. 14: 601-614.

Passioura, J.B. (1972). Effects of root geometry on the yield of wheat growing on stored water. Aust. J. Agric. Res. 23: 745-752.

Passioura, J.B. (1984). Hydraulic resistance of plants. I. Constant or variable? Aust. J. Plant Physiol. 11: 333-339.

Passioura, J.B. (1988). Water transport in and to roots. Annual Review of Plant Physiology and Plant Molecular Biology 39: 245-256.

Paw U, K.T. and Meyers, T.P. (1989). Investigations with higher order canopy turbulence model into mean source-sink levels and bulk canopy resistances. Agric. For. Meteorol. 47: 259-271.

Paw U, K.T., Qie, J., Su, H.B., Watanabe, T., and Brunet, Y. (1995). Surface renewal analysis: a new method to obtain scalar fluxes. Agric. For. Meteorol. 74: 119-137.

Perelyot, N.A. (1970). Trudy Ukrainskogo nauchno-issleedovatelskogo gidro-meteorologicheskogo instituta (Proc. Ukrainian Inst. Hydrometeorol.) 94: 46-51.

Philip, J.R. (1966). Plant water relations: Some physical aspects. Ann. Rev. Pl. Physiol. 17: 245-268.

Pinker, R.T., Thompson, O.E., and Eck, T.F. (1980). The albedo of a tropical evergreen forest. Q.J.R. Meteorol. Soc. 106: 551-558.

Pregitxer, K.S., Kubiske, M.E., Yu, C.K., and Hendrick, R.L. (1997). Relationships among root branch order, carbon, and nitrogen in four temperate species. Oecologia 111: 302-308.

Pukkala, T., Beaker, P., Kuuluvainen, T., and Oker-Blom, P. (1991). Predicting spatial distribution of direct radiation below forest canopies. Agric. For. Meteorol. 55: 295-307.

Pukkala, T., Kuuluvainen, T., and Stenberg, P. (1993). Below-canopy distribution of photosynthetically active radiation and its relation to seeding growth in a boreal *Pinus sylvestris* stand: A simulation approach. Scand. J. For. Res. 8: 313-325.

Ranatunga, Kemachandra, Nation, E.R., Barratt, D.G. (2008) Review of soil water models and their applications in Australia Environmental Modelling & Software, Volume 23, Issue 9, Pages 1182-1206

Ranatunga, Kemachandra., and Murty, V.V.N. (1992). Modelling irrigation deliveries for tertiary units in large irrigation systems. Agric. Water Manage. 21: 197-214.

Raupach, M.R. (1989). Canopy transport processes. In: W.L. Steffen and O.T. Denmead (Eds.), Flow and Transport in the Natural Environment: Advances and applications. Springer-Verlag, Berlin, pp. 1-33.

Raupach, M.R. (1991). Vegetation-atmosphere interaction in homogeneous and heterogeneous terrain: some implications of mixed-layer dynamics. In: A. Henderson-Sellers and A.J. Pitman (Eds.), Vegetation and Climatic Interactions in Semi-arid Regions. Kluwer Academic Publishers, The Netherlands, pp. 105-120.

Raupach, M.R. (1995). Vegetation-atmosphere interaction and surface conductance at leaf, canopy and regional scales: discussion. Aust. J. Plant Physiol. 15: 705-716.

Raupach, M.R., and Finnigan, J.J. (1988). Single-layer models of evaporation from plant canopies are incorrect but useful, whereas multilayer models are correct but useless: Discuss. Aust. J. Plant Physiol. 15: 705-716.

Richards, L.A. (1931). Capillary conductivity of liquid through porous media. Physics 1: 318-333.

Richards, P.W. (1983). The tree-dimensional structure of tropical rain forest. In: S.L. Sutton, T.C. Whitmore, and A.C. Chadwick (Eds.), Tropical rain forest: Ecology and Management. Blackwell, Oxford, pp. 3-10.

Robinson, D., Linehan, D.J., and Caul, P. (1991). What limits nitrate uptake from soil? Plant Cell Environ. 14: 77-85.

Ross, J. (1975). Radiative transfer in plant communities. In: J.L. Monteith (Ed.), Vegetation and the Atmosphere, Vol. 1: Principles. Academic Press, New York, pp. 332-365.

Ross, J., (1981). The Radiation Regime and Architecture of Plant Stands. Dr. W. Junk Publishers, The Hague, The Netherlands, pp 391-443.

Rotenberg, E., Mamane, Y., Joseph, J.H. (1999). Long wave radiation regime in vegetation-parameterisations for climate research Environmental Modelling & Software 13 (3-4), 361-371.

Running, S.W., and Coughlan, J.C. (1988). A general model for forest ecosystem processes for regional applications. 1. Hydrologic balance, canopy gas exchange and primary production processes. Ecol. Model. 42: 125-154.

Ryel, R.J., Beyschlag, W., and Caldwell, M.M. (1993). Foliage orientation and carbon gain in two tussock grasses as assessed with a new whole-plant gas-exchange model. Functional Ecology 7: 115-124.

Salby, M.L. (1996). Fundamentals of Atmospheric Physics. Academic Press, London, 627 pp.

Santantonio, D., Hermann, R.K. and Overton, W.S. (1977). Root biomass studies in forest ecosystems. Pedobiologia 17: 1-31.

Sattelmacher, B., Gerendas, J., Thoms, K., Bruck, H., and Bagdady, N.H. (1990). Interaction between root growth and mineral nutrition. Environ. Exp. Bot. 33: 63-73.

Saugier, B. (1996). The evapotranspiration of grasslands and crops. Comptes Rendus de l'Academie d'Agriculture de France 82: 133-153.

Schreuder, H.T., Swank, W.T. (1974). Coniferous stands characterized with the Weibull distribution. Can. J. For. Res. 4: 518-523.

Sellers, P.J., 1985. Canopy reflectance, photosynthesis and transpiration. Int. J. Remote Sen. 8:1335-1372.

Shibusawa, S. (1994). Modelling the branching growth fractal pattern of the maize root system. Plant Soil 165: 339-347.

Shinozaki, K., Yoda, K., Orava, P.J., and Kira, T. (1964). A quantitative analysis of plant form-the pipe model theory. 1. Basic analysis, Jap. J. Ecol. 14: 97-105.

Sinoquet, H., and Bonhomme, R. (1992). Modelling radiative transfer in mixed and row intercropping systems. Agric. For. Meteorol. 62: 219-240.

Soumar, A., Groot, J.J.R., Kon_, D., and Radersma, S. (1994). Structure spatiale du systeme racinaire de deux arbres du Sahel: Acacia seyal and Sclerocaryea birrea. Rapport PSS No. 5. Wageningen, 45 pp.

St Aubin, G., Canny, M.J., and McCully, M.E. (1986). Living vessel elements in late metaxylem of sheathed maize roots. Ann. Bot. 58: 145-157.

Steudle, E., and Brinckmann, E. (1989). The osmometer model of the root: Water and solute relations of roots of *Phaseolus coccineus*. Botanica Acta. 102: 85-95.

Stewart, J.B. (1971). The albedo of a pine forest. Q.J.R. Meteorol. Soc. 97: 561-564.

Stewart, J.B. (1988). Modelling surface conductance of pine forest. Agric. For. Meteor. 43: 19-35.

Strandman, H., Väisänen, H., Kellomäki, S. (1993). A procedure for generating synthetic weather records in conjunction of climatic scenario for modelling of ecological impacts of changing climate in boreal conditions. Ecol. Model. 70: 195-220.

Tardieu, F. (1988). Analysis of the spatial variability of maize root density: II Distance between roots. Plant Soil 107: 267-272.

Thornley, J.H.M., and Johnson, I.R (1990). Plant and Crop Modelling: A Mathematical Approach to Plant and Crop Physiology. Clarendon Press, Oxford, England, 669 pp.

Tyree, M.T. (1988). A dynamic model for water flow in a single tree: evidence that models must account for hydraulic architecture. Tree Physiol. 4: 195-217.

van Engelen, A.F., and Guerts, H.A.M. (1983). Een Rekenmodel dat het Verloop van de Temperatuur over een Etmaal Berekent uit Drie Termijnmetingen van de Temperatuur. KNMI, De Bilt, The Netherlands, 44 pp.

Van Noordwijk, M., and Brouwer, G. (1995). Roots as sinks and sources of carbon and nutrients in agricultural systems. Advances in Agroecology 23: 234-239.

Van Noordwijk, M., Spek, L.Y., and De Willigen, P. (1994). Proximal root diameters as predictors of total root system size for fractal branching models. 1. Theory. Plant Soil 164: 107-118.

Waggoner, P.E., and Reifsnyder, W.E. (1968). Simulation of temperature, humidity and evaporation profiles in a leaf canopy. J. Appl. Meteorol. 7: 400-409.

Walker, B. H., and Langridge, J.L. (1996). Modelling plant and soil water dynamics in semi-arid ecosystems with limited site data. Ecol. Model. 87: 153-167.

Wang, Y., and Baldocchi, D.D. (1989). A numerical model for simulating the radiation regime within a deciduous canopy. Agric. For. Meteorol. 46: 313-337.

Wang, Y.P., and Jarvis, P.G. (1990). Description and validation of an array model-MAESTRO. Agric. For. Meteorol. 51: 257-280.

Weatherley, P.E. (1970). Some aspects of water relations. Adv. Bot. Res. 3: 171-186.

Weaver, J.E. (1968). Classification of root systems of forbs of grassland and a consideration of their significance. Ecology 39: 393-401.

Weiss, A., and Norman, J.M. (1985). Partitioning solar radiation into direct and diffuse, visible and near-infrared components. Agric. For. Meteorol. 34: 205-213.

West, P.W., and Wells, K.F. (1992). Method of application of a model to predict the light environment of individual tree crowns and its use in a eucalypt forest. Ecol. Model. 60: 199-231.

White, D.A., Beadle, C.L., Sands, P.J., Worledge, D., and Honeysett, J.L. (1999). Quantifying the effect of cumulative water stress on stomatal conductance of *Eucalyptus globulus* and *Eucalyptus nitens*: a phenomenological approach. Aust. J. Plant Physiol. 26: 17-27.

Whitehead, D., Grace, J.C., and Godfrey, M.J.S. (1990). Architectural distribution of foliage in individual *Pinus radiata* D: Crowns and the effects of clumping on radiation interception. Tree Physiol. 7: 135-155.

Williams, M., Rastetter, E.B., Fernades, D.N., Goulden, M.L., Wofsy, S.C., Shaver, G.R., Meillo, J.M., Munger, J.W., Fan, S.M., and Nadelhoffer, K.J. (1996). Modelling thesoil-plant-atmosphere continuum in a Quercus-Acer stand at Harvard forest: the regulation of stomatal conductance by light, nitrogen and soil-plant hydraulic properties. Plant Cell Environ. 19: 911-927.

Worner, S.P. (1988). Evaluation of diurnal temperature models and thermal summation in New Zealand. J. Econ. Entomol. 81: 9-13.

Wu, J. (1990). Modelling the energy exchange processes between plant communities and environment. Ecol. Model. 51: 233-250.

Yang, S., Liu, X., and Tyree, M.T. (1998) A model of stomatal conductance in sugar maple (*Acer saccharum* Marsh). J. Theoretical Biol. 191: 197-211.

Yong, X., Miller, D.R, and Montgomery, M.E. (1993). Vertical distribution of canopy foliage and biologically active radiation in a defoliated/defoliated hardwood forest. Agric. For. Meteorol. 67: 129-146.

Zamir, M. 2001. Fractal dimensions and multifractility in vascular branching. J. Theor. Biol. 212: 183–190.

Zeng, X., Decker, M. (2009) Improving the Numerical Solution of Soil Moisture–Based Richards Equation for Land Models with a Deep or Shallow Water Table. J. Hydrometeor, 10, 308–319.

Zermeno, G.A., Hipps, L.E. (1997). Downwind evolution of surface fluxes over a vegetated surface during local advection of heat and saturation deficit. J. Hydrol. (Amsterdam) 192: 189-210.

9

Solar Radiation Modeling and Simulation of Multispectral Satellite Data

Fouzia Houma[1] and Nour El Islam Bachari[2]
[1]National School for Marine Sciences and Coastal Management (ENSSMAL), Campus Dely Ibrahim Bois des Cars, Algiers Laboratory Marine and Coastal Ecosystems,
[2]Faculty of Biological Sciences, University of Science and Technology Houari Boumediene, USTHB, BP 32 El Alia, Bab Ezzouar Algiers Laboratory analysis and application of radiation (LAAR) USTO, Oran Algeria

1. Introduction

All bodies emit and reflect the flow of energy in the form of electromagnetic radiation. The relative variation of the energy reflected or emitted as a function of wavelength, is the spectral signature of the object considered in a given state. The spectrum can be used to identify and determine its status. For a satellite, making measurements in a number of spectral bands, the spectral signature of an object will correspond to different levels of radioactivity recorded in each of them.

The principle of remote sensing is the detection of electromagnetic radiation that carries information from the soil-atmosphere either by reflection or by transmission from a radiometer on board the satellite. The signal received by the radiometer is the result of physical, biological and geometrical objects on the ground. For a better use of satellite measurements, we must answer the following questions: At what point on the earth's surface so far is it? What is the value of measuring that?

Answering these questions requires the definition: What exactly are the physical quantities measured by the measurement system? What disturbs the measurement system does what it is supposed to measure? Which model can you describe the disturbances? How does one characterize the quality of measurement?

To understand this complex phenomenon, we have developed an analytic model (SDDS) of radiatif transfer simulation in water coupled to an atmospheric model in order to simulate measure by satellite. This direct model permits to follow the solar radiance in his trajectory Sun-Atmosphere - Sea - Depth of sea- sensor. The goal of this simulation is to show for every satellite of observation (SPOT, Landsat MSS, Landsat TM) possibilities that can offer in domain of oceanography. (Bachari,1997)

An interaction model of the solar spectrum with the Earth-atmosphere system is developed to calculate the various components of solar radiation at ground and upper atmosphere. (Bachari, 1999; Houma and al.,2010)

In this research, we are interested in applying the model to simulate the radiative transfer through the atmosphere under realistic conditions for assessing the significance of the effects of the atmosphere and conditions on shooting satellite images. The main objective of this application is the analysis of satellite measurements, along with their variations atmospheric parameters. The spectral signature of water is used here to simulate the action of the satellites. (Gordon, 1974)

2. Modelling the interaction of the solar spectrum with the atmosphere

An analytical model for simulating radiative transfer in water coupled with an atmospheric model can adequately simulate the signal of a body of water to the attitude of the satellite to analyze the effects of atmospheric parameters and shallow water. This model determines the scattered radiation by a body of water in the software simulation of satellite data *SDDS* (Bachari, 1997), its main function is the calculation of the spectral radiance reflected from the sea water level sensor.

The purpose of modeling is to understand how different components of the measurement system combine to make a measurement. The form and content of a model depends on their purpose. The model is constructed to describe and characterize the measurement system to understand the phenomena which he is registered and to predict their behavior under the effect of an external action or as a result of a partial modification of the system itself same. The model developed is to break the middle-ground atmosphere into subsets in interaction with the solar spectrum and the sensor onboard the satellite.

The source irradiates the object and the latter reflects the radiation in all directions, some of this radiation is captured. The radiation received by a radiometer on board the satellite, is composed of two main terms: brightness caused by the surface in the field of vision sensor and a brightness that is not caused by the surface in the field of vision. The first term is useful information, it is due to the direct and indirect solar radiation. The second term, considered the noise is due to the light scattered by the atmosphere. (Gordon & Clark, 1981; Becker & Rffy, 1990)

3. Spectral irradiance on the ground

The sun is the source of energy in passive remote sensing; solar radiation carries the information of the natural environment for its intrinsic properties (wavelength, polarization, phase shift). Knowledge of the spectral distribution of the radiation that reaches the high atmosphere is very important for various applications. (Chadin, 1988).The solar spectrum has been the subject of several measures on the ground, air and satellite

Assuming that the atmosphere is transparent, the solar spectrum $E_{0\lambda}(w.cm^{-2}\mu m^{-1})$ reaching the soil does not undergo any change in its trajectory.

The spectral irradiance in the upper atmosphere depends on the latitude of the location (latitude=φ), declination of the axis of rotation of the earth (δ) and time (h)

The solar spectrum on the ground is given by the following equation (Ratto, 1986):

$$E_{\lambda} = E_{0\lambda}\,(1+f)\cos(\theta_z) \tag{1}$$

$E_{0\lambda}$: is the solar spectrum outside Earth's atmosphere, λ: wavelength of emission of radiation, 1: Astronomical Unit (1UA= 1.496x 10^8 km), f: is the correction factor of distance sun-soil (this factor depends on the number of days and cos (θz) is the zenith angle).

In clear sky atmosphere, the concentration of gases and aerosols varies with the changing weather conditions and geographical position. Gases and aerosols absorb and scatter solar radiation on a selective basis throughout the optical path. Gases, principally ozone, carbon dioxide and water vapor are the bodies responsible for absorption of the solar spectrum. Air molecules and aerosols are the body responsible for the dissemination of solar radiation in all directions. (Prieur & Morel, 1975). The effects of absorption and scattering functions are presented by the transmittance according to Bouguer's law (Bouguer, 1953):

$$T_\lambda = I_\lambda / I_{0\lambda} \quad (2)$$

I_λ is the spectral radiation output and $I_{0\lambda}$ is the radiation spectral λ input.

Diffusion occurs during the interaction between the incident radiation and particles or large gas molecules in the atmosphere (water droplets, dust, smoke ...). Where the suspended particles are negligible compared to the wavelength, the phenomenon that occurs is Rayleigh scattering. (Fröhlich & Brusa, 1981)

The diffusion of a particle occurs independently of other particles. The radiation will be distributed in all directions, the forward scattered radiation is equal to the radiation scattered backward.

3.1 Model description of irradiance

This model is to calculate the solar spectral irradiance (irradiance) direct normal and horizontal diffuse to the conditions of a cloudy sky not. This code calculates a range of 0.3 and 4.0 microns with a pitch of 10 nm. This code introduces a number of parameters such as solar zenith angle, the angle of inclination, atmospheric turbulence, the amount of water vapor precipitated amount of ozone, pressure and albedo (Guyot & Fagu ,1992).

Monochromatic distribution of a direct solar beam can be computed as a function of a number of variables, including optical mass and a wide variety of atmospheric parameters-for exemple, water-vapor content, ozone layer thickness, and turbidity parameters.

In the ultraviolet and visible region, it is essentially ozone absorption, Rayleigh scattering, and aerosols that control attenuation of the direct beam. The transmittance by aerosols is minimum at the short wavelengths and increases slowly as the wavelength increases. (Morel & Gentili, 1993)

3.2 Direct spectral irradiance on the ground

The equation of the light arriving directly from the sun at ground level for a wavelength λ is as follows:

$$I_{d\lambda} = H_{o\lambda} D T_{r\lambda} T_{a\lambda} T_{w\lambda} T_{o\lambda} T_{u\lambda} \quad (3)$$

Where

- $H_{o\lambda}$: represents the irradiance in the upper atmosphere of Earth-Sun distance an average wavelength λ.
- D is the correction factor for Earth-Sun distance.
- $T_{r\lambda}$, $T_{a\lambda}$, $T_{w\lambda}$,$T_{o\lambda}$ et $T_{u\lambda}$ are respectively the functions of the transmittance of the atmosphere for a wavelength λ of molecular diffusion (Rayleigh), mitigation of aerosols, the absorption of water vapor, the absorption of ozone and gas absorption

The direct irradiation on a horizontal surface is obtained from the equation multiplied by cos (Z), or Z is the solar zenith angle:

$$I_{d\lambda} = I_{d\lambda} \cos(Z) \tag{4}$$

Spectral diffuse irradiance

a. The diffuse irradiance on a horizontal surface

The diffuse irradiance on a horizontal surface is based on three components:

- Component of Rayleigh scattering $I_{r\lambda}$
- Release component aerosol $I_{a\lambda}$
- Component that takes into account multiple reflections of light between the ground and air

The total $I_{s\lambda}$ diffuse illumination is given by the sum.

$$I_{s\lambda} = I_{r\lambda} + I_{a\lambda} + I_{g\lambda} \tag{5}$$

The diffuse illumination on an inclined surface

The global spectral irradiance on an inclined surface is represented by:

$$\begin{aligned} I_{T\lambda}(t) = I_{d\lambda} \cos(\theta) + I_{s\lambda} \langle \{ I_{d\lambda} \cos(\theta) / [H_{o\lambda} D \cos(Z)] \} \\ + 0.5[1+\cos(t)][1 - I_{d\lambda} / (H_{o\lambda} D)] \rangle \end{aligned} \tag{6}$$

where

- θ is the angle of incidence of direct beam on an inclined surface
- t is the angle of the inclined surface

The angle of inclination to a horizontal surface is 0 ° and 90 ° to a vertical surface. The global irradiance on a horizontal surface is given by:

$$I_{T\lambda} = I_{d\lambda} \cos(Z) + I_{s\lambda} \tag{7}$$

Figures 1, 2, 3 shows the specter of global illumination DTOT, diffuse illumination DIF and direct illumination DIR.

The input parameters are:

- Optical thickness = 0.51.
- Turbulence ALPHA = 1.14.
- The amount of ozone O_3 = 0.53 atm.cm.

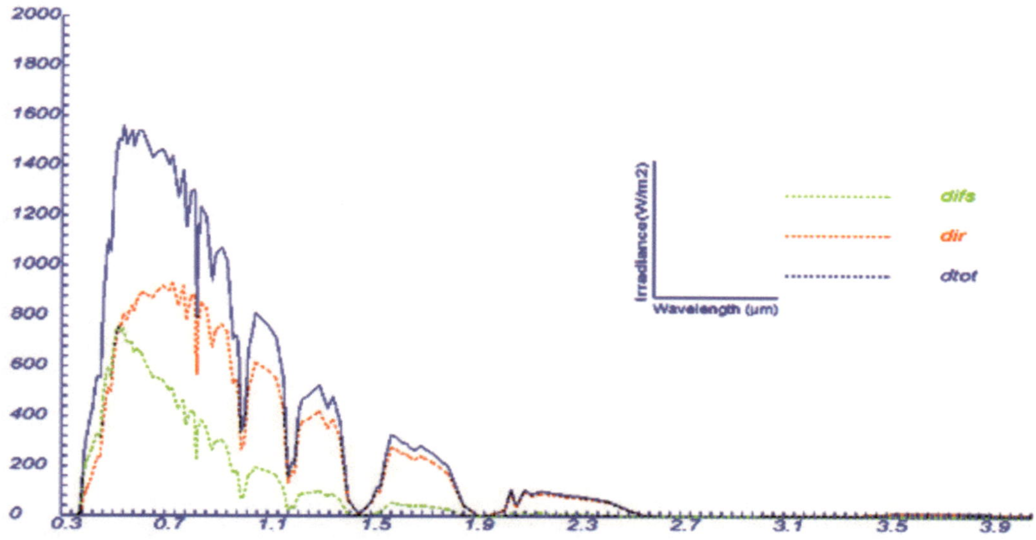

Fig. 1. Solar illumination for an area generally horizontal

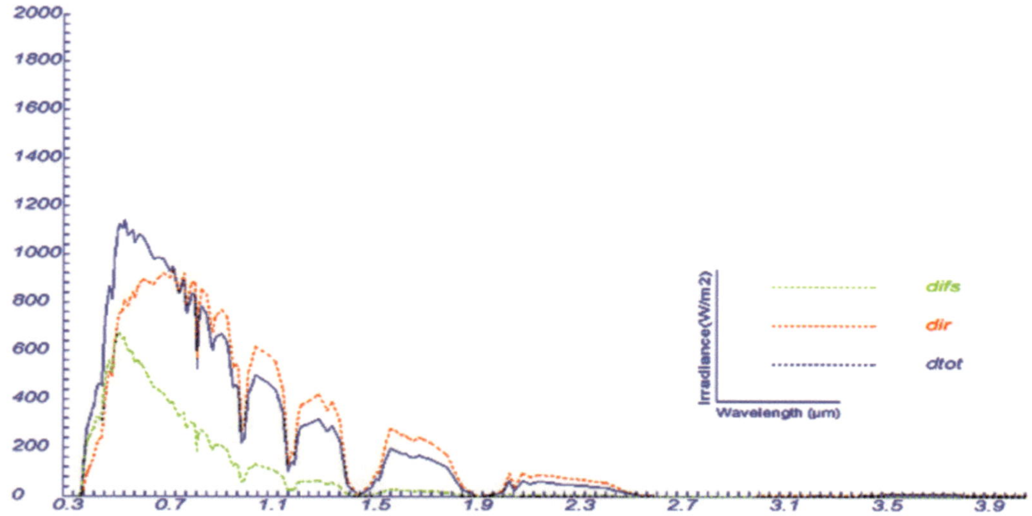

Fig. 2. The solar irradiance for an area generally vertical

- Precipitable water W = 1.42 cm.
- Incline angle for a surface generally vertical or horizontal : TILT = 0.0°; in the case of an inclined surface we take the angle TILT = 60.0°;
- Surface pressure 840 millibars.
- The number of days in the year 96 (6 April 2009)
- The number of wavelength 122
- The solar zenith angle Z = 53 °

Figure 4 shows that the maximum solar irradiance is with a solar zenith angle of 0 °, which explains when the sun is overhead (the sun is at noon) solar intensity is high and second illumination decreases with increasing solar zenith angle.

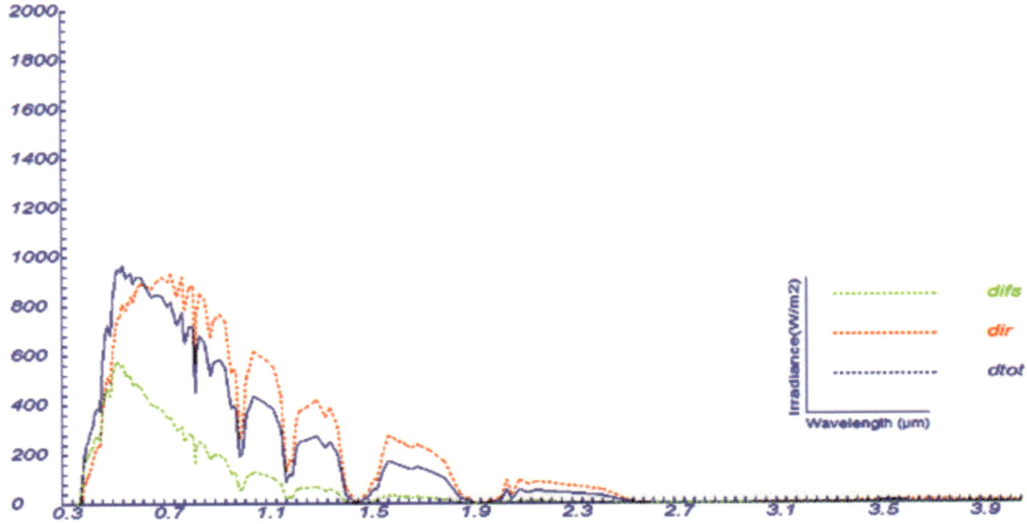

Fig. 3. The solar irradiance for an area generally inclined

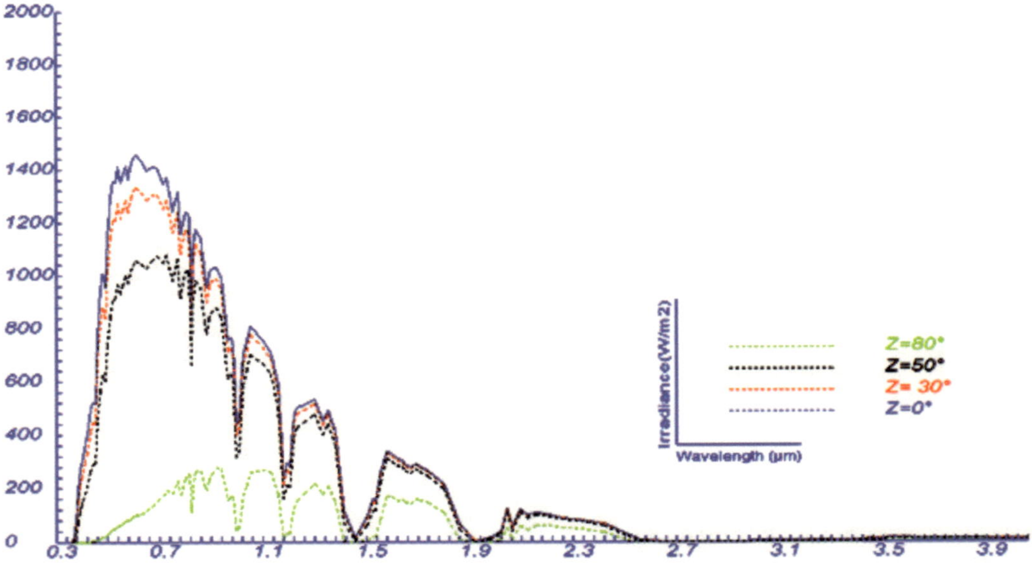

Fig. 4. Variation of solar irradiance depending on the angle of incidence.

Figure 5 shows a small optical thickness results in an intense light while a high optical thickness shows a low light, in other words, unlike the light varies with the optical thickness. (Kaufmann, and Sendra , 1992).

Figure 6 shows the variation of water vapor only affect the light weakly, but it is very important as if we compare it with the influence of the number of days in the year of the illumination.

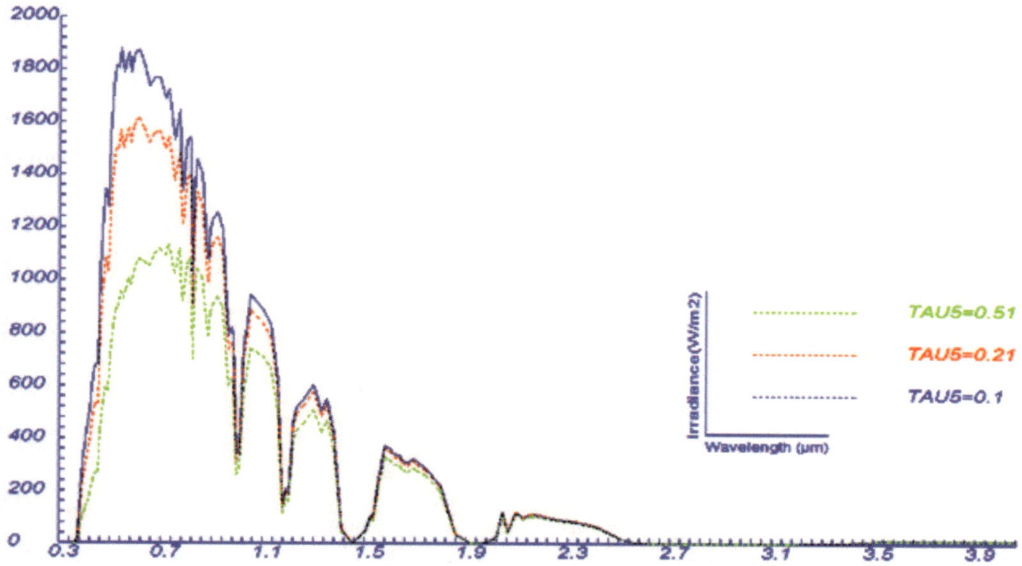

Fig. 5. Variation of solar Irradiance according to the optical thickness

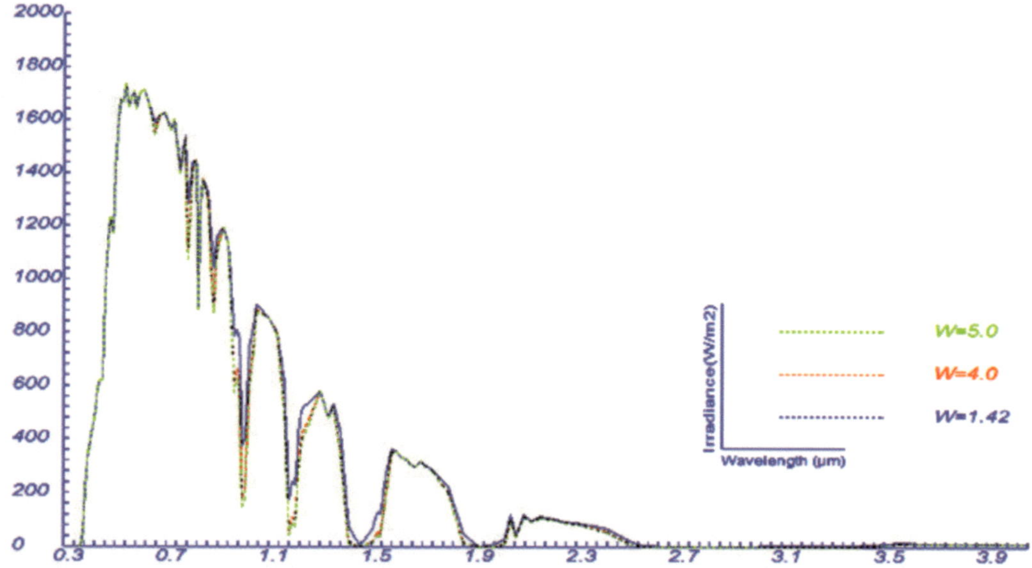

Fig. 6. Variation of illumination depending on the water vapor precipitated.

4. Modeling the radiation reflected by the ground

In this work, we determine the physical quantity measured by the system of shooting (sensor) which is sunlight reflected by the soil-atmosphere averaged in some way in the spectral band considered the sensor. For this, we described the various factors affecting a satellite measurement.

Thus, after characterization data on the spectrum of electromagnetic radiation, reflection, emission and atmospheric transmission is determined and developed by the optical properties of the elements of natural surfaces, radiation reflected by the surface water and radiation captured by satellites. (Houma and *al.*,2004)

The methods used in atmospheric modeling can be divided into direct method and indirect method. Generally, direct methods are represented by the development of a model of interaction of the solar spectrum with the various elements that are in the path of solar radiation the sun-ground and ground sensor. The radiance L_{sat} captured by the satellite is the sum of the three luminances:

1. The luminance of the system from the ground - atmosphere considering the ground as a black body
2. L_{sol} luminance reflected from the ground toward the sensor
3. L_v luminance reflected from nearby objects but observed in the direction of the ground

$$\mathbf{L_{sat} = L_a + L_{sol} + L_v} \tag{8}$$

Either a pixel image coordinate (x, y) and a spectral band b, the radiation that excites the sensor λ K (x, y, b) (w.cm^{-2} µm^{-1} sr^{-1}) according to Teillet (Teillet , 1986) is written as follows:

$$\mathbf{K_\lambda(x,y,b) = C(x,y)\, R_\lambda(x,y,\, b) + H_\lambda(b) + \Delta_\lambda(x,y,b)} \tag{9}$$

C(x,y) is a multiplicative constant that describes the form factor that is the orientation of the surface topography relative to the sun's position during the shooting; R(x, y, b) is the radiation reflected by the ground is proportional to the average reflectance of the pixel (x, y) in the spectral band b; H_λ (b) is the distribution of soil-atmosphere (noise) by considering the earth as a black body and Δ_λ (x, y, b) represents a residual variable that creates the effect of neighborhood.

R (x, y, b) is given by the following equation:

$$\mathbf{R_\lambda(x,y,b)=S_\lambda(b)T_\lambda(b)G_\lambda(b)\, \rho_\lambda(x,y)} \tag{10}$$

S_λ(b) : represents the gain factor of the system in the channel b (sensor sensitivity),

T_λ(b) is the atmospheric transmittance of the earth to the satellite in channel b,

G_λ(b) : is the global radiation and ρ_λ(x,y,b) : reflectance of the pixel (x, y) (assumed Lambertian) in the channel b.

Using a database of spectral signatures and spectral extinction coefficients to model parameters. A numerical code to track the signal in the solar sun-trip ground and ground sensor.

The energy quantity R_λ(x,y,b) is transformed into a numbered account, which includes all information about the Earth-atmosphere system, the geometric conditions of shooting and optical properties of the sensor.

It is obvious that the atmospheric absorption and scattering vary across an image due to three effects:

- Change in weather conditions across an image.
- Change of observation relative to the position of the sun.
- Variation in the average radiation in the area surrounding the pixel observed at all times.

It is therefore necessary to analyze different types of information in order to quantify and qualify. (Becker,1978). To do this, we should dissect the process of taking an image, estimate its multiple components and determine at what level the various categories of information can be determined follow atmospheric correction models . (Bukata et *al.*,1995)

4.1 Simulation analysis of the reflectance of sea water

This model is followed by a detailed study of factors affecting the optical properties of sea water. To correctly interpret satellite data, we must solve the equation of radiative transfer soil-atmosphere. Solving the transfer equation is based on atmospheric models at several levels that require a considerable mass of meteorological data generally not available.

The first test is performed to explain the blue sky, was made by Lord Rayleigh that the assumptions of his theory are: the particles are small compared to the wavelength, the scattering particles and the medium does not contain free charges (not conductive), therefore, the dielectric constant of the particles is almost the same as that of the medium.

In vertical viewing, then the reflectance is lower than when the sun is at its zenith . The set of simulated data depends on the reflectance and the spectral amplitude of the radiation that reaches the ground is maximum at the zenith, so the measure is more affected by radiation than because of the dependence of reflectance of the zenith angle.

The zenith angle determines the illumination received by the target surface and is involved in all elements of calculating the various transmittances and radiation. Radiation received, for all channels, decreases if the solar zenith angle tends to a horizontal position; it is maximum when the sun is at its zenith . The zenith angle is involved in all elements of calculating the various transmittances and radiation, it depends on the latitude, the inclination of the sun and time.

The information spectral radiometers are determined by the wavelengths recorded by the sensor. The width of each spectral band radiometer defined spectral resolution. We consider that the observation is made in a plane perpendicular to the direction of the grooves.

Solar radiation travels through space as electromagnetic waves. In the case where the wave propagates in a medium refractive index and suddenly she meets any other medium characterized by a different index of refraction, part of the wave is then transmitted into the second medium and the other part is reflected in the first medium. The amplitude of the reflected wave depends on the nature of the medium, shape and lighting conditions.

Part of the global radiation reaching the ground is reflected to the sensor by the coefficient of reflectance. The major problem in determining the reflected radiation is the development of a model that generates all the soil properties affecting in a direct spectral signature (lighting condition, roughness, soil type,....) or indirect (color , salinity, humidity, etc.).

4.2 Total radiation reaching the satellite

The luminance level of the satellite is the sum of the intrinsic brightness of the atmosphere and the luminance of the target which represents the sea water in our case. The radiation recorded at the satellite is given by the relation

$$B= (1/\pi) \int[(C(x,y)T_\lambda(b)G_\lambda(b)\rho_\lambda(x,y)(1-s\rho_\lambda(x,y))+H_\lambda(b))S_\lambda(b)d\lambda/\int S_\lambda(b)\, d\lambda \quad (11)$$

$(1/\pi)$ is a normalization factor, $T_\lambda(\theta_v)$ transmittance of direct radiation toward the sensor, s : spherical albedo of the atmosphere et $S(\lambda)$sensitivity function optical sensor (Sturm, 1980).

The sensor has a spectral response δ_λ ,the recorded signal at the sensor is the luminance:

$$L = \int_0^\infty \left(\frac{I_\lambda \rho_\lambda}{\pi} T_{atm} + \frac{I_{0\lambda} \cos\theta_z}{\pi} \rho_{a\lambda} \right) \delta_\lambda \, d\lambda \quad (12)$$

The luminance level of the satellite is the sum of the intrinsic brightness of the atmosphere and the luminance of the target which represents the sea water in our case. (Deschamps et al., 1983)

The radiation reflected from the water surface to the satellite passes through the atmosphere in a direct way with an angle θ_v and undergoes attenuation before being captured by the satellite.

$$G_{\lambda(sat)} = I_{e\lambda}.\tau'_\lambda \quad (13)$$

$I_{e\lambda}$: radiation reflected from the surface of the water; τ'_λ : total spectral transmittance

The amount of energy that reaches the satellite sensor is the sum of that from the ground and scattered by the atmosphere. The radiation emitted by the sea water that reaches the sensor is:

$$R_{(sol-atm)} = \left[I_{e\lambda} \left(\tau'_\lambda + \tau'_{da\lambda} + \tau'_{dr\lambda} \right) \right] / \pi \quad (14)$$

The radiation scattered by the atmosphere:

$$R_{(atm-sat)} = \frac{I_{o\lambda}.\cos\vartheta_z}{\pi} \rho_{a\lambda} \quad (15)$$

Radiation reaching the satellite is composed of spectral global radiation reflected from the sea water passing through the atmosphere $R_{(sol-atm)}$) and part of the radiation scattered by the atmosphere $R_{(atm-sat)}$. (Bricaud ,1988)

So the radiation that reaches the sensor is expressed:

$$R_\lambda = R_{(sol-atm)} + R_{(atm-sat)} \quad (16)$$

The signal of sea water recorded at the sensor is:

$$L_\lambda = \int_{\lambda_1}^{\lambda_2} \left(\frac{I_{e\lambda} \tau'_{atm}}{\pi} + \frac{I_{o\lambda} . \cos \vartheta_z}{\pi} \rho_{a\lambda} \right) \delta_\lambda d\lambda \tag{17}$$

with : L_λ : Radiation calculated in sea water for the canal λ.

$\Delta\lambda$: Spectral band of the channel.

δ_λ : Sensitivity of the channel.

5. Simulation of satellite data for SDDS

Based on this physical model, we developed a simulation system of satellite data to correct the scattered radiation of atmospheric effects.

A library of spectral signatures is introduced, it covers the main ground objects that have a reflectance in the bands of the electromagnetic spectrum. The combination of spectral signatures and different radiances allows us to calculate the spectral radiance reflected from the surfaces. The simulation results depend on the choice of input parameters. The software allows to show the influence of the effects of various parameters and geometric characteristics of the structures on the signal reaching the sensors onboard the satellite SPOT, LANDSAT and IRS1C.

To highlight the effect of a given parameter on the satellite measurement, are assigned fixed data for all variables in the case of a clear sky and for geometrically well defined.

The second part is from satellite SPOT, LANDSAT and IRS1C, applying the method of covariance matrix (a method that can provide a correction locally specific in the sense that it relates to the pixels of a given region of the image), one can estimate the atmospheric noise through a program to input data images from different channels and outputs the atmospheric noise of these channels.The physical quantity measured by the shooting system (sensor) is a solar radiation reflected by the soil-atmosphere averaged in some way in the spectral band considered the sensor. It depends on angle of illumination and shooting.

To determine the different radiation received at the satellite, the data input parameters are astronomical, geographical and atmospheric.

5.1 Atmospheric correction of remotely sensed data

Atmospheric correction is a major issue in visible or near-infrared remote sensing because the presence of the atmosphere always influences the radiation from the ground to the sensor.

As introduced before, the atmosphere has severe effects on the visible and near-infrared radiance.

First, it modifies the spectral and spatial distribution of the radiation incident on the surface.

Second, radiance being reflected is attenuated.

Third, atmospheric scattered radiance, called path radiance, is added to the transmitted radiance.

The atmosphere transmittance is:

$$\boldsymbol{T_{\theta}} = \boldsymbol{Exp}(-\boldsymbol{\tau}/\boldsymbol{Cos}(\boldsymbol{\theta})) \tag{18}$$

Where τ is the atmospheric optical thickness and θ can be sonar zenith angle or satellite angle view.

The optical thickness is composed of:

$$\tau_{\lambda} = \tau_{a\lambda} + \tau_{R\lambda} + \tau_{d\lambda} \tag{19}$$

$\tau_{a\lambda}$: Thickness selective absorption

$\tau_{R\lambda}$: Thickness scattering by small particle (molecular) is named Rayleigh diffuse

$\tau_{d\lambda}$: Thickness scattering by medium particle is named Mie diffuse

For a given spectral interval, the solar irradiance reaching the earth's surface is

$$E_g = \int_{\lambda_1}^{\lambda_2} (E_S T_i \text{Cos}\,(i) + E_d) d\lambda \tag{20}$$

The scattering is dominated by aerosols while back scattering is mainly due to Rayleigh scattering. A number of path radiance determination algorithms exists. For a nadir view as Landsat MSS, TM and SPOT HRV are usually used. In this section, we only tried to introduce some basic concepts of this complex topic. This is only a single-scattering correction algorithm for nadir viewing condition. More sophisticated algorithms which count multiple-scattering do exist. Some examples of these algorithms are LOWTRAN7, 5S (Simulation of the Satellite Signal in the Solar Spectrum 5S) and 6S (Second Simulation - aircraft, altitude of target).

There are FORTRAN codes available for these algorithms. The 5S and 6S are proposed by Tanre (Tanre et al., 1990).

For small wavelengths the atmospheric contribution is very important, we wish to point out that the impact angle is also important he was noticed by a degradation of the signal that reaches the sensor signals for all contributors to the signal exciting the radiometer. The simulation analysis shows the need to correct the satellite images of atmospheric effects in order to identify objects on the ground. (Figure7)

The composition of the atmosphere disturbs the path of electromagnetic radiation between the source on the one hand, between the earth and the satellite on the other. Atmospheric effects resulting absorption and diffusion, performed jointly by the two major components: gases and aerosols.

When a signal is recorded, it is the part of the spectral radiation scattered by air molecules and aerosols to the outside of the system soil - atmosphere (Popp, 1994).

This signal $H_{\lambda}(b)$ is called atmospheric noise :

$$H_{\lambda}(b) = E(\lambda) T_{at\text{-}sat}(\lambda) \tag{21}$$

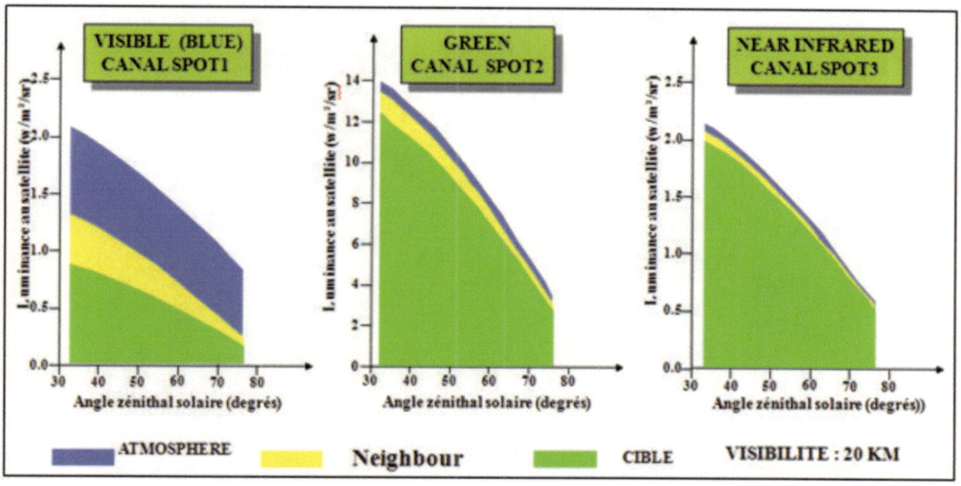

Fig. 7. Contribution of the atmosphere on the multispectral channels SPOTXS.

$T_{at\text{-}sat}(\lambda)$ is the spectral transmittance of the atmosphere, such that its optical path is calculated by replacing the zenith angle from the viewing angle.

To dissect the effects of the various elements contributing to satellite measurements, we developed a software simulation of satellite data (SDDS) using the visual language Basic.6. The tool is based on modeling of radiation and atmospheric effects. The monitoring of the solar spectrum as a double drive-ground and ground sun-sensor is implemented based on the concepts of codes 5S, 6S, to simulate different radiances reaching the sensor. For the operation of the system we used the extinction coefficients of the solar spectrum in the developed Lowtran.6 and a bank of spectral signatures extracted from the software ENVI.4.3 (2007).

6. Analysis of variation in luminance

The physical quantity measured in the system shooting (sensor) is a solar radiation reflected by the soil-atmosphere averaged in some way in the spectral band considered the sensor. It depends on the angle of illumination and shooting. (Bachari et al., 1997)

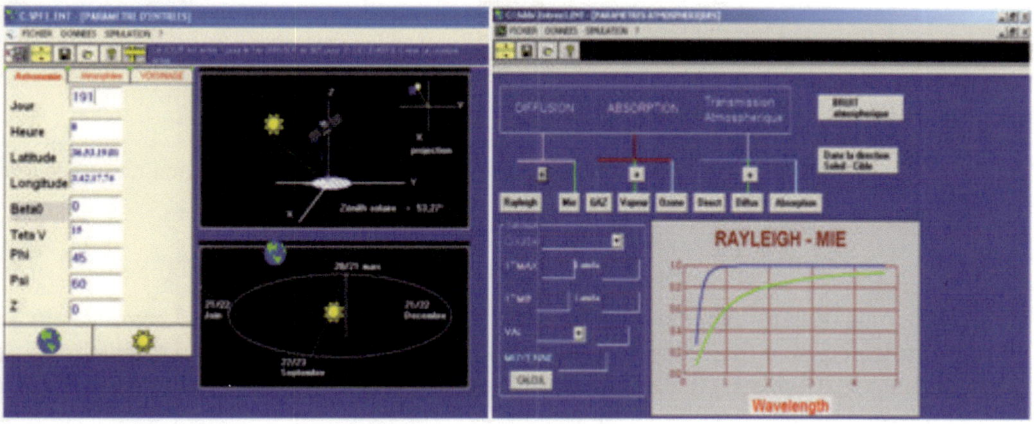

Fig. 8. Simulation of the scattered radiation as SDDS

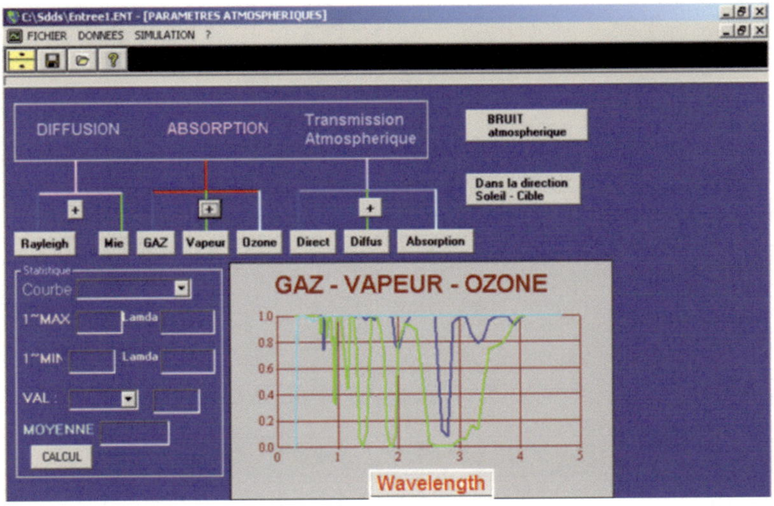

Fig. 9. Simulation of the spectral irradiance of water, ozone and gas as SDDS

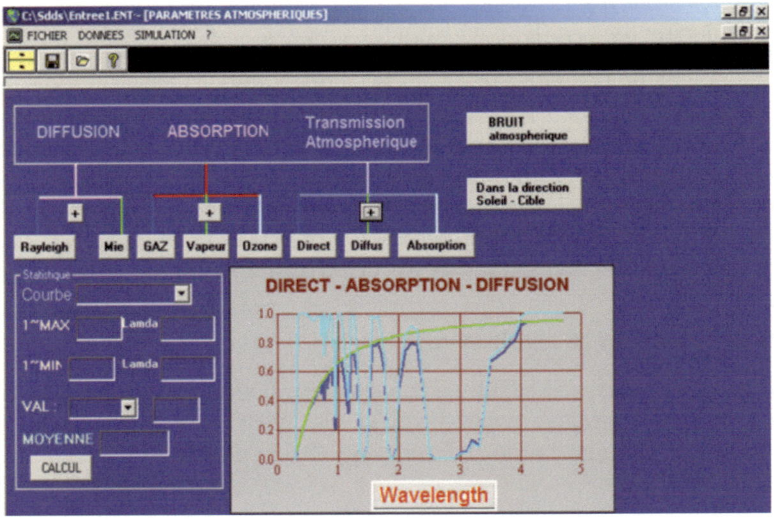

Fig. 10. Simulation of the spectral irradiance on the ground as SDDS.

6.1 Effect of solar zenith angle

The zenith angle is involved in all aspects of calculation of the various transmittances and radiation, it depends on the latitude, the inclination of the sun and time.

$$\cos(\theta_z) = \cos(\delta).\cos(\text{latitude}).\cos(15(12\text{-heure})) + \sin(\text{latitude})\sin(\delta) \tag{22}$$

The zenith angle determines the illumination received by the ground and involved in all aspects of calculation of the various transmittances and radiation. Radiation received, for all satellite channels, decreases if the solar zenith angle tends to a horizontal position, and is maximum when the sun is at its zenith (Figure 11). The following figure shows the contrast angle (zenith-sun) radiances:

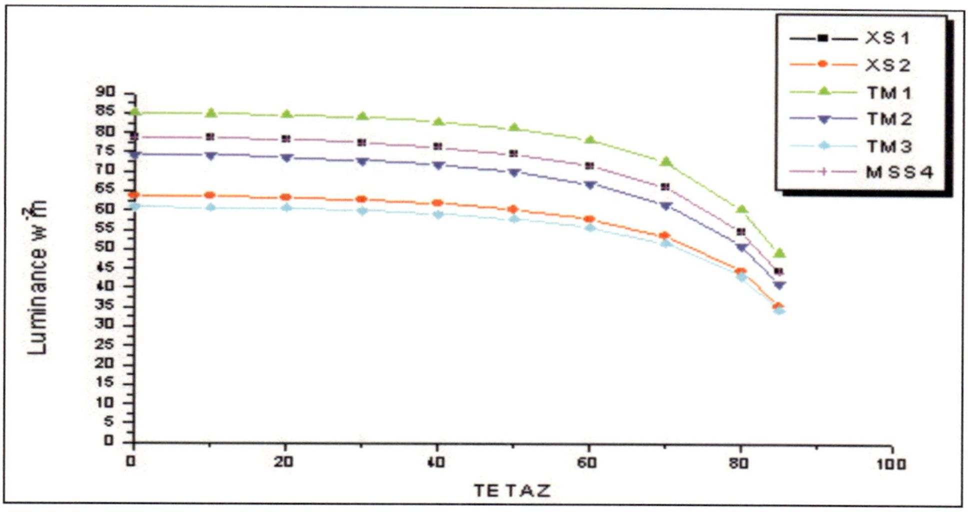

Fig. 11. The effect of solar zenith angle on luminance.

6.2 Effect of the zenith angle of observation

The zenith angle of observation determines the length of the journey made by atmospheric radiation. The simulation results show that the radiation collected is small if the angle of observation tends to a horizontal position. The growth of the zenith angle of observation leads to an increase in air mass and a decrease in transmittance. The following figure shows the contrast of the luminance for different sensors. In the case of surface spectral signature weak as water, atmospheric noise becomes important to the reflected radiation if the angle of observation believed to a horizontal position. The atmospheric contribution increases, therefore the luminance level of the sensor increases. The brightness peaks at a zenith angle of observation [60°, 70°] and then begins to decrease if θ_V differs from 70°.

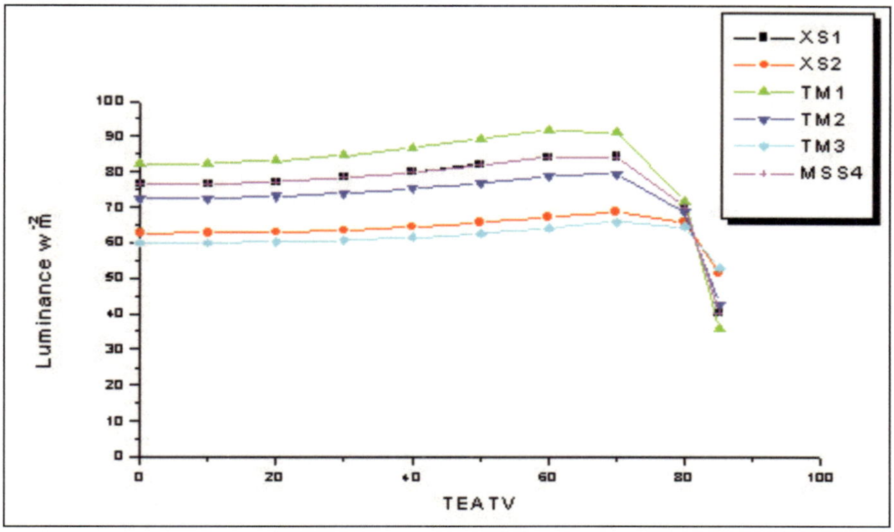

Fig. 12. The effect of the zenith angle of observation on the luminance.

6.3 Effect of relative humidity

The presence of water vapor in the atmosphere depends on the location and altitude above the ground. Analysis of simulated data shows the insensitivity of the channels XS1, XS2, TM1, TM2, TM3, MSS4 to increase or decrease of water vapor in the atmosphere.The transition from a dry to a humid atmosphere causes a slight decrease in the extent to channel XS3 and TM4. Constructors radiometers onboard satellites SPOT and LANDSAT have avoided the windows of absorption of radiation by water vapor.

The following figure shows the change in apparent radiance between the two extreme amounts of water vapor. (Figure 13)

The apparent radiance level of the system SPOT, LANDSAT is practically independent of the wet state of the atmosphere because the spectral bands used do not contain the total absorption window of radiation by water vapor.

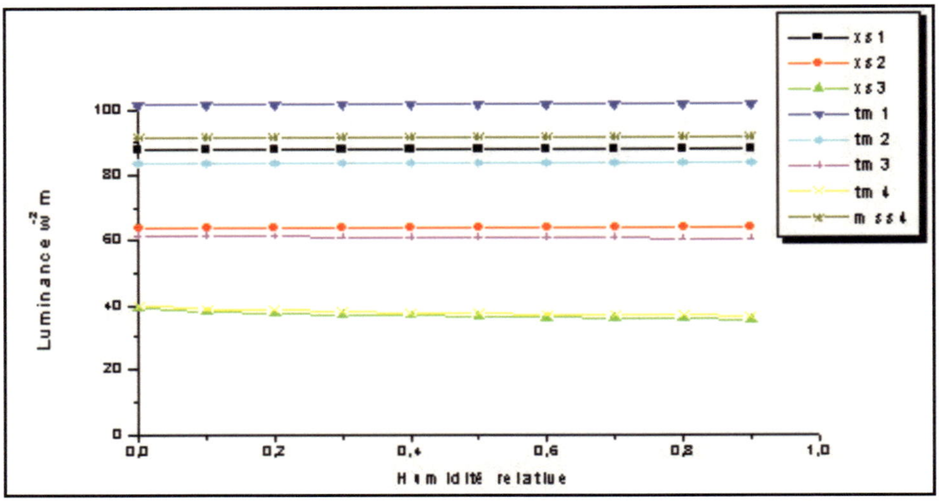

Fig. 13. The effect of relative humidity on the radiance.

6.4 Effect of diffusion parameter Fc

Mie scattering has a significant influence on the measured signal at the sensor described by the diffusion parameter Fc. The atmosphere absorbs in the field of short wavelength and becoming more transparent. The information is degraded in the channels depending on the diffusion parameter Fc. (Figure14). The following figure shows that for low surface reflectance as the case of water degradation in the channel becomes more comparable, especially for channels XS1 and MSS4.

7. Correction of satellite images

Each pixel of an image is a digital count from 0 to 255, which translates into a color using an editable pre-selected distribution in image processing. Generally the relationship between digital count and the luminance is linear:

$$K(x,y) = a_1 \,.\, CN + a_0 \tag{23}$$

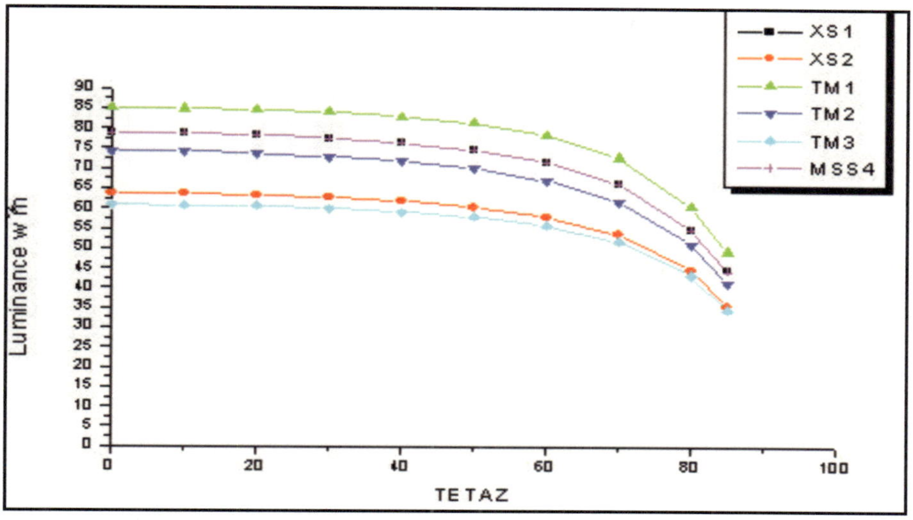

Fig. 14. The effect of atmospheric turbidity parameter Fc on the radiance.

with factors a_1 and a_0 are calibration coefficients.

For the same conditions of image capture, simulation of the observed brightness is determined by the relationship between apparent brightness and simulated the account corresponding digital images processed.

For the three channels of the HRV sensor, radiometric conversion is given by the relations:

$$CN_{XS1} = 1.23\ L_{XS1} + 0.22 \quad (24)$$

$$CN_{XS2} = 1.24 L_{XS2} - 0.08 \quad (25)$$

$$CN_{XS3} = 1.32\ L_{XS3} - 0.59. \quad (26)$$

Application to images

The same method was applied directly to the accounts of digital Landsat 2003 and Spot 2004. The images are processed using the software to process satellite images PCSATWIN developed by (Bachari et al, 1997)

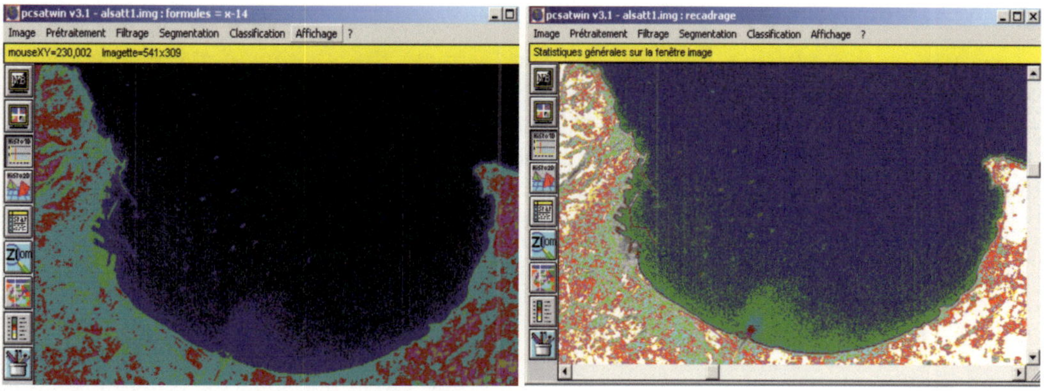

Fig. 15. SPOT XS1 image before and after radiometric correction.

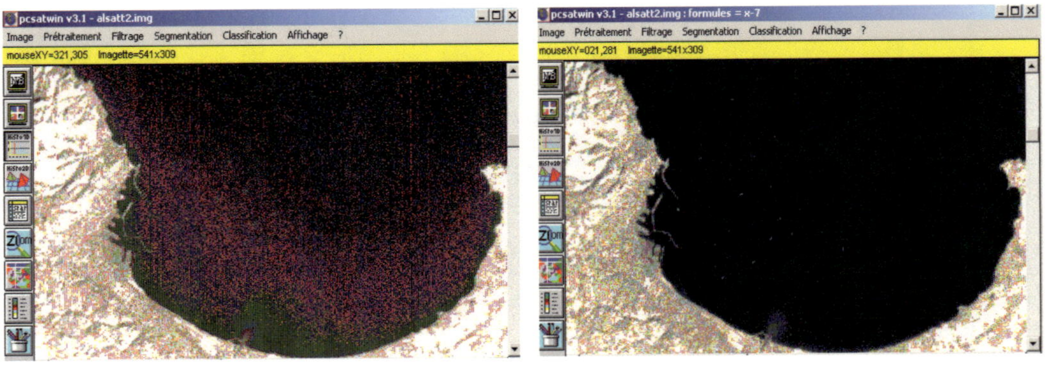

Fig. 16. Landsat TM1 image before and after radiometric correction.

Calculation of the reflectance

The luminance is happening to global satellite is expressed by the relation:

$$L = a\bar{\rho} + b\,. \tag{27}$$

With ρ_λ is the reflectance at the sea surface, E is the total illumination received by the surface. The average reflectance is connected to the luminance by the following equation:

$$\bar{\rho} = a\,CN + b \tag{28}$$

The conversion factors obtained by modeling of radiation on SPOT satellite channels (XS1, XS2, XS3 and) and Landsat (TM1, TM2, TM3 and TM4) are given in the table below: (Bachari,2006)

Channel	**XS1**	**XS 2**	**XS 3**	**TM_1**	**TM_2**	**TM_3**	**TM_4**
a	0. 0024	0,0025	0,0031	0,0017	0,0033	0,0026	0,0036
b	– 0,05	–0,0433	–0,0217	– 0,099	– 0,0723	– 0,0416	– 0,0295

Table 1. Conversion factors accounts simulated digital reflectance.

Quality of radiometric corrections

According to the criterion Rouquet, we can estimate the quality of the correction by comparing the properties of the raw images and corrected. (Morel & Prieur, 1977) For a given image, the atmospheric effect is minimum if the contrast and the ratio of standard deviation is the average maximum, this quantitative criterion is also applied systematically for the selection of good quality data and is also a primary method of atmospheric correction, which tends to minimize atmospheric effects.

There is also a primary method of atmospheric correction, which tends to minimize atmospheric effects. The properties of the images are combined and corrected in the following table (Bachari,1999)

For a simple analysis of the results expressed in Table 2, we note that the criterion used is justified for the corrected images. The application of atmospheric corrections is shown in the

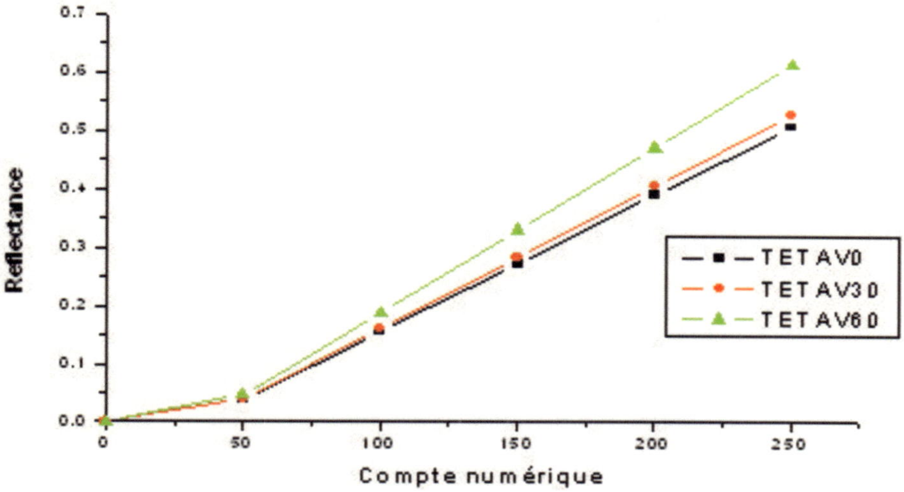

Fig. 17. The digital count -reflectance at ground level

Image	CN_{min}	CN_{max}	CN_{max}- CN_{min}	Ecart-type (σ)	Moyenne (m)	V=(σ/m)*100
XS1 Brute	38	254	216	18,23	65,10	28
XS1 Corrigée	29	255	226	24,99	38,76	64
XS2 Brute	25	235	210	21,04	55,26	38
XS2 Corrigée	11	255	244	32,80	58,58	56
XS3 Brute	17	213	206	26,40	57,85	46
XS3 Corrigée	14	255	241	37,28	67,39	55

Table 2. Statistical data of radiometric channels SPOT HRV.

graph defining the linear relationship between the reflectance calibration and account for two-channel digital SPOT XS1 and XS2 for images corrected and uncorrected images (Figure18).

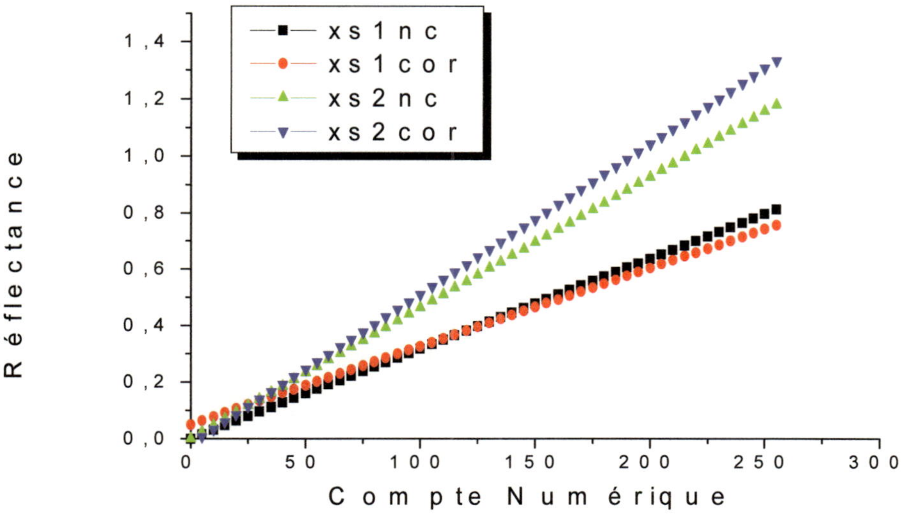

Fig. 18. Linear calibration reflectance before and after atmospheric correction.

This simple method of correction amounts to replacing the linear calibration another linear relationship applied to both SPOT images, this new relationship calibration results to correct the measured reflectances.

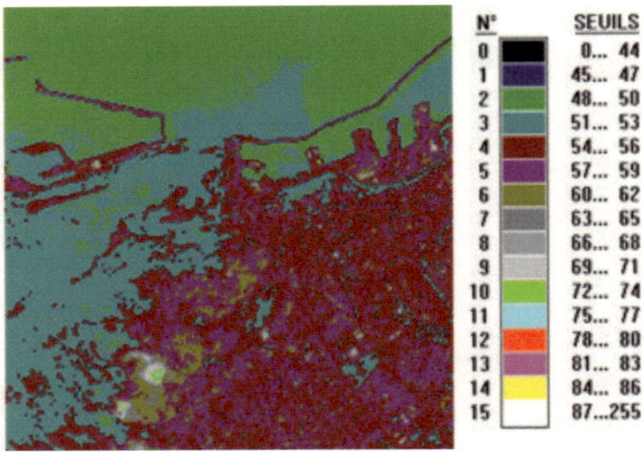

Fig. 19. XS1 image of the Bay of Oran

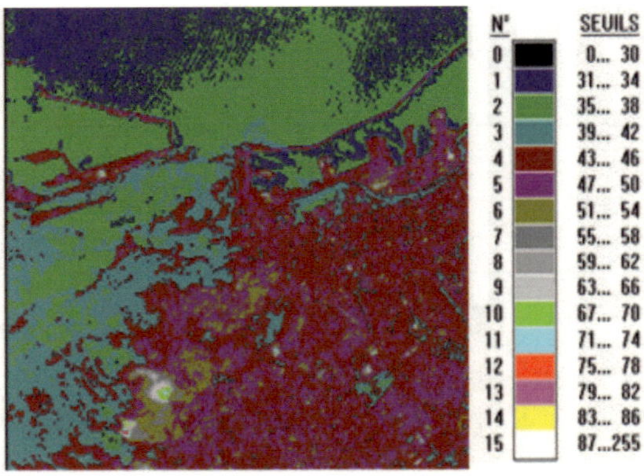

Fig. 20. XS1 corrected image of the Bay of Oran

The statistical properties of the corrected images are presented in the following histograms:

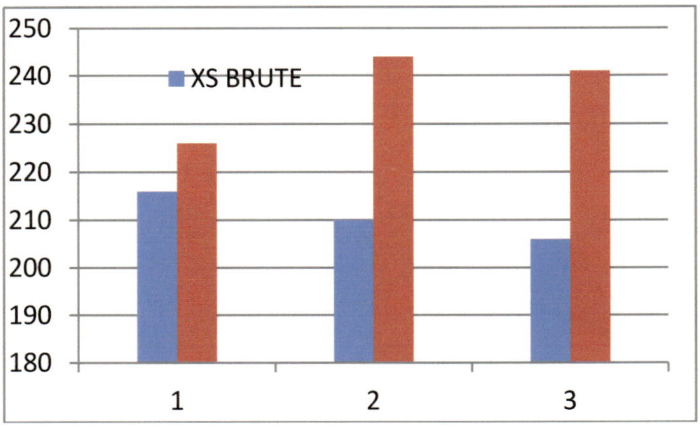

Fig. 21. Histogram spread between raw images and images corrected SPOT.

Corrected images and spectra of radiation have a greater spread and therefore a high contrast, this justifies the first condition of Rouquet.

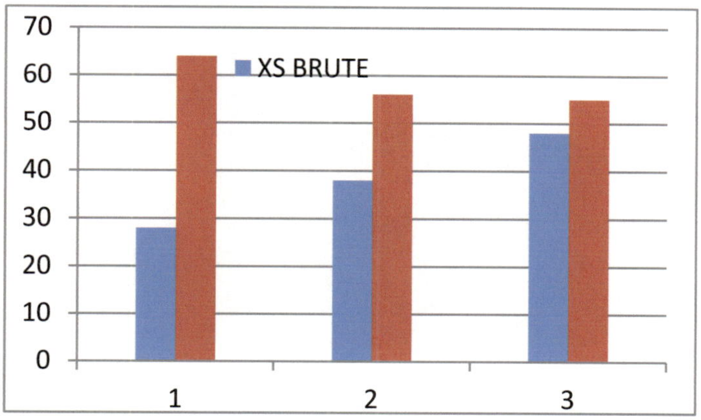

Fig. 22. Coefficient of variation of the three images corrected for atmospheric effects.

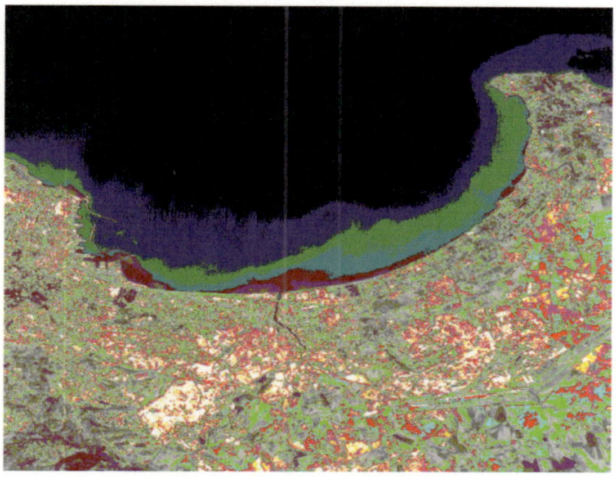

Fig. 23. Radiometric correction in the SPOTXS1 channel of the bay of Algiers.

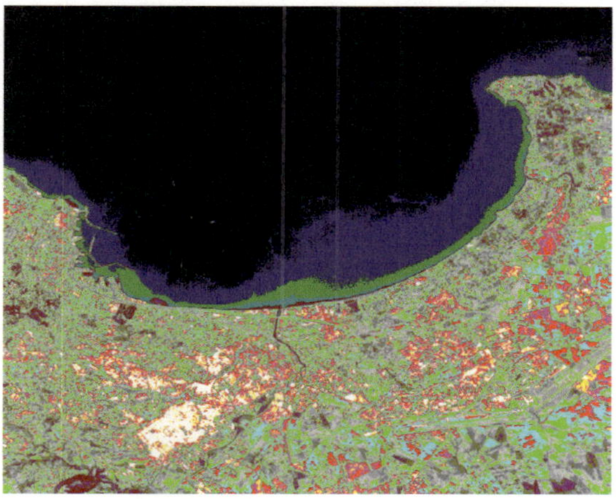

Fig. 24. Radiometric correction in the SPOTXS2 channel of the bay of Algiers.

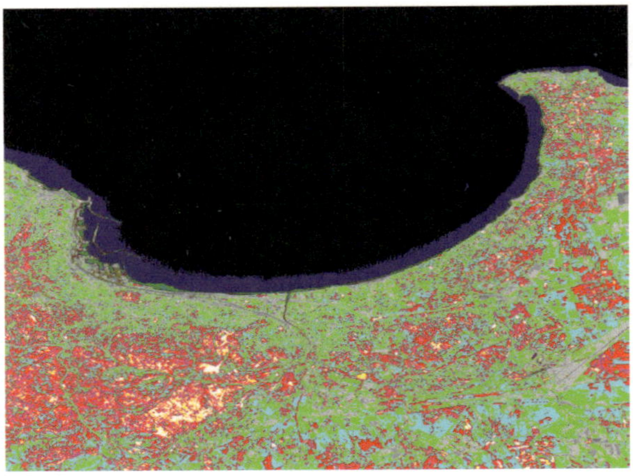

Fig. 25. Radiometric correction in the SPOT XS3 channel of the bay of Algiers.

In the histogram of the coefficient of variation we wish to point out that the first channel has a high coefficient corrected this can be explained as the effect of correction is more experienced in this channel than in the other two channels. In the third channel (NIR), Mie scattering and Rayleigh are less experienced.

Images are acquired by satellite sensors (HRV pour SPOT 2) (B1 (green: 0.50 - 0.59 microns), B2 (red: 0.61 - 0.68 microns), and B3 (near infrared: 0.78 - 0.89 microns).

8. Modelling of satellite measure under sea water

The knowledge of the topography of the seafloor is important for several applications. The principle of measure of bathymetry necessarily takes this model of reflectance joining the intensity of radiometric signal measured by the satellite to the depth as a basis; it can call on the physical method that requires the knowledge of all parameters governing this model (optic properties of water, coefficient of reflection of the bottom, transmittance of the atmosphere (Minghelli-Roman et *al.*,2007). The model provides of image mono channel where each pixel of the maritime domain is represented either by a radiometry in-situ but rather by a calculated depth. In general the use of hybrid multiple SPOT band regression algorithms are superior to the exclusive use of any single band. (Bachari & Houma 2008, Houma et *al.*,2010)

The spectral distribution of the submarine radiance varies in a complex way with the depth, in relation with the selective character of the attenuation.

The total signal received by a sensor operating at high altitude water above can be decomposed in a first time, in two terms:

$$S_{t_\lambda} = S_{a\lambda} + S_{e_\lambda} \tag{29}$$

with $S_{s\lambda}$ is an atmospheric component and $S_{e\lambda}$ is a water component

In a second time, it is possible to analyze the composing water measured near the surface:

$$S_{e_\lambda} = S_{s\lambda} + S_{d\lambda} + S_{f\lambda} \tag{30}$$

with $S_{s\lambda}$ a specular reflection at the surface, $S_{f\lambda}$ is a reflectance of the bottom in shallow waters, $S_{d\lambda}$ a component owed to the diffuse reflection by volume water.

$$S_{e\lambda} = G_\lambda \rho_s + G_\lambda \rho_s.\omega_0 \left(\frac{1}{\rho_s} - 1 - \left(\frac{1}{\rho_s} - 1 \right) R_a \exp(-kz(\cos\vartheta_z + \cos\vartheta_v) \right) \\ + G_\lambda (1 - \rho_s) R_a \exp - kz(\cos\vartheta_z + \cos\vartheta_v) \quad (31)$$

with ρ_λ a reflectance of the sea water, Ra, a reflectance of the bottom, k, is attenuation coefficient, z, a depth, ω_0 albedo of diffusion of water molecules, θ_z a zenith angle and θ_v a viewer angle of the sensor.

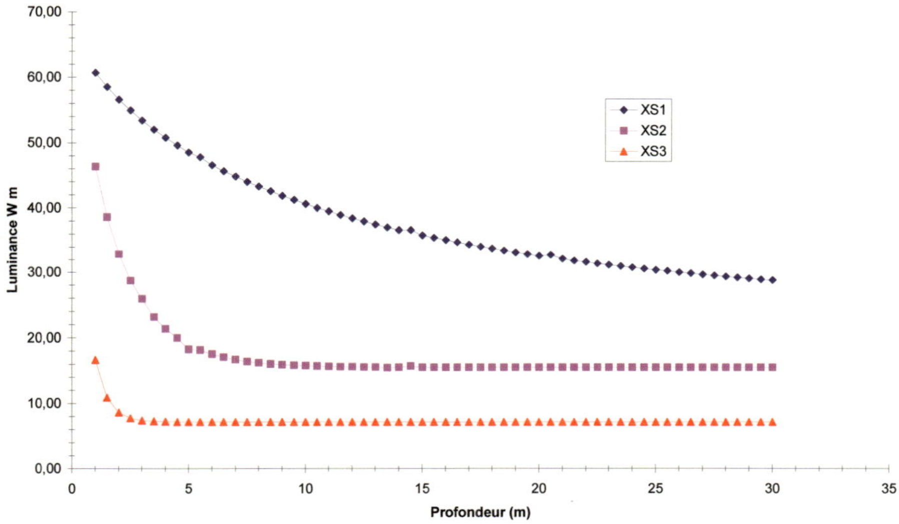

Fig. 26. A variation a luminance's of a SPOT XS with a depth

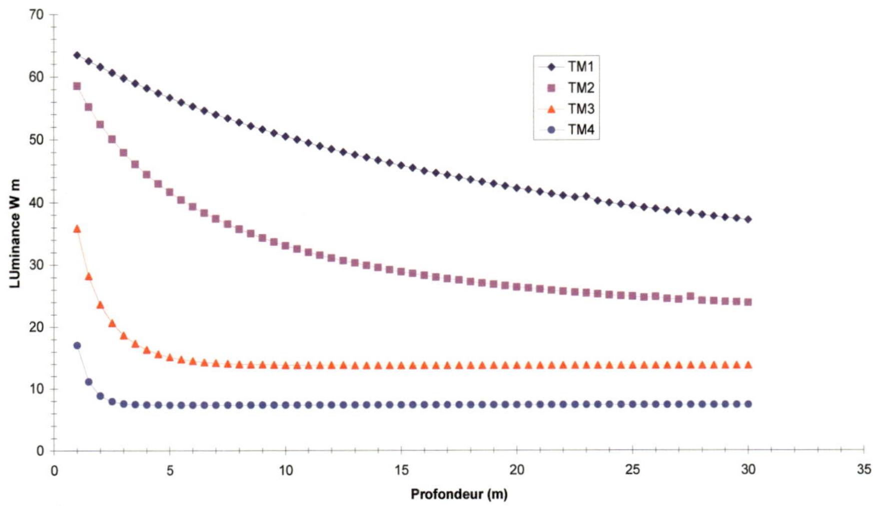

Fig. 27. A variation a luminance's of a LANDSAT TM with a depth.

In the case of the channel 2 of the spot the sensitivity of this channel to effects of the bottom can reach 10 meters. For the channel 1 of the spot the effect of the bottom can reach funds that pass the 30 meters. For the TM1 the effect of the bottom can reach funds of 40 to 50 meters.

The CN2 quantities and CN1 are luminance's corrected of the atmospheric effects. As in this case we removed the point that present a maximum of SM. This singular point presents an anomaly that indicates the streamlined convergence of currents.

Notices the variable Z is just to the middle in relation to the two luminance's. We removed data that correspond to depths superior to 60ms. Since for depths that pass the 50 meters the effect of the bottom on measure satellite is non-existent. The set of the corrected data are analyzed by an exponential regression. Results in this case are better presented and the curve of adjustment is representative of the cloud of points since the coefficient of interrelationship reaches the 88%. (Bachari & Houma, 2005)

Correlative analysis

1. Monoband model

Results of studies led on the bathymetry are bound directly to features of satellites. On curves of reflectance's we notice that more the depth is important, more the radiance is absorbed and more the level of radiometry is weak. The spectral resolution permits to observe in the light waters objects as far as 40 ms on the XS1 channel, and as far as 15m on the XS2 channel. The XS3 channel when to him doesn't bring any bathymetric information since the infrared is absorbed by water.

The reflectance of the bottom, during his ascension toward the surface, sudden a selective attenuation. All it has for effect a bruising of the spectral answer of the bottom that returns discriminations noise in that the depth increases.

$$\textbf{XS1} \qquad Z = -0{,}27823 + 2{,}68400.\exp(-(XS1-47)/5{,}8886) \qquad (32)$$

$$\textbf{XS2} \qquad Z = -0{,}2579 + 5{,}83395.\exp(-(XS2-25)/4{,}24134) \qquad (33)$$

2. Several channels'

For this reason one tried to achieve some multiple and polynomial interrelationships between these three variables:

$$Z = -148.8 - 2.76\,XS1 + 17.93\,XS2 - 0.014\,XS1^2 + 0.57 * XS1 * XS2 - 2.04 * XS2^2 \qquad (34)$$

The developed equation permits us to achieve a direct extraction of the bathymetries while combining the two pictures satellites one hold in the channel1 and the other in the channel 2. To achieve the extraction of the bathymetry from pictures satellites we used the software *PCSATWIN*. (Bachari,1997)

9. Conclusion

A methodology was developed to solve the problem caused by the effect of the atmosphere that usually results in a signal noise. To make this work, we first followed the path of the solar spectrum as a double drive-ground and ground sun-sensor. To highlight the contribution of different elements making up the signal that reaches the sensor. Simulation software satellite data is developed.

The proposed radiometric correction method is simple because it is based on pixels that are known to support their radiometric images. The luminances are simulated using the software (SDDS) which allows us to establish rules between reported digital luminance and luminance reflectance. The techniques of normalization of images are to correct the atmospheric degradation, the effects of illumination and variations in the responses of the sensors in the multitemporal and multispectral imaging. Thus the methods developed in this section may be modified or combined as required by the user.

10. References

Bachari N.E.I., Belbachir A.H ., et Benbadji N.,1997 .Numerical Methods for Satellite Imagery Analysis, AMSE.,J Volume.38, N°1,2, pp 49-60.

Bachari N.E.I., et Houma F., 2005. Combination of data soils and data extracted of satellites images for the survey of the bay of Algiers, MS'05 Rouen, 6-8 Juillet 2005,France.

Bachari N.E.I., et Houma F., 2008.Contribution of satellites visible and infrared images for the follow-up of the inshore water quality; International Conference on "Monitoring & Modeling of Marine pollution" (INCOMP 2008); KISH 1-3 /12 2008

Bachari N.E.I.,1999. *« Méthodologie d'analyse des données satellitaires en utilisant des données multi-sources » Thèse de Doctorat d'état* en Physique; Rayonnement-Matière, 11 avril 1999, Oran USTO, Algérie.222p.

Becker F., 1978. Physiques fondamentales de la télédétection. Ecole d'été de physique spatiale C.N.E,S.

Becker F., et Rffy M., 1990. Modèles et modélisation en télédétection, Télédétection spatiale: Aspects physiques et modélisation., pp 37-162, Toulouse, Cepadues-édition, 1032 pages.

Bougeur P., 1953. Essai d'optique sur la graduation de la lumière Ann.Pys.8

Bukata R.P., Jérome J.H., Kondratyev C., and Poszdnyakov D.V., 1995. *Optical properties and remote sensing of inland and coastal waters*. CRC Press, Boca Raton, Florida

Chadin A.,1988. Les modèles d'interaction rayonnement-atmosphère et détemination de paramètres météorologiques et climatologiques à partir d'observation satellitaires. Télélédétection spatiale : Aspects Physiques et Modélisation, Cepadues Editions, p 1031.

Deschamps P.Y., Herman M., and Tanré D., 1983. Modeling of the atmospheric effects and its application to the remote sensing of ocean color. *Appl. Opt.*, 22: 3751-3758

Gordon H.R., et Clark D.K., 1981. Clear water radiances for atmospheric correction of Coastal Zone Color Scanner imagery », *Appl. Opt.*, no 20, p. 4175-4180.

Guyot G., and Fagu X.,1992. Radiometric corrections for quantitative analysis of multispectral, multitemporal and multisystem satellite data, Int, J, Remo.Sens .

Houma F., Abdellaoui A., Bachari N.E.I., and Belkessa R., 2010.Contribution of Multispectral Satellite Imagery to the Bathymetric Analysis of Coastal sea bottom. Application to Algiers bay, Algeria. *Journal Physical Chemical News, volume 53, P57-61.*

Houma F., R. Belkessa, Khouider A. Bashar NIS, and Z. Derriche, 2004. Correlative Study of Physico-chemical parameters and satellite data to characterize IRS1C water pollution. Application to the Bay of Oran, Algeria. Water Science , volume 17 / 4, 429-446.

Kaufmann Y.J., and Sendra C., 1992. Algorithm for automatic corrections to visible and near-IR satellite imagery *Int.J .Rem.Sens.*, 9, 1357-1381.

Minghelli-Roman A., Polidori L., Mathieu-Blanc S., Loubersac L, and François Cauneau ., 2007. Bathymetric Estimation Using MeRIS Images in Coastal Sea waters .IEEE Geoscience and Remote Sensing Letters, Vol. 4, No. 2, April 2007

Morel A. and Gentili B., 1993. Diffuse reflectance of oceanic waters; II: Bidirectional aspects. Applied Optics, vol. 32,p. 6864-6879.

Morel A., and Prieur L.,1977. Analysis of variations in ocean color. Limnology and Oceanography, vol. 22, no 4, p. 709-722.

Popp T., 1994. Atmospheric correction of satellite images in the solar spectral range, J. Research. Centre.

Prieur L., Morel A., 1975. Relations théoriques entre le facteur de réflexion diffuse de l'eau de mer, à diverses profondeurs, et les caractéristiques optiques. UGGI. XVIe Ass. Gle. Grenoble. Août 1975. Symposium Interdisciplinaire d'Optique Océanique.I.S.30, n°13 : 250-251.

Ratto C.F., 1986. Sun-earth astronomical relationships and the extraterrestrial solar radiation. Université di Genova.

Sturm B.,1980. "Optical properties of water applications of remote sensing to water quality determination", dans G. Fraysse (dir), *Remote Sensing Applications in Agriculture and Hydrology*. Balkema, Rotterdam, p.471-495.

Tanre D., Deroo C., Duhaut P., Herman M., Morcette J.J., Perbos J., Dechamps P.Y.,1990. Technical note, Description of a computer code to simulate the satellite signal in the solar spectrum: The SSSSS, Int, J. Rem. Sens, 11, 659-668.

Teillet P.M., 1986. Image correction for radiometric effects in remote sensing. Int. J. Remote sensing, Vol. 7. No.12, pp 1637-1651.

10

Reactivity Trends in Radical-Molecule Tropospheric Reactions – A Quantum Chemistry and Computational Kinetics Approach

Cristina Iuga[1], Annia Galano[2], Raúl Alvarez-Idaboy[3], Ignacio Sainz-Dìaz[4], Víctor Hugo Uc[1] and Annik Vivier-Bunge[2]

[1]Universidad Autónoma Metropolitana, Azcapotzalco
[2]Universidad Autónoma Metropolitana, Iztapalapa
[3]Facultad de Química, Universidad Nacional Autónoma de México
[4]Instituto Andaluz de Ciencias de la Tierra, CSIC-Universidad de Granada
[1,2,3]México
[4]Spain

1. Introduction

The most relevant chemical reactions that take place in the atmosphere involve free radicals and volatile organic compounds (usually termed VOC). In the troposphere, the main sink of volatile organic compounds (VOC) is oxidation, initiated typically by reaction with hydroxyl (OH) free radicals. Many of these processes give rise to the formation of new radicals, which ultimately cause higher OH radical levels and thus higher rates of reactions of the other VOC present.

Kinetic investigations of the OH radical reaction with VOC's are essential for the evaluation of their significance in air pollution. Reaction rate coefficients are used, for example, in estimating their tropospheric lifetimes, or in atmospheric chemical model calculations which are used to generate distribution maps of air pollutants under given meteorological and topographical conditions. Furthermore, the use of temperature dependent reaction rate coefficients in model calculations increases their accuracy, since the temperature gradient of the troposphere and the seasonal temperature variations can be taken into consideration.

As a result of almost three decades of research, the rate constants and mechanisms of the initial reactions of OH and NO_3 radicals with VOCs are now reliably known or can be estimated. Significant advances have been made in our understanding of the mechanisms of the reactions subsequent to the initial OH and NO_3 radical attack on selected VOCs and of first-generation products formed from these reactions. Extensive and comprehensive reviews on the current state of knowledge of atmospheric reactions of VOCs have been written periodically over the years. Modern rate constant measurements are often precise, and individual values are known fairly well. In addition, methods exist for estimating rate constants for the reactions of VOCs with OH and NO_3 radicals which can be used when data are not available. A realistic uncertainty estimate for most VOCs is a factor of 2. Data concerning rate constants for the reactions of the radical intermediates are much more limited and are usually restricted to the simplest cases. It has been assumed that the higher molecular weight radicals react with the same rate constant as their low molecular weight counterparts.

Theoretical calculations provide the data needed to support application of thermochemically-based estimation methods and to evaluate proposed reaction sequences. Estimates of reaction heats and free energies are also used to rule out chemically unreasonable reaction schemes. In recent years the development of theoretical methods for the calculation of potential energy surfaces has led to the direct computation of rate constants. These computational techniques have been tested by comparison with available data for known reactions, and then applied to understand and predict mechanisms and reaction rates. In addition, computational methods provide a useful tool for reducing the time required in the laboratory by suggesting specific product compounds for analysis in chamber experiments. For example, for many combustion systems, detailed kinetic models often consist of up to several thousands of elementary reactions whose kinetic parameters are mainly estimated from those available for similar reactions. Computational kinetics affords one of the simplest and most cost-effective methods for calculating thermal rate constants, by applying the conventional transition state theory (TST), which requires only structural, energetic, and vibrational frequency information for reactants and transition states. On the basis of the TST framework, much progress has been made in developing direct ab-initio methods for calculating rate constants from first principles.

For the last decade the main emphasis of our research in atmospheric chemistry has been to investigate mechanisms and kinetics of many different atmospherically important reactions of selected VOCs and of their reaction products. In particular, we have studied the OH and NO_3 initiated oxidation of several groups of organic compounds (alkanes, alkenes, dienes, aromatics, aldehydes, carboxylic acids, alcohols, ethers, etc.) under tropospheric conditions, as well as water-assisted reactions and heterogeneous processes in the presence of mineral aerosol surfaces.

One of our major findings has been the correct explanation of the anti-Arrhenius behavior that is common to many radical-molecule reactions. Indeed, many of these reactions present negative activation energies, i. e. their rate constant decreases with increasing temperature. Following the suggestion of Singleton and Cvetanovic proposed in 1976 (Singleton & Cvetanovic, 1976) and using quantum chemistry methods, we showed that the existence of a stable Van der Waals pre-reactive complex in the entrance of the reaction channel explains satisfactorily the observed kinetic data. Thus, radical-molecule reactions must be seen as complex reactions consisting of more than one elementary step. This mechanism and the resulting rate constant expression are derived in the next Section.

Heterogeneous processes that involve the impact of aerosols and trace vapors on the VOC atmospheric chemistry is a fairly new field of research, and its importance has steadily grown due to its environmental importance. Laboratory studies, together with field observations and modeling calculations, have clearly demonstrated the importance of heterogeneous processes in the atmosphere. In this sense, it is very important to understand the role of particulate matter and the extent to which heterogeneous reactions on solids as well as multiphase reactions in liquid droplets contribute to the atmospheric chemistry. For example, the potentially reactive surface of mineral aerosols may be a significant sink for many volatile organic compounds in the atmosphere, and consequently it could influence the global photooxidant budget. In addition, the special nature of H_2O as a third body may need to be taken into account.

In this work, we summarize the theoretical methodology employed in our work to study the mechanisms and kinetics of the reaction of a variety of COV with OH radicals, in the gas phase and in the presence of mineral aerosols. Also, the possibility of single-water molecule catalysis of OH reactions with volatile organic compounds is discussed.

2. Theoretical methodology

Ab initio quantum chemical calculations can provide results approaching benchmark accuracy for small molecules in the gas phase (Martin & de Oliveira, 1999) and they have proven to be very useful to complement experimental studies. Small molecules in the gas phase are typically addressed by high-level methods such as CCSD(T), QCISD(T) and MRCI, which in many cases are as accurate as experiments (Friesner, 2005). A wide variety of properties such as: structures (Thomas, 1993); thermochemistry (Guner, 2003); spectroscopic quantities (Stanton & Bartlett (1993); and kinetics (Fernandez-Ramos, 2006) can be effectively computed.

In this section, a summary of the theoretical methodology employed in our work since 1994, and a discussion of the various methods in connection with the kinetics calculations of tropospheric reactions, is presented.

Electronic structure calculations have been performed using MP2 and DFT methods. The reliability of DFT methods to properly describe chemical reactions has been discussed elsewhere (Siegbahn & Blomberg, 1991; Fernandez-Ramos, 2006).

Although the calculations performed in the atmospheric chemistry related to the VOC oxidation cover a list of different quantum chemistry methods that have been used over the years, our latest work has been performed using the M052X density functional method (Zhao, 2006) developed specifically for kinetic calculations by Professor Truhlar and his group. The M05-2X functional has been parametrized to take into account dispersion forces. This functional has previously been tested to model complex reactions, and it has been shown that it provides excellent structures, energies and kinetics results at a reasonable computational cost, thus allowing treatment of large systems (Vega-Rodriguez & Alvarez-Idaboy (2009).

2.1 B//A approach

We have used a procedure that has become common in the study of the stationary points of chemical reactions of polyatomic systems because it is relatively inexpensive from a computational point of view and it usually reproduces correctly the main features of the reaction path. It is known as B//A approach, and it consists of geometry optimizations at a given level (A) followed by single point calculations, without optimization, at a higher level (B). Based on our previous experience, the use of B//A approach at CCSD(T)//BHandHLYP level of theory properly describes the energetic and kinetics features of VOCs + OH hydrogen abstraction reactions. In addition, for this kind of reactions it has been proved that the differences in geometries between several DFT methods compared to CCSD and QCISD are minimal for BHandHLYP (Szori, 2006). However in some of our earlier works other methods were used. More recently in the Truhlar group (University of Minnesota) new DFT methods have been developed that can achieve the required accuracy for kinetic and thermodynamic calculations without the need of refining energies with post Hartree-Fock methods. This is the case of the M05-2X functional that we have adopted for some of our recent studies (Zavala-Oseguera et al., 2009; Galano et al. 2011, Galano & Alvarez-Idaboy, 2009; Pérez-González & Galano 2011; Leon-Carmona & Galano, 2011; Galano 2011; Gao et al., 2010; Iuga, et al. , 2010, Iuga et al. 2011).

In some cases the results obtained with this functional are closer to experimental results than the ones obtained at the CCSD(T)//BHandHLYP level.

2.2 Free radical kinetics

We present here a brief overview on the methodology used to compute the rate constants in the examples that are going to be discussed later. For a chemical reaction that occurs through a stepwise mechanism involving the formation of a reactant complex in the entrance channel, at least two steps must be considered in the kinetics, namely: (1) the formation or the reactant complex from the isolated reactants, and (2) the formation of the products from the reactant complex. For VOCs + OH reactions that mechanism can be written as:

$$VOC + \mathrm{OH}\bullet \underset{k_{-1}}{\overset{k_1}{\rightleftarrows}} [VOC \text{ ---- } \mathrm{OH}]\bullet \xrightarrow{k_2} Products$$

Provided that $k_{-1} + k_2 > k_1$, i.e, the complex rapidly disappears, a steady-state analysis leads to a rate coefficient for each overall reaction channel which can be written as:

$$k = \frac{k_1 k_2}{k_{-1} + k_2} \tag{1}$$

In general the energy barrier for k_{-1} is about the same size as that for k_2, in terms of enthalpy. However, the entropy change is much larger in the reverse reaction than in the formation of the products. The activation entropy ΔS_2 is small and negative because the transition state structure is tighter than the reactant complex, while ΔS_{-1} is large and positive because six vibrational degrees of freedom are converted into three translational plus three rotational degrees of freedom. This leads to k_{-1} values that are much larger than k_2. Based on this assumption, first considered by Singleton and Cvetanovic, the overall rate coefficient (k) can be rewritten as:

$$k = \frac{k_1 k_2}{k_{-1}} = K_{eq} \cdot k_2 \tag{2}$$

where k_2 is the rate constant corresponding to the second step of the mechanism, i.e., transformation of the reactant complex into products; and K_{eq} is the equilibrium constant between the isolated reactants and the reactant complex. Applying basic statistical thermodynamic principles the equilibrium constant (k_1/k_{-1}) of the fast pre-equilibrium between the reactants and the reactant complex may be obtained as:

$$K_{eq} = \frac{Q_{RC}}{Q_R} \exp\left[(E_R - E_{RC}) / RT\right] \tag{3}$$

where Q_{RC} and Q_R represent the partition functions corresponding to the reactant complex and the isolated reactants, respectively.

Let us emphasize that when it is claimed that the reactant complexes are lower in energy than the isolated reactant, this statement is made in terms of enthalpy or ZPE corrected electronic energies. On the other hand, the Gibbs free energy associated with step 1, i.e, the formation of the reactant complex ($G_1 = G_{RC} - G_R$), is always positive in a wide range of temperatures around 300 K. The expression for the equilibrium constant can also be written as:

$$K_{eq} = \frac{k_1}{k_{-1}} = \exp\left(-\Delta G_1 / RT\right) \tag{4}$$

From this expression, it is evident that $K_{eq} < 1$ for endergonic processes ($\Delta G_1 > 0$), i.e. $k_{-1} > k_1$, which validates the steady state hypothesis.

On the other hand k_2 can be calculated within the frame of the Transition State Theory. A typical reaction profile for this kind of reactions, in terms of Gibbs free energy and enthalpy, is shown in Figure1.

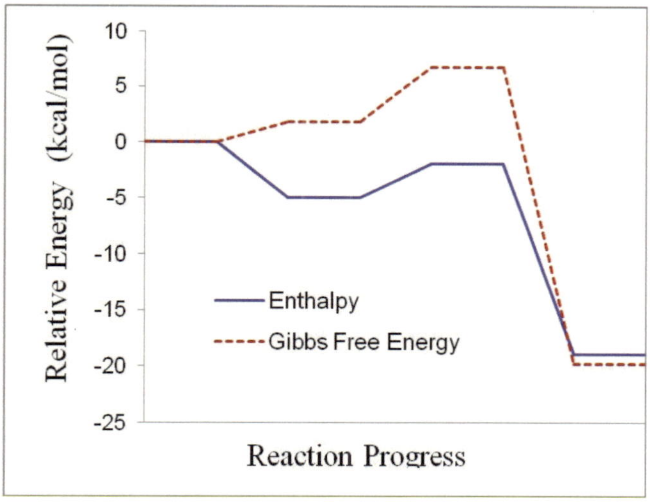

Fig. 1. Typical energy profiles, at 298.15 K, in terms of enthalpies and Gibbs free energies, of an atmospheric reaction with negative activation energy..

2.3 Conventional transition state theory (CTST)

From a phenomenological point of view, numerous experiments have shown that the variation of the rate constant with temperature can be described by the Arrhenius equation:

$$k = A\exp\left(-Ea / RT\right) \tag{5}$$

where Ea is the activation energy and A is the pre-exponential or frequency factor, which may have a weak dependence on temperature. If a reaction obeys the Arrhenius equation, then the Arrhenius plot ($ln\ k$ versus $1/T$) should be a straight line with the slope and the intercept being $-Ea/R$ and A, respectively.

In a unimolecular process, under high-pressure conditions, an equilibrium distribution of reactants is established and the Transition State Theory formula can be applied to calculate k_2:

$$k_2 = \sigma\kappa_2 \frac{k_B T}{h} \frac{Q_{TS}}{Q_{RC}} \exp\left[\left(E_{RC} - E_{TS}\right) / RT\right] \tag{6}$$

where σ is the symmetry factor, which accounts for the number of equivalent reaction paths, κ is the tunneling factor, k_B and h are the Boltzmann and Planck constants, respectively, Q_{TS} is the transition state partition function, and the energy difference includes the ZPE corrections. In this approach the reactant complex is assumed to be in its vibrational ground state.

In a classical treatment the influence of the complex exactly cancels in equation 2 and the overall rate coefficient depends only on the properties of OH, VOCs, and the transition states. However, when there is a possibility of quantum mechanical tunneling, the existence of the complex means that there are extra energy levels from where tunneling may occur so that the tunneling factor increases. Since it has been assumed that a thermal equilibrium distribution of energy levels is maintained, the energy levels from the bottom of the well of the complex up to the barrier might contribute to tunneling.

2.4 On the Basis Set Superposition Error (BSSE) dilemma

One issue of concern when modeling weakly bonded systems is the BSSE, which is caused by the truncation of the basis set. From a theoretical point of view its existence has been well established (Mayer, 1983; Vargas et al., 2000; Mayer, 1996). It is especially important in weakly bound complexes such as Van der Waals and hydrogen-bonded complexes and transition states, because for these systems the BSSE and the binding energies are of the same order of magnitude. The most widely used and simplest way to correct BSSE is the counterpoise procedure (CP) (Jansen & Ros, 1969; Boys & Bernardi, 1979). In CP, for a dimer system formed by two interacting monomers: CP^2, the BSSE is corrected by calculating each monomer with the basis functions of the other one (but without its nuclei or electrons), using so-called "ghost orbitals". However, the use of this method is polemic since several authors have proposed that it overestimates the BSSE (Frisch et al., 1996; Schwenke & Truhlar, 1985; Morokuma & Kitaura, 1981; López et al., 1999; Hunt & Leopold, 2001; Valdés & Sordo, 2002a, 2002b).

It has been pointed out by Dunning (Dunning, 2000) that "It is quite possible and even probable that the binding energies computed without the counterpoise correction are closer to the complete basis set limit than the uncorrected values. This situation is due to the fact that BSSE and basis set convergence error are often of opposite sign." Since it cannot be established *a priori* if that is the case, we have modeled the closest system to our reactant complexes with known experimental binding energy: the water dimer. The experimental value for the electronic dissociation energy (*DE*) of water dimer is 5.4 kcal/mol. We have modeled the water dimer at the CCSD(T)//BHandHLYP/6-311++G(d,p) level and *DE* values of 5.2 and 3.9 kcal/mol are obtained without and with CP^2 corrections, respectively. The CP uncorrected value is closer to the experimental one, suggesting that for the water dimer, at this level of calculation, the BSSE and Basis set truncation errors cancel each other. Since the reactant complexes between oxygenated VOCs and OH radicals are formed in an equivalent chemical way, i.e they are formed trough the interaction of the H atom in the OH radical and an oxygen atom in the VOCs, and as it will be discussed later they also have electronic binding energies that are very similar in magnitude to that of the water dimer, it is reasonable to assume that the same cancellation of errors occurs in our systems.

Recently we have performed a detailed investigation on the role of the cancelation of BSSE and the Basis Set Truncation Error (Alvarez-Idaboy & Galano, 2010). We have found that the inclusion of CP corrections systematically leads to results that differ from the CBS-extrapolated ones to a larger extent than the uncorrected ones. Contrary to the general belief, this effect is more significant if only weak interactions are analyzed. Accordingly, from a practical point of view, when relatively small basis sets are used, we do not recommend the inclusion of such corrections in the calculation of interaction energies, since it may lead to values with larger discrepancies with the accurate ones. We think that the best way of dealing with BSSE, for large-sized systems that make CBS extrapolations computationally unfeasible, is not to use CP corrections, but instead to make a computational effort for increasing the basis set. This approach does not eliminate BSSE but it significantly decreases it and, more importantly, it proportionally decreases all the errors arising from the basis set truncation. This does not necessarily mean that one should use a very large basis set, as it has been interpreted by several authors, but rather only to increase the basis set. For example, instead of using 6-31G(d,p) with counterpoise correction it is better to perform 6-31++G(d,p) calculations without CP. The interaction energies obtained from the latter methodology should be closer to the experimental values than those arising from using the first one.

3. Gas phase reactions

In this section, we will discuss the most important trends in the reactivity of VOCs towards OH free radicals in the gas phase.

In most radical–molecule reactions, a complex two-step reaction mechanism is assumed, which involves the formation of reactant complexes that are in equilibrium with the separated reactants, followed by an irreversible step that leads to the products:

$$\text{Step 1: } VOC + \text{OH}\bullet \underset{k_{-1}}{\overset{k_1}{\rightleftarrows}} [VOC \text{ ---- HO}]$$

$$\text{Step 2: } [VOC \text{ --- HO}]^{\cdot} \xrightarrow{k_2} Products$$

In H-abstraction reactions, the corresponding radical and one water molecule are formed, while in the OH-addition reactions, an adduct is formed. In general, in saturated and oxygenated VOCs oxidations, H-abstraction is the most important mechanism, while the predominant mechanism in unsaturated VOCs oxidation is OH-addition. In general, branching ratios may change as a function of temperature.

3.1 H-abstraction

Hydrogen abstraction of saturated hydrocarbons by the hydroxyl radical is one of the most important classes of reactions in combustion chemistry. This is clear from the more than 70 experiments and theoretical studies that are available in the literature for the smallest reaction in this class, namely, the $OH\cdot + CH_4 = H_2O + \cdot CH_3$ reaction. However, kinetic information for reactions involving larger hydrocarbons, for example, larger than C5, is limited.

In urban air, oxygenated compounds represent about 20% of the emitted VOCs. They are also formed in situ in the atmosphere from the oxidation of other VOCs.

The reactions of OH radicals with oxygenated compounds in the gas phase seem to proceed via formation of complexes that involve one or two hydrogen bonds. Several theoretical studies on these complexes show binding energies large enough to overcome any inaccuracy of the calculation method. In particular, the relevance of including such complex in the modeling of oxygenated VOCs + OH reactions has been discussed in detail (Galano & Alvarez-Idaboy, 2008). Taking them into account, together with any possible intramolecular interaction in the transition states structures, leads to very reliable theoretical values of the

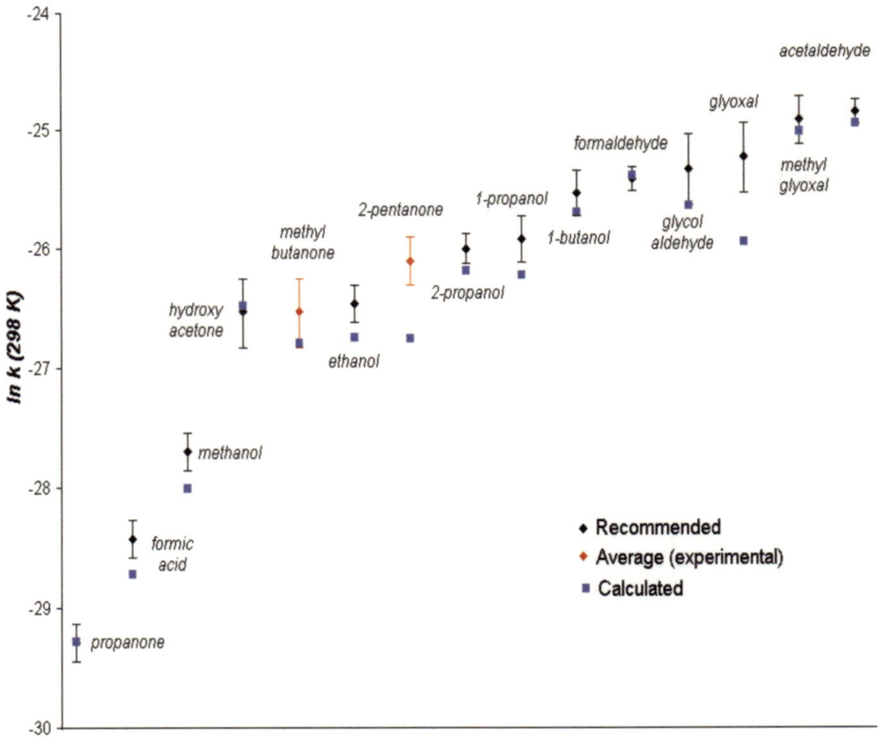

Fig. 2. Comparison between recommended experimental, and calculated rate coefficients, at 298 K.

rate coefficients. To prove this statement, Figure 2 shows a comparison between calculated rate coefficients, at 298 K (Alvarez-Idaboy, 2001; Galano et al. 2002a; Alvarez-Idaboy et al., 2004; Galano et al. 2002b; Galano et al., 2004; Galano et al., 2005; Galano, 2006) for several VOCs + OH reactions and recommended values (Atkinson et al., 1999). For 2-pentanone and methylbutanone there are no recommendations, therefore the average from the experimental values was used instead for 2-pentanone, and the only experimental value available for methylbutanone. As this figure shows, the calculated rate coefficients are in excellent agreement with the available experimental data. The calculated values are within the error range from experimental determinations in most of the studied cases. The largest discrepancies were found for glyoxal and 2-pentanone, which differ from the recommended values in 51 and 48 %, respectively. Even for them the agreement is very good, since an error of 1 kcal/mol in energies represent about one order of error in the rate constant, and in all the cases the discrepancies are smaller than that. Accordingly, the calculated results in Figure 2 can be considered within the error inherent to the most accurate quantum mechanical calculations, or even smaller.

All the main conclusions from these articles are now accepted in recent works (Tyndall et al., 2002) and it is well known that hydrogen abstraction is the main reaction channel if not unique (Butkovskaya et al., 2004).

3.2 OH-addition

OH-addition channel is the dominant pathway in unsaturated VOCs oxidation. In the troposphere, alkenes participate in a sequence of reactions which ultimately lead to their breakdown into highly toxic aldehydes, at the same time altering the equilibrium ratio of nitrogen oxides and indirectly producing ozone.

The mechanism of the reaction has been studied in our group for several alkenes from a theoretical point of view. Initially we obtained the potential energy surface of the propene + OH reaction, both in the presence (Dıaz-Acosta et al., 1999) and in the absence of oxygen, (Alvarez-Idaboy et al., 1998), and the preferred site for addition of OH and O_2. A complex mechanism involving the formation of a reactant complex was postulated to explain the observed negative activation energy. An interesting finding was the elucidation of the branching ratios in the propene–OH reaction. According to the results, the apparent discrepancy between thermochemical data (which favor addition of the OH radical to the central carbon atom of propene) and experimental results obtained indirectly from a product study (which finds that 65% of the final products are consistent with addition of OH to the terminal carbon) can be explained if one also considers the subsequent reaction (OH–propene) + O_2. At atmospheric pressure, addition at the central carbon atom is energetically favored, both in terms of barrier heights and of heats of formation of the products, although differences in energy between the two pathways are quite small (less than 2.5 kcal/mol), implying, in fact, that both channels should occur. However, a crossing occurs between the energy profiles when O_2 is added, favoring, by as much as 4.0 kcal/mol, addition of O_2 to the less stable OH–propene adduct. The branching ratio of both reactions, with the predominance of one isomer over the other, is expected to be both pressure and temperature dependent. In addition, a study of the temperature dependence of the rate constant for the O_2 addition reaction should yield a different slope for reactions to both initially formed adducts.

Later, in a more detailed study that included rate constant calculations (Alvarez-Idaboy et al., 2000), it was possible to explain the origin of the negative activation energy observed in experiments. We concluded that, when the OH + alkene reaction occurs at atmospheric pressure, the following hold:

- The reaction is not elemental.
- The overall addition of OH is irreversible, due to the large thermal effect of reaction that is usually larger than 30 kcal/mol.
- The overall rate depends on the rates of two competitive reactions: the reverse of the first step and the second step, the former one being more affected by temperature than the latter.

We showed that only Singleton and Cvetanovic's[6] hypothesis is compatible with the above three points, i.e., the existence of a relatively stable pre-reactive complex, in equilibrium with the reactants, and from which the addition adduct is formed irreversibly. However, the activation energy of the OH-alkene reactions at high pressures can be calculated as the difference between the energy of the TS and that of the reactants, without having to obtain the pre-reactive complex. This hypothesis and methodology proposed for the first time in OH + alkene reactions was later used to explain the mechanisms and kinetics of many other addition reactions and was the basis for the study of other atmospheric reactions (Francisco-Márquez et al., 2003; Francisco-Márquez, et al., 2004; Francisco-Márquez et al. 2008; Vega-Rodriguez & Alvarez-Idaboy, 2009).

In a series of studies on OH reactions with aromatic hydrocarbons (Uc et al. 2000; Uc et al.,2001; Uc et al. 2002; Uc et al. 2004; Uc et al. 2006; Andino & Vivier-Bunge, 2008) we showed that pathways involving the initial formation of *ipso* adducts might also play a non-negligible role in tropospheric chemistry (Figure 3). In general, only addition to *ortho, meta,*

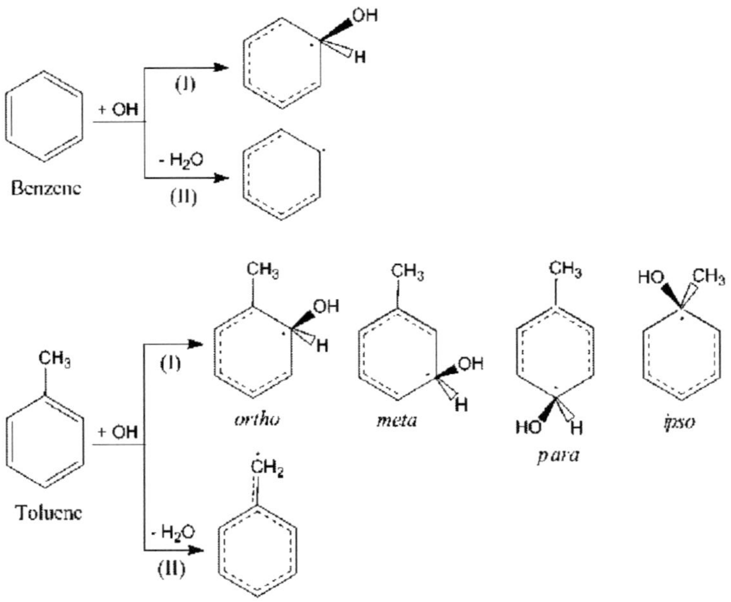

Scheme I

Fig. 3. Possible channels in OH addition to benzene and toluene.

and *para* positions have been considered in laboratory studies for OH/aromatic systems. However, only about 70% of the reacted carbon has been fully accounted for. Trayham (Trayham,1979) described numerous examples of ipso free radical substitution and he emphasized the importance of the ipso position in free radical reactions with aromatic hydrocarbons. Tiecco (Tiecco, 1980) showed that radical additions to ipso sites could be important, or even more important, than addition to the nonsubstituted carbon sites of aromatic compounds. Rather than a steric hindrance, stabilization due to interactions between the lone pair on the oxygen atom and two methyl hydrogen atoms has been described in some of these studies. In the case of toluene reaction with OH, we estimated that ipso addition contributes 13% to the overall reaction. Results have been summarized in (Andino & Vivier-Bunge, 2008).

3.3 Competing mechanisms

In some reactions, competing channels occur, and different mechanisms are favored depending on temperature. This is the case for benzene and toluene, where three distinct regions can be observed in the Arrhenius plots of log k versus $1000/T$ (Finlayson-Pitts & Pitts, 1986). At low temperatures, below 325 K, the plot is linear but with a slope that is negative for benzene and positive for toluene. At temperatures above 380 K, a linear plot is obtained with a negative slope, typical of a normal Arrhenius behavior. Between these two temperatures, a non-exponential decay is observed (Perry et al. 1977; Tully et al., 1981) and the Arrhenius plot presents an abrupt discontinuity. There are no recommended values for the rate constants in this region. The Arrhenius plot for toluene +OH is shown in Figure 5.

The existence of the positive slope in the low-temperature region in the toluene reaction, which implies that the observed activation energy is negative, can be explained in terms of a complex mechanism that has been discussed in Section 2. The non-exponential OH decay in the intermediate temperature region and the unusual Arrhenius plots have been rationalized on the basis of two types of reaction paths, addition and abstraction, occurring simultaneously. It has been postulated that the thermalized OH-aromatic adduct formed by the OH radical addition to the ring decomposes back to reactants (Tully et al. 1981), thus decreasing the importance of the addition channel and leading to bimolecular reaction rate-constant values significantly lower than those measured near room temperature.

Computational kinetics calculations (Uc et al. 2008) were performed, using quantum chemistry data for these reactions under pseudo-first-order conditions, in order to explain the observed Arrhenius plots. We employed a theoretical approach that is in line with both experimental and environmental conditions and that takes into account the fact that aromatic concentrations are in large excess compared to OH concentrations. Thus, pseudo-first-order kinetics was used to describe the addition reactions, and the possibility of the reverse reaction was explicitly introduced. Branching ratios were obtained for all channels.

The study used pseudo-first-order conditions, with aromatic concentrations in large excess compared to OH concentrations as is the case both in previous experiments and in the atmosphere. Our results are in excellent agreement with the experimental data in the whole 200-600 K temperature range. They reproduce the observed non-exponential OH decay. The suggestion that the non-exponential OH decays observed in the 300 K < T < 400 K range are caused by the decomposition of thermalized OH-aromatic adducts back to reactants and the competition between abstraction and addition channels was confirmed. Moreover, we showed that the low-temperature onset of the non-exponential decay depends on the concentration of the aromatic compounds: the lower the concentration, the lower the temperature onset. This finding could have atmospheric implications since, under atmospheric conditions, the non-exponential decay occurs in the 275-325 K range, which corresponds to temperatures of importance in tropospheric chemistry. The temperature dependence of benzene + OH rate constant at different times, for an initial aromatic concentration of $[arom]_0 = 1 \times 10^{13}$ molecule/cm³ is shown in Figure 4.

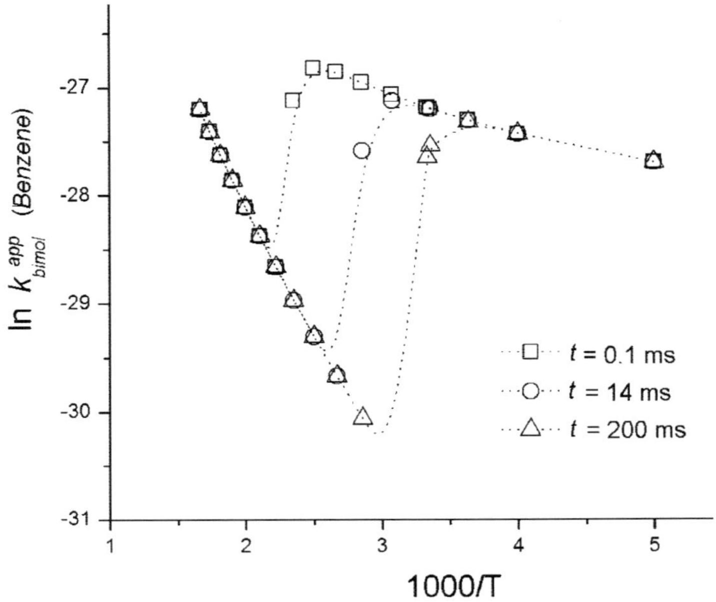

Fig. 4. Temperature dependence of benzene + OH rate constant at t = 0.1, 14 and 200 ms, for pseudo-first-order conditions with $[benzene]_0 = 1 \times 10^{13}$ molecule/cm³

4. Water-assisted gas phase reactions

A 2007 publication in Science (Vöhringer-Martinez et al., 2007) suggested the possible role of a single water molecule in the oxidation reaction of acetaldehyde by a hydroxyl radical. The authors used a wind-tunnel apparatus with detector lasers to study the reaction between gaseous hydroxide radicals and acetaldehyde molecules ($OH + CH_3CHO \longrightarrow H_2O + CH_3O$), in the 60-300 K temperature range. They detected a small but detectable increase in the rate constant at low temperatures. Accompanying theoretical calculations showed that the complex formed between an acetaldehyde molecule and a single water molecule is more vulnerable to OH radical attack than free acetaldehyde. Indeed, the energy barrier for the complex + OH reaction is considerably smaller than for acetaldehyde + OH. A comment by I. Smith in the Perspectives Section in the same issue of Science (Smith, 2007) acknowledged the interest of this work, but questioned its relevance to atmospheric chemistry in terms of the negligibly small amount of existent water-acetaldehyde complex.

In fact, the kinetic effect of the presence of water had already been studied (Canneaux et al., 2004) in a theoretical study of the reaction of acetone with OH radicals in the presence of a single water molecule. Their conclusion was that the tropospheric concentration of OH-acetone-water complexes would be very small, but that they might still be detected experimentally. Since then, several articles have been published suggesting possible one-molecule catalysis in reactions of OH radicals with other organic volatile compounds (Chen et al., 2008; Buszek & Francisco, 2009). All of them involve oxygenated VOC's, since, in these cases, the OH reaction transition state barriers may be considerably decreased by hydrogen bonds involving a water molecule.

There has also been considerable speculation about the role of water complexes and other hydrogen-bonded molecular complexes on the kinetics and dynamics of gas-phase free radical reactions (Luo et al. 2009; Long et al. 2010). A field measurement in the near-infrared has revealed a relatively high concentration of 6 x 10^{14} molecule cm^{-3} of water dimer in the atmosphere (Pfeilsticker et al. 2003). This finding has been reproduced very well by a theoretical study (Dunn et al., 2004). However, although detectable, the water dimer concentration represents only about 1% of the total water present in air in tropospheric conditions. The concentration of VOC-water complexes in air is expected to be even smaller in most cases (Galano et al. 2010), and the catalytic effect of water is therefore questionable.

In a series of articles on the reactions of acetaldehyde (Iuga et al., 2010a), glyoxal (Iuga et al., 2010b), acetone (Iuga et al., 2011a) and formic acid (Iuga et al., 2011b) with OH radicals in the presence of water, we showed that consideration of the initial water complexation step is essential in the rate constant calculation. Even if the VOC-H_2O complex + OH rate constant is often orders of magnitude larger than the VOC + OH rate constant, if very little complex is formed, the resulting rate constant will not be significantly modified.

Since the simultaneous collision of the three molecules involved (VOC + H_2O + OH) is very improbable, the termolecular mechanism is ruled out. Hence, the most probable mechanism consists of two consecutive bimolecular elementary steps, presumably complexation of the VOC with water, followed by reaction of the complex with an OH radical. Reaction with OH, in turn, has been shown to involve the initial formation of a pre-reactive complex, as discussed in Section 2. Thus, the mechanism can be written as:

$$\text{Step 0: } VOC + \mathrm{H_2O} \underset{k_{-0}}{\overset{k_e}{\rightleftharpoons}} [VOC \text{ ---- } \mathrm{H_2O}]$$

$$\text{Step 1: } [VOC \text{ ---- } \mathrm{H_2O}] + \mathrm{OH}\bullet \underset{k_{-1}}{\overset{k_1}{\rightleftharpoons}} [VOC \text{ ---- } \mathrm{H_2O} \text{ --- } \mathrm{HO}]^{\cdot}$$

$$\text{Step 2: } [VOC \text{---- } \mathrm{H_2O} \text{ --- } \mathrm{HO}]^{\cdot} \xrightarrow{k_2} Products$$

The most stable water complex structures formed in Step 0 with acetaldehyde, glyoxal, acetone and formic acid are shown in Figure 5, and the percent of complex formed at 298K is indicated under each structure.

When these complexes react with OH radicals, in the entrance reaction channel a termolecular pre-reactive complex is formed, which presents multiple H-bonds. In all cases the reaction path depends on the initial molecular arrangement, since various combinations are possible. These termolecular complexes are in general very stable because they involve

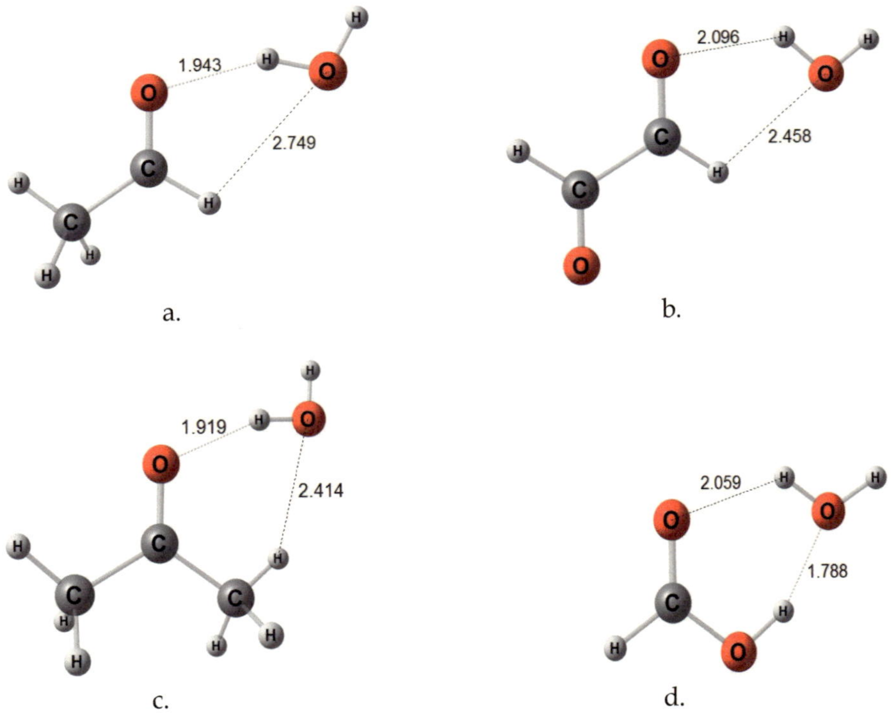

Fig. 5. Water complexes and percent of complex formed at 298K, with acetaldehyde, glyoxal, acetone and formic acid.

strong H bonds between water, a polar molecule and the free radical. Some of these complexes are shown in Fig. 6 and relevant distances are given in Å.

The concentration of atmospheric water, although variable, is always in large excess with respect to both OH and COV's. This, together with the fact that, in the above mechanism, water is a catalyst and therefore regenerates, allows us to use fixed water concentrations to obtain pseudo second order rate constants that depend parametrically on water concentration. Thus, the rate constant should be calculated as:

$$k = \sigma\, K_{eq0}\, K_{eq1}\, k_2\, \kappa_2\, [water]$$

where K_{eq0} and K_{eq1} are the equilibrium constants for Steps 0 and 1 respectively, k_2 and □2 are the unimolecular rate constant and the tunneling factor for Step 2, and□□ is the reaction symmetry number.

An alternative (but exactly equivalent) way to interpret the pseudo second order rate constant is to multiply the second order rate constant of the water-VOC complex + OH reaction by the molar fraction (X) of VOC that is complexed: this fraction is simply K_{eq0} [water]. The latter approach has the advantage of allowing for a direct comparison between rate constants obtained with the present methodology and similar calculations in which Step 0 was ignored. In this approach, the rate constant is then simply obtained as:

$$k(\text{overall}) = (1 - X)\, k(\text{VOC}) + X\, k(\text{VOC-water complex})$$

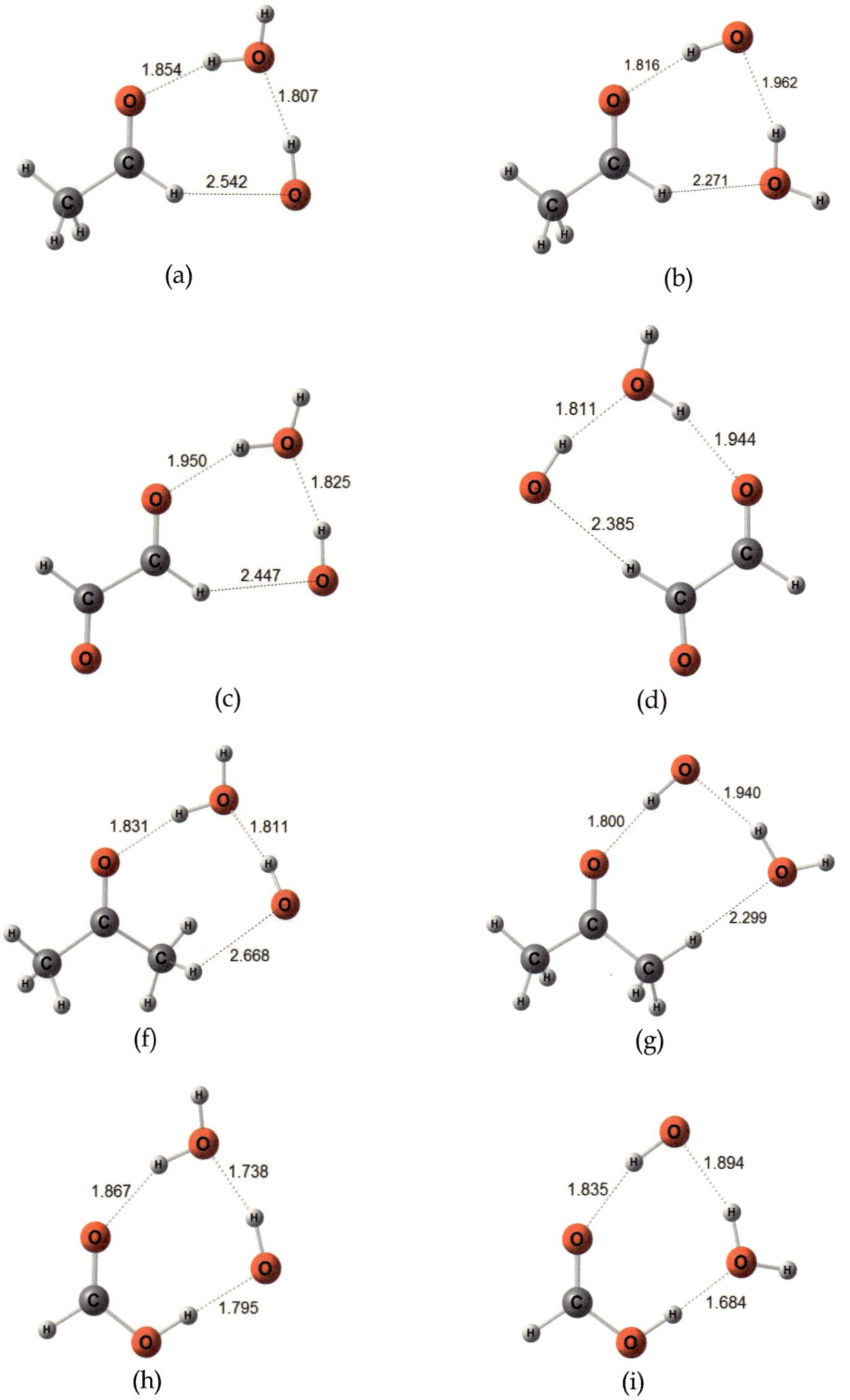

Fig. 6. Termolecular pre-reactive complexes for acetaldehyde, glyoxal, acetone and formic acid.

In Table 1, we have collected the results for acetaldehyde, glyoxal, acetone and formic acid, at 298K. The calculated percent amount of complexed VOC is reported in the second column. Rate constants for the VOC + OH reaction in the absence of water are given in the third column, those for the complex are shown in the fourth column, and the overall result is reported in the last column.

VOC	% complex	k(VOC)	k(VOC-water) complex	k(overall)
acetaldehyde	0.01	1.58×10^{-11}	1.32×10^{-9}	1.58×10^{-11}
glyoxal	0.01	1.95×10^{-11}	4.25×10^{-10}	1.95×10^{-11}
acetone	0.02	1.04×10^{-13}	3.29×10^{-12}	1.04×10^{-13}
formic acid	12.0	4.39×10^{-13}	1.48×10^{-13}	4.04×10^{-13}

Table 1. Calculated percent amount of complexed VOC and rate constants at 298K.

It is interesting to discuss the specific reaction mechanism of each of the studied VOC individually, both in the presence and in the absence of water.

Reaction of acetaldehyde with OH occurs mainly by abstraction of the formyl hydrogen atom. Although the rate constant for the complex is several orders of magnitude larger than in the absence of water, the amount of complex formed is so small that the overall rate constant is practically unaffected even in a troposphere with 100% humidity (water concentration of 7.95×10^{17} molecules cm^{-3}, at 298.15 K). The variation with temperature was also analyzed. Slight changes are noticeable only below 220 K, i. e. at temperatures that are unattainable under usual atmospheric conditions. The experiment of Vöhringer-Martinez et al. was performed under severe laboratory conditions, and a large acceleration was observed at very low temperatures (50 K).

The reaction of free glyoxal with OH occurs by H-abstraction. When a water complex is formed, one of the formyl hydrogen atoms binds to water while the other one is still available for abstraction. Reaction of the glyoxal-water complex with OH is about 27 times faster than the corresponding reaction for free glyoxal + OH. However, the amount of water complex formed is less than 0.01%, and a reaction rate enhancement by single-water-molecule catalysis does not occur.

Reaction of both free and water-complexed acetone with OH occurs by hydrogen abstraction from a methyl group. Again, the amount of complex formed is only about 0.02 % at 298 K, and although the step 2 barrier is considerably lowered in the presence of a water molecule, no observable rate increase is expected.

The case of formic acid with OH is different than the preceding ones in the sense that the amount of water complex formed is small but relevant under atmospheric conditions (about 12% at 298 K), and it could in principle be large enough to produce a measurable increase in the overall rate constant. However, in the presence of water, the reaction mechanism is different from the one in a dry atmosphere. While in the water-free case the favored path is a PCET carboxylic hydrogen abstraction, in the water-assisted reaction the carboxylic hydrogen is strongly involved in the formic acid-water complex and it is therefore less available for reaction with an OH radical. Free energy values point to a preference for a formyl hydrogen abstraction channel, which is considerably slower. Thus, once more, no single-water-molecule catalysis is observed.

In conclusion, it is safe to state that one water molecule *does not* accelerate the reaction of OH radicals with most volatile organic compounds under atmospheric conditions. The apparent rate coefficient taking into account atmospheric water concentration is considerably smaller than the one in the absence of water, independently of the method of calculation. The apparent disagreement between the present results and the experimentally observed rate coefficients in Vöhringer-Martinez et al. could be due to a possible catalytic effect of water molecules associations or because the experimental conditions do not match the atmospheric ones. Such reactions may still occur in molecular complexes consisting of several water molecules. These molecular complexes are known to be abundant prior to atmospheric aerosol nucleation events (Zhang et al., 2002), and have been shown to catalyze organic (including glyoxal) reactions in nanoparticles (Zhao et al., 2009; Wang et al., 2010). The study of these very interesting reactions is, however, outside of the scope of the present work.

5. Adsorption and reactivity of vocs on a mineral aerosol surface

Clay particles are present in large quantities in mineral dust in atmospheric aerosols and their interaction with organic species may, in principle, influence atmospheric reactions and the prediction of environmental risks. Thus, it is interesting to study the kinetics and mechanisms of adsorption and reaction of atmospheric volatile organic compounds bound to a model clay surface in order to understand and describe, at the molecular level, its effect on the reaction.

In this section, mechanistic and kinetic data for OH-initiated oxidation reactions of VOC's adsorbed on mineral aerosols models are discussed.

The rigid tetrahedron SiO_4 is the building block of all siliceous materials, from zeolites to quartz and amorphous silica. Clay minerals, or phyllosilicates, are formed by sheets of SiO_4 tetrahedrons joined to a sheet of Al oxide octahedrons (Figure 1). The ideal surface of a phyllosilicate is characterized by the presence of a large number of siloxane Si-O-Si bridges, forming hexagonal rings. However, a natural clay surface presents many structural defects and fractures, and its chemical properties are largely due to the presence of active sites on the surface. These are mainly acid sites (Brönsted sites, associated to aluminol and silanol groups) and Lewis sites (such as in four-coordinated Al phyllosilicates) that have large specific surfaces and catalytic properties. Therefore, their presence in aerosols can be expected to play an important role in the heterogeneous chemistry of the troposphere.

Within the quantum mechanical methodology, small clusters of silicate groups are used to model silicate surfaces (Sauer et al., 1994). The basic premise behind the cluster models is that reactions and adsorption are local phenomena that are primarily affected by the nearby surface structure. The active site is described explicitly by the interactions between the local molecular orbitals of the adsorbate and the adsorbent. Clearly, this methodology has the disadvantage that the electronic system is represented only partially, due to the small size and the discrete nature of the cluster employed. However, it has been shown to be adequate to represent a mineral aerosol surface and to allow the use of high accuracy quantum methods to describe the energetics of adsorption, reaction paths, and intermediates formation.

On the natural clay surface, silanol groups are the most reactive sites for adsorption of organic molecules and for their reactivity. Orthosilicic acid $Si(OH)_4$ has been validated as a good model for both isolated and geminal silanol hydroxyl groups and we have used it to mimic the OH reaction with formaldehyde and formic acid on a silicate surface (Iuga et al., 2008a). Vicinal silanol groups may be studied by means of the $(HO)_3Si\text{-}O\text{-}Si(OH)_3$ dimer. In this model, substitution of a Si cation by a tetrahedral Al is useful to represent a wide range of silicates with Lewis or Brönsted acid properties. The silicate hexagonal six-member ring characteristic of a clay surface may be studied by means of a cyclic hexamer model, with and without an OH surface group. In Figure 7 we show the silicate models employed.

The adsorption energy is defined as the difference between the total electronic energy of the surface-adsorbate complex and the sum of those of the isolated molecule and the model surface, including ZPE corrections:

$$\Delta E_{\text{adsorption}} = E_{\text{adsorption complex}} - (E_{\text{molecule}} + E_{\text{surface}}) + \Delta(\text{ZPE})$$

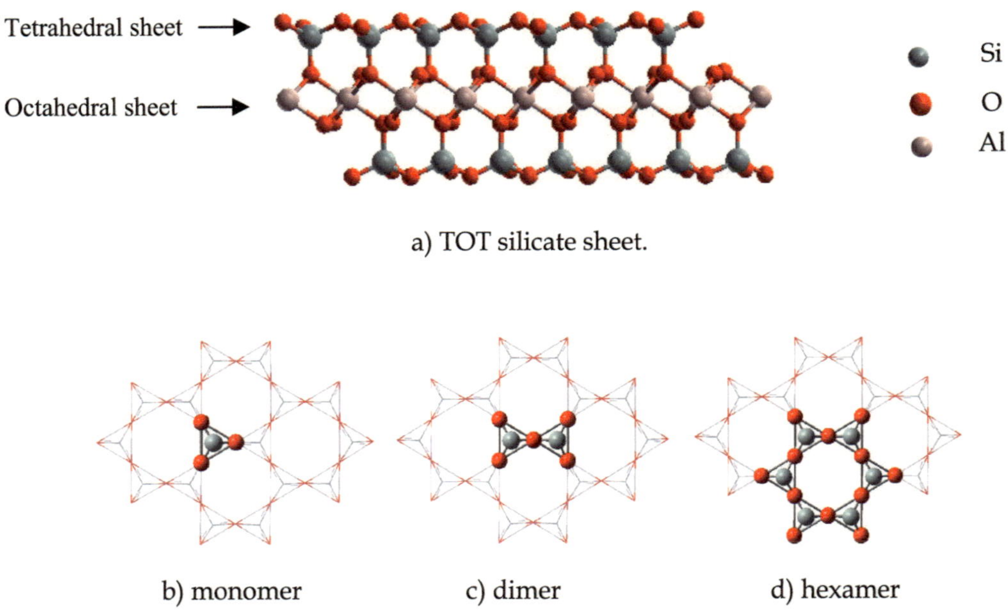

Fig. 7. Silicate cluster models.

The most stable adsorption complex in the reaction of OH· radicals with formaldehyde adsorbed on an $Si(OH)_4$ monomer is shown in Fig. 8. In this complex, the formaldehyde molecule is clearly oriented perpendicularly to the surface and it attaches to the model surface by two hydrogen bonds. Surface hydroxyls play the role of a weak hydrogen donor and the adsorbed molecule assumes the role of the base.

The vibrational properties of silanol groups can be easily studied by infrared spectroscopy. On highly dehydrated surfaces, a single, well-defined band due to the OH stretch is measured at about 3742 cm^{-1} on an otherwise featureless spectral region extending from 3730 to 2000 cm^{-1}. It is then easy to study the perturbations of the OH stretching frequency upon adsorption of molecules from the gas phase. As an example, the formaldehyde

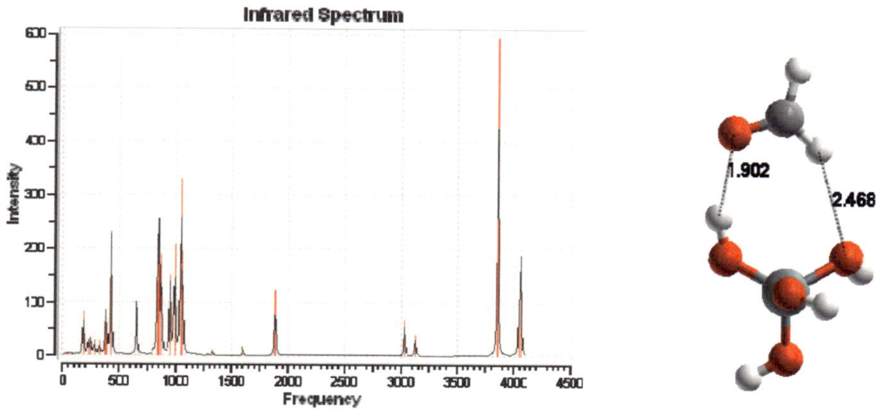

υ_{OH} $Si(OH)_4$	υ_{OH} Complex	Δ υ_{OH}
4062 cm^{-1}	3858 cm^{-1}	204 cm^{-1}

Fig. 8. Adsorption complex calculated infrared spectrum.

adsorption complex calculated infrared spectrum is shown in Fig. 8. The observed shift in the frequency of the O-H stretching involved in the main interaction is 204 cm^{-1}.

In the main reaction channel, formaldehyde reacts with OH radicals by hydrogen abstraction, to form a water molecule and a bound formyl radical. We showed that the rate constant for the H-abstraction reaction is an order of magnitude smaller when formaldehyde is bound to $Si(OH)_4$ than in the gas phase. Thus, on the basis of the calculated rate constants one can conclude that, when the OH-formaldehyde reaction occurs in the presence of dust, at atmospheric pressure, it is *slower*, by a factor of about ten, than the reaction in the gas phase. If one takes into account the fact that silicate aerosols are known to trap some of the OH radicals in the troposphere, the resulting decrease in the formaldehyde reaction rate with OH could be significant.

On a perfect silicate surface with no defects, there are no OH groups and adsorption is much weaker. This case can be modelled with a simple hexamer with no OH groups on the surface, as shown in Figure 9.

Results obtained for the formaldehyde + OH reaction using larger cluster models (Iuga et al., 2008b) with OH groups on the surface yield results that are equivalent to those of the $Si(OH)_4$ model and suggest that, at least for small molecules, the monomer model is sufficient to qualitatively predict the main effects of mineral aerosols.

Adsorption of larger (C2-C5) aliphatic aldehydes on silicate clusters active sites, and their subsequent reaction with OH radicals was also studied (Iuga et al. 2010). Different adsorption complexes were found, and it is clear that the adsorption complex structure determines the subsequent path of its reaction with OH. The initial step in the OH reaction with acetaldehyde occurs according to the same mechanism as in the gas phase, i.e., the aldehydic hydrogen abstraction. Starting from propanal, another abstraction channel becomes increasingly important, involving mainly the abstraction of a hydrogen of the aliphatic chain. In the presence of a silica monomer model surface, the reaction rate of acetaldehyde with OH is found to be about one seventh its value in the gas phase. Because

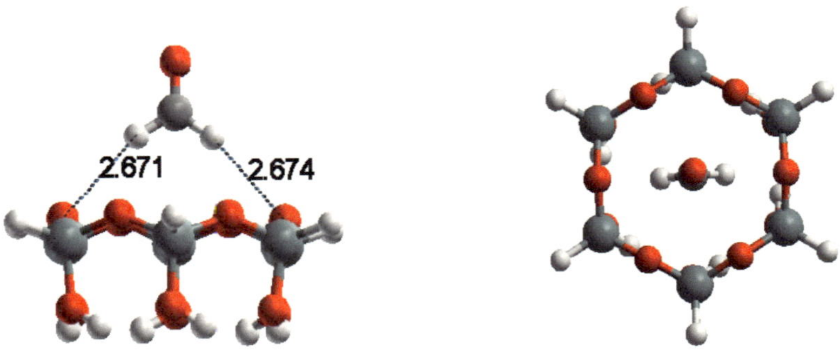

E_{ADS} = 3.05 kcal/mol

Fig. 9. Formaldehyde adsorbed complex.

of the abundance of this contaminant, our results may have implications in tropospheric chemistry. For larger aldehydes, the rate constant is also consistently smaller than in the gas phase. Results are summarized in Table 2.

VOC	*Alkyl H-abstraction*		*Aldehydic H-abstraction*		$k_{overalll}$ (x 10^{-11})	$k_{experimental}$ (x 10^{-11})
	k_{eff} (x 10^{-11})	%	k_{eff} (x 10^{-11})	%		
Acetaldehyde	0.0068	3.00	0.22	97.00	0.23	1.50, 1.50, 1.44
Propanal	0.30	42.86	0.40	57.14	0.70	1.90, 2.06, 1.99
Butanal	1.37	65.87	0.71	34.13	2.08	2.88, 2.38, 2.38
Pentanal	0.38	35.18	0.70	64.82	1.08	2.48, 2.61, 2.76

NIST data base (http://kinetics.nist.gov/kinetics/index.jsp).

Table 2. Calculated total rate constants (in cm^3 $molecule^{-1}$ s^{-1}) and branching ratios at 298 K, for OH H-abstraction reactions of C2-C5 aldehydes on the $Si(OH)_4$ model. Experimental values correspond to the gas phase reaction have been taken from the NIST data base.

Adsorption of other contaminants, such as carboxylic acids and several polyaromatic heterocycles have also been studied on silicate model clusters. In general, these compounds use their most reactive groups to add on the silanol groups, and consequently their OH reactivity is smaller than in the gas phase.

6. Conclusions

In this article, we have reviewed our work on the mechanisms and kinetics of selected VOCs towards OH free radicals.

Quantum chemistry and computational kinetics methods have been used to model mechanisms and kinetics of the reactions of OH radicals with several groups of organic compounds (alkanes, alkenes, dienes, aromatics, aldehydes, carboxylic acids, alcohols, ethers, etc.) under tropospheric conditions. We have calculated reaction profiles, rate constants and branching ratios for numerous volatile organic compounds with OH and other radicals. Insight into reactivity trends, both in the gas phase and in the presence of mineral aerosol models has been obtained. The relative site reactivity of the studied compounds towards OH radicals has been shown to be strongly influenced by intramolecular hydrogen-bond-like interactions that arise in the transition states. The usefulness of quantum chemical calculations to elucidate the detailed mechanisms of OH radical reactions with oxygenated VOCs has been proven. We show that the theoretical methodology employed provides accurate kinetic data that reproduce well the available experimental results and provide new data for a large number of tropospheric reactions. Although the calculations performed in the references cover a list of different quantum chemistry methods that have been used over the years, our latest work has been performed using the M05-2X density functional method developed specifically for kinetic studies by the Truhlar group.

The correct explanation of the anti-Arrhenius behavior common to many radical-molecule reactions has been elucidated. The importance of including reactant complexes in the modeling in order to obtain accurate values of the rate coefficients, has been shown. The best results are those obtained when it is assumed that such complexes are in their vibrational ground state.

The possibility of one-water-molecule catalysis in OH reactions with volatile organic compounds is discussed, and it is shown that it does not occur.

Since the main daytime tropospheric sink of oxygenated VOCs is their reactions with OH radicals, the mechanistic and kinetic information discussed in this work is essential in order to fully understand their tropospheric chemistry as well as their subsequent fate. Hopefully, the large amount of experimental and theoretical work that has been revisited here, which has been devoted to chemical reactions of environmental significance, could contribute in some extent to act in the right direction and prevent more damage to the atmosphere.

7. Acknowledgment

This work is a result of the FONCICYT Mexico-EU 'RMAYS' network, Project N° 94666. We gratefully acknowledge the Laboratorio de Visualización y Cómputo Paralelo at Universidad Autónoma Metropolitana-Iztapalapa and the Dirección General de Cómputo y de Tecnologías de Información y Comunicación (DGCTIC) at Universidad Nacional Autónoma de México for computer time.

8. References

Alvarez-Idaboy J. R. & Galano A. *Counterpoise corrected interaction energies are not systematically better than uncorrected ones: comparison with CCSD(T)CBS extrapolated valuesTheor Chem Acc* 2010, *126*, 75.

Alvarez-Idaboy, J. R., Cruz-Torres, A., Galano, A. & Ruiz-Santoyo M. E. *Structure-Reactivity Relationship in Ketones + OH Reactions: A Quantum Mechanical and TST Approach J. Phys. Chem. A*, 2004, *108*, 2740.

Alvarez-Idaboy, J. R., Diaz-Acosta, I. & Vivier-Bunge, A. *Energetics of the mechanism of the OH-propene reaction at low pressures in an inert atmosphere*, J. Comput. Chem. 1998, 88, 811.

Alvarez-Idaboy, J. R.; Mora-Diez, N.; & Vivier-Bunge, A. *A Quantum Chemical and Classical Transition State Theory Explanation of Negative Activation Energies in OH Addition To Substituted Ethenes* J. Am. Chem. Soc. 2000, *122*, 3715.

Andino J. & Vivier-Bunge, A. *Tropospheric Chemistry of Aromatic Compounds Emitted from Anthropogenic Sources,* in Advances in Quantum Chemistry Special Issue: Applications of Quantum Chemistry to Atmospheric Science, Vol. 55, edited by M.E. Goodsite and M.S. Johnson, Elsevier, (2008), p. 297-310.

Atkinson, R., Baulch, D. L., Cox, R. A., Hampson, R. F. Jr., Kerr, J. A., Rossi, M. J. & Troe, J. *Evaluated Kinetic and Photochemical Data for Atmospheric Chemistry, Organic Species: Supplement VII, J. Phys. Chem. Ref. Data* 1999, *28*, 191.

Boys, S. F.& Bernardi F. *The calculation of small molecular interactions by the differences of separate total energies. Some procedures with reduced error,* Mol. Phys. 1970, *19*, 553; Buszek, R. J. & Francisco, J. S. *The Gas-Phase Decomposition of CF3OH with Water: A Radical-Catalyzed Mechanism, J. Phys. Chem. A*, 2009, *113*, 5333.

Canneaux, S., Sokolowski-Gomez, N., Bohr, F. & Dobe, S. *Theoretical study of the reaction OH + acetone: a possible kinetic effect of the presence of water? Phys. Chem. Chem. Phys.* 2004, 6, 5172.

Chen H. T.; Chang J. G.; Chen H. L. *A Computational Study on the Decomposition of Formic Acid Catalyzed by (H2O)x, x)0-3: Comparison of the Gas-Phase and Aqueous-Phase Results, J. Phys. Chem. A* 2008, 112, 8093.

Díaz-Acosta, I., Alvarez-Idaboy, J. R. &Vivier-Bunge, A. *Mechanism of the OH-propene-O_2 reaction: An ab-initio study,* Int. J. Chem. Kinet. 1999, *31*, 29.

Dunn, M. E., Pokon E. K. & Shields, G. C. *Thermodynamics of Forming Water Clusters at Various Temperatures and Pressures by Gaussian-2, Gaussian-3, Complete Basis Set-QB3, and Complete Basis Set-APNO Model Chemistries; Implications for Atmospheric Chemistry, J. Am. Chem. Soc.* 2004, *126*, 2647.

Dunning, T. H. Jr. *A Road Map for the Calculation of Molecular Binding Energies, J. Phys. Chem. A* 2000, *104*, 9062.

Fernandez-Ramos, A., Miller A., Klippenstein, S. J. & Truhlar, D. G. *Modeling the Kinetics of Bimolecular Reactions, Chem. Rev.* 2006, *106*, 4518, and references therein.

Finlayson-Pitts, B. J. & Pitts, N. *Atmospheric Chemistry: Fundamentals and Experimental Techniques*; Wiley-Interscience, New York, 1986.

Francisco-Márquez, M.; Alvarez-Idaboy, J. R.; Galano, A.; Vivier-Bunge, A. *Theoretical study of the initial reaction between OH and isoprene in tropospheric conditions,* Phys. Chem. Chem. Phys. 2003, *5*, 1392.

Francisco-Márquez, M.; Alvarez-Idaboy, J. R.; Galano, A.; Vivier-Bunge, A. *On the role of cis conformers in the reaction of dienes with hydroxyl radicals,* Phys. Chem. Chem. Phys. 2004, *6*, 2237.

Francisco-Márquez, M.; Alvarez-Idaboy, J. R.; Galano, A.; Vivier-Bunge, A. *Theoretical study of the tropospheric reaction of dienes with mercapto radicals,* Chem. Phys. 2008, *344*, 3, 273.

Friesner,R. A., *Ab initio quantum chemistry: Methodology and applications, PNAS* 2005, *102*, 6648.

Frisch, M. J., Del Bene, J. E., Binkley, J. S. & H. F. Schaefer III, *Extensive Theoretical Studies of the Hydrogen-Bonded Complexes $(H_20)_2$, $(H_20)_2H^+$, $(HF)_2$, $(HF)_2H^+$, F_2H^- and $(NH_3)_2$, J. Chem. Phys.* 1986, *84*, 2279.

Galano A., & Alvarez-Idaboy Atmospheric Reactions of Oxygenated Volatile Organic Compounds + OH Radicals: Role of Hydrogen-Bonded Intermediates and Transition StatesJ. R. in Advances in Quantum Chemistry Special Issue: Applications of Quantum Chemistry to Atmospheric Science, Vol. 55, edited by M.E. Goodsite and M.S. Johnson, Elsevier, (2008), p. 245.

Galano, A. Alvarez-Idaboy, J. R., Ruiz-Santoyo, M. E. & Vivier-Bunge, A. *Glycolaldehyde + OH gas phase reaction: a quantum chemistry + CVT/SCT approach* J. Phys. Chem. A. 2005, *109*, 169.

Galano, A. *On the direct scavenging activity of melatonin towards hydroxyl and a series of peroxyl radicals Phys. Chem. Chem. Phys.* 2011, *13*, 7147

Galano, A. Theoretical Study on the Reaction of Tropospheric Interest: Hydroxyacetone + OH. Mechanism and Kinetics *J. Phys. Chem. A.* 2006, *110*, 9153.

Galano, A., Alvarez-Idaboy J. R., Bravo-Perez, G. & Ruiz-Santoyo, M. E. *Gas phase reactions of C1–C4 alcohols with the OH radical: A quantum mechanical approach, Phys. Chem. Chem. Phys.* 2002, *4*, 4648

Galano, A., Alvarez-Idaboy J. R., Ruiz-Santoyo, M. E. & Vivier-Bunge, A. *Mechanism and kinetics of the reaction of OH radicals with glyoxal and methylglyoxal: a quantum chemistry + CVT/SCT approach*, ChemPhysChem. 2004, *5*, 1379.

Galano, A., Alvarez-Idaboy, J. R., Ruiz-Santoyo, M. E. & Vivier-Bunge, A. *Rate Coefficient and Mechanism of the Gas Phase OH Hydrogen Abstraction Reaction from Formic Acid: A Quantum Mechanical Approach, J. Phys. Chem. A*, 2002, *106*, 9520.

Galano, A., Narciso-López, M. & Francisco-Márquez, M. Water *Complexes of Important Air Pollutants: Geometries, Complexation Energies, Concentrations, Infrared Spectra, and Intrinsic Reactivity J. Phys. Chem. A*, 2010, *114*, 5796.

Galano, A.; Alvarez-Idaboy, J. R. *Guanosine + OH Radical Reaction in Aqueous Solution: A Reinterpretation of the UV-vis Data Based on Thermodynamic and Kinetic Calculations Org. Lett.* 2009, *11*, 5114

Galano, A.; Francisco-Marquez, M.; Alvarez-Idaboy, J. R. *Mechanism and kinetics studies on the antioxidant activity of sinapinic acid Phys. Chem. Chem. Phys.* 2011, *13*, 11199.

Gao, T.; Andino, J. M.; Alvarez-Idaboy, J. R. *Computational and experimental study of the interactions between ionic liquids and volatile organic compounds Phys. Chem. Chem. Phys.* 2010, *12*, 9830.

Gauss, J. & Stanton, J. F. *Perturbative treatment of triple excitations in coupled cluster calculations of nuclear magnetic shielding constants, J. Chem. Phys.* 1996, *104*, 2574.

Guner, V., Khuong, K. S., Leach, A. G., Lee, P. S., Bartberger, M. D. & Houk, K. N. *A Standard Set of Pericyclic Reactions of Hydrocarbons for the Benchmarking of Computational Methods: The Performance of ab Initio, Density Functional, CASSCF, CASPT2, and CBS-QB3 Methods for the Prediction of Activation Barriers, Reaction, Energetics, and Transition State Geometries, J. Phys. Chem. A* 2003, *107*, 11445.

Halasz, G., Vibok, A ., Valiron, P. & Mayer, I. *BSSE-Free SCF Algorithm for Treating Several Weakly Interacting Systems*

Hunt, S. W. & Leopold, K. R. *Molecular and Electronic Structure of C5H5N-SO3: Correlation of Ground State Physical. Properties with Orbital Energy Gaps in Partially Bound Lewis Acid-Base Complexes, J. Phys. Chem. A* 2001, *105*, 5498.

Iuga, C., Alvarez-Idaboy, J. R. & Vivier-Bunge, A. *Can a Single Water Molecule Really Catalyze the Acetaldehyde + OH Reaction in Tropospheric Conditions?*, J. Phys. Chem. Lett. 2010, *1*, 3112.

Iuga, C., Alvarez-Idaboy, J. R. & Vivier-Bunge, A. *Mechanism and Kinetics of the Water-assisted Formic Acid+OH Reaction under Tropospheric Conditions*, J. Phys. Chem. A 2011, *115*, 5138.

Iuga, C., Alvarez-Idaboy, J. R. & Vivier-Bunge, A. *On the Possible Catalytic Role of a Single Water Molecule in the Acetone + OH Gas Phase Reaction: A Theoretical Pseudo Second-order Kinetics Study*, Theoretical Chemistry Accounts 2011, *129*, 209.

Iuga, C., Alvarez-Idaboy, J. R. & Vivier-Bunge, A. *Single Water-Molecule Catalysis in the Glyoxal + OH Reaction in Tropospheric Conditions: fact or fiction? A Quantum Chemistry and Pseudo-Second Order Computational Kinetic Study, Chemical Physics Letters.* 2010, *501*, 11.

Iuga, C., Esquivel Olea, R. & Vivier-Bunge A., *Mechanism and Kinetics of the OH· Radical Reaction with Formaldehyde Bound to an $Si(OH)_4$ Monomer*, J. Mex. Chem. Soc. 2008a *51*, 36.

Iuga, C., Sainz-Díaz, C. I. & Vivier-Bunge, A. *On the OH Initiated Oxidation of C2-C5 Aliphatic Aldehydes in the Presence of Mineral Aerosols,* Geochimica et Cosmochimica Acta, 2010b *74*, 12, 3587.

Iuga, C., Vivier-Bunge A., Hernández-Laguna, A. & Sainz-Díaz, C.I. *Quantum Chemistry and Computational Kinetics of the Reaction between OH Radicals and Formaldehyde Adsorbed on Small Silica Aerosol Models*, J. Phys. Chem. C 2008b, *112*, 4590.

Jansen, H. B. & Ros, P. *Non-Empirical Molecular Orbital Calculations on the Protonation of Carbon Monoxide, Chem. Phys. Lett.* 1969, *3*, 140.

Leon-Carmona, J. R.; Galano, A. Is Caffeine a Good Scavenger of Oxygenated Free Radicals? *J. Phys. Chem. B* 2011, *115*, 4538.

Long, B., Tan, X. F., Ren, D. S. & Zhang, W. J. *Theoretical study on the water-catalyzed reaction of glyoxal with OH radical, J. Mol. Struct*, 2010, 956, 30.

López, J. C., Alonso, J. L., Lorenzo, F. J., Rayon, V. M. & Sordo, J. A. *The tetrahydrofuranhydrogen chloride complex: Rotational spectrum and theoretical analysis, J. Chem. Phys.* 1999, *111*, 6363.

Martin, J. M. L.& de Oliveira, G. *Towards standard methods for benchmark quality ab initio thermochemistry – W1 and W2 theory, J. Chem. Phys.* 1999, *111*, 1843.

Mayer, I. *J. Phys. Chem.* 1996, *100*, 6332.

Mayer, I. *Towards a "Chemical" Hamiltonian, Int. J. Quantum Chem.* 1983, *23*, 341.

Morokuma, K. & Kitaura, K. *Chemical Application of Atomic and Molecular Electronic Potentials*, Politzer, P. (Ed.) 1981, Plenum: New York,.

Pérez-González, A.; Galano, A. *OH Radical Scavenging Activity of Edaravone: Mechanism and Kinetics J. Phys. Chem. B* 2011, *115*, 1306.

Perry, R. A., Atkinson, R. & Pitts, J. N. *Kinetics and mechanism of the gas phase reaction of hydroxyl radicals with aromatic hydrocarbons over the temperature range 296-473 K, J. Phys. Chem.* 1977, *81*, 296.

Pfeilsticker, K., Lotter, A., Peters C. & Bösch, *Atmospheric Detection of Water Dimers via Near-Infrared Absorption, H. Science* 2003, *300*, 2078.

Sauer, J., Ugliengo, P., Garrone, E., & Saunders, V.R., *Theoretical Study of van der Waals Complexes at Surface Sites in Comparison with the Experiment, Chem. Rev.* 1994, *94*, 2095.

Schwenke, D. W. & Truhlar, D. G. *Systematic Study of Basis Set superposition Errors in the Calculated Interaction Energy of two HF molecules. J. Chem. Phys.* 1985, *82*, 2418.

Siegbahn and Blomberg, *Density Functional Theory of Biologically Relevant Metal Centers,* 1999

Singleton, D. L. & Cvetanovic, R. J. *Temperature dependence of the reaction of oxygen atoms with olefins, J. Am. Chem. Soc.* 1976, *98*, 6812.

Smith I. *Single-Molecule Catalysis, Science* 2007, 315, 470.

Stanton, J. F. & Bartlett, R. J. *The equation of motion coupled-cluster method. A systematic biorthogonal approach to molecular excitation energies, transition probabilities, and excited state properties, J. Chem. Phys.* 1993, *98*, 7029 .

Szori, M., Fittschen, C., Csizmadia, I. G. & Viskolcz, B. *Allylic H-Abstraction Mechanism: The Potential Energy Surface of the Reaction of Propene with OH Radical, J. Chem. Theory Comput.* 2006, 2, 1575 .

Thomas, J. R., Deeleeuw, B. J., Vacek, G., Crawford, T. D., Yamaguchi, Y. & Schaefer III, H. F. *The balance between theoretical method and basis set quality: A systematic study of equilibrium geometries, dipole moments, harmonic vibrational frequencies, and infrared intensities, J. Chem. Phys.* 1993, *99*, 403.

Tiecco, M. *Radical ipso attack and ipso substitution in aromatic compounds Acc. Chem. Res.* 1980, *13*, 51.

Trayham, J. G. *Ipso substitution in free-radical aromatic substitution reactions Chem. Rev.* 1979, *79*, 323.

Tully, F. P., Ravishankara, A. R., Thompson, R. L., Nicovlch, J. M., Shah, R. C., Kreutter, N. M. & Wine, P. H. *Kinetics of the reactions of hydroxyl radical with benzene and toluene J. Phys. Chem.* 1981, *85*, 2262.

Uc, V. H., Galano, A., Alvarez-Idaboy J. R., García-Cruz, I. & Vivier-Bunge, A. *A theoretical determination of the rate constant for OH hydrogen abstraction from toluene* J. Phys. Chem. A. 2006, 110, 10155.

Uc, V. H., García-Cruz, I., Grand A. & Vivier-Bunge, A. *Theoretical prediction of EPR coupling constants for the determination of the selectivity in the OH addition to toluene,* J. Phys. Chem. A 2001, *105*, 6226.

Uc, V. H., García-Cruz, I., Hernández-Laguna A. & Vivier-Bunge, A. *New channels in the reaction mechanism of the atmospheric oxidation of toluene,* J. Phys. Chem. A 2000, *104*, 7847.

Uc, V. H., Grand, A., Hernández-Laguna A. & Vivier-Bunge, A. *Isomeric adduct stability in the addition of atomic radicals to toluene: H, O(^{3}P), F and Cl,* Phys. Chem. Chem. Phys. 2002, *4*, 5730.

Uc, V. H., Grand, A., Hernández-Laguna A. &. Vivier-Bunge, A. *Stability and selectivity trends in the addition of atomic radicals to xylenes,* J. Mol. Struct. (Theochem) 2004, *684*, 171.

Uc, V. H.; Alvarez-Idaboy, J. R.; Galano, A. & Vivier-Bunge, A. *A theoretical explanation of non-exponential OH decay in reactions with Benzene and Toluene using pseudo-first order conditions* J. Phys. Chem. A. 2008, *112*, 7608.

Valdés, H. & Sordo, J. A. *Ab initio and DFT studies on van der Waals trimers: the OCS.(CO2)2 complexes. J. Comput. Chem.* 2002, *23*, 444.

Valdés, H. & Sordo, J. A. *Ab Initio Study on the (OCS)2.CO2 van der Waals Trimers J. Phys. Chem. A* 2002, *106*, 3690.

Vargas, R., Garza, J., Dixon, D. A. & Hay, B. P. *How Strong Is the C-H…O=C Hydrogen Bond? J. Am. Chem. Soc.* 2000, *122*, 4750.

Vega-Rodriguez A. & Alvarez-Idaboy, J. R. *Quantum chemistry and TST study of the mechanisms and branching ratios for the reactions of OH with unsaturated aldehydes Phys. Chem. Chem. Phys.,* 2009, *11*, 7649.

Vöhringer-Martinez E., Hansmann, B., Hernandez, H., Francisco J. S., Troe, J. & Abel, B. *Water Catalysis of a Radical-Molecule Gas-Phase Reaction Science* 2007, *315*, 497.

Wang, L., Khalizov, A. F., Zheng, J., Xu, W., Ma, Y., Lal, V. & Zhang, *Atmospheric nanoparticles formed from heterogeneous reactions of organics, R. Nature Geosci* 2010, *3*, 238.

Zavala-Oseguera, C.; Alvarez-Idaboy, J. R.; Merino, G.; Galano, A. *OH Radical Gas Phase Reactions with Aliphatic Ethers: A Variational Transition State Theory Study J. Phys. Chem. A* 2009, *113*, 13913

Zhang, R. & Park, J. & North S. W. *Hydroxy Peroxy Nitrites and Nitrates from OH Initiated Reactions of Isoprene J.Am. Chem. Soc.* 2002,*124*, 9600.

Zhao, Y., Schultz, N. E. & Truhlar, D. G. *Design of Density Functionals by Combining the Method of Constraint Satisfaction with Parametrization for Thermochemistry,*

Thermochemical Kinetics, and Noncovalent Interactions J. Chem. Theory Comput. 2006, 2, 364.

Zhao,J., Khalizov, A., Zhang, R. *Hydrogen-Bonding Interaction in Molecular Complexes and Clusters of Aerosol Nucleation Precursors, J. Phys. Chem.* A 2009, 113, 680.

11

Fourier Transform Spectroscopy with Partially Scanned Interferograms as a Tool to Retrieve Atmospheric Gases Concentrations from High Spectral Resolution Satellite Observations – Methodological Aspects and Application to IASI

Carmine Serio[1], Guido Masiello[1] and Giuseppe Grieco[2]
[1]*CNISM, Unitá di Ricerca di Potenza, Universitá della Basilicata, Potenza*
[2]*DIFA, Universitá della Basilicata, Potenza*
Italy

1. Introduction

Fourier Transform Spectroscopy with partially scanned interferograms (FTS*PSI) is a technique to obtain the difference between spectra of atmospheric radiance at two diverse spectral resolutions (Grieco et al., 2010; 2011; Kyle, 1977; Smith et al., 1979).

In the context of infrared remote sensing, the idea of using partially scanned interferograms for the retrieval of atmospheric parameters dates back to Kyle (1977) who argued that large portions of the spectrum (the Fourier transform of the interferogram and vice versa) could bring poor or no information for a given atmospheric parameter, whereas small ranges in the interferogram domain could concentrate much information about the parameter at hand. Kyle (1977) exemplified the technique for temperature, whereas a correlation interferometer was proposed for the observation of atmospheric trace gases by Goldstein et al. (1978). The direct inversion of small segments of interferometric radiances for the purpose of temperature retrieval was further analyzed and exemplified in Smith et al. (1979).

In some circumstances, according to Kyle (1977) the interferogram domain can provide a data space in which information about the atmospheric thermodynamical state can be encoded much more efficiently than in the spectral domain. Exploiting this idea, Grieco et al. (2010) has shown how to perform a dimensionality reduction of high spectral resolution infrared observations, which preserves the spectral coverage of the original spectrum . Once compared to the usual way of reducing a high amount of spectral data by simply considering a sparse (optimal) selection of the spectral channels, the methodology has shown a better performance mostly for the retrieval of water vapor.

Unlike the previous works by Grieco et al. (2010); Kyle (1977); Smith et al. (1979), which concentrated on the direct use of inteferogram radiances, in this study, we consider the point of transforming back to the spectral domain the partial scanned interferogram, in such a way to obtain a difference spectrum with improved signal-to-noise ratio. This difference spectrum, instead of the partial interferogram, is considered for the retrieval of atmospheric gases.

Depending on the interferogram range, the difference spectrum can isolate emission features of atmospheric gases from, e.g., the strong surface emission. In this respect the technique is mostly suited for nadir looking radiometers/spectrometers on board of satellites, since for this instrumentation the observed infrared radiance is made up by the atmospheric component plus that of surface emission. The issue of properly choosing the partial interferogram has to do with correlation interferometry. Because of the Wiener-Khinchin-Einstein theorem (e.g. see Bell (1972)), the interferogram is the *auto*-correlation function of the light spectrum, which means that periodic or almost periodic features present in the spectrum can yield sharp peaks (constructive interference). in the interferogram signal. As an example for the specific case of CO_2, rotation transitions yield a spectrum characterized by a periodic pattern with a period of about 1.5 cm^{-1}. This periodic pattern determines a strong signature (coherent interference) in the interferogram domain at about 1/1.5 cm=0.66 cm and overtones. Thus, for the case of CO_2, molecular spectroscopy fundamentals tell us how to exactly choose the proper partial interferogram.

In this study, FTS*PSI will be exemplified to show how we can handle and partly separate from the spectrum the surface emission in order to develop and implement suitable schemes for the retrieval of the columnar load of CO_2, CO, CH_4 and N_2O. The methodology will be applied to the Infrared Atmospheric Sounding Interferometer (IASI), which is flying on board the Metop-A (Meteorological Operational Satellite) platform. IASI was developed in France by the Centre National d'Etudes Spatiales (CNES) and is the first of three satellites of the European Organization for the Exploitation of Meteorological Satellites (EUMETSAT) European Polar System (EPS). The instrument spectral bandwidth covers the range from 645 to 2760 cm^{-1} (3.62 to 15.50 μm), with a sampling interval $\Delta\sigma = 0.25$ cm^{-1}. Thus, each single spectrum yields 8461 data points or channels. The calibrated IASI interferogram extends from 0 to a maximum OPD of 2 cm.

As already outlined, the main objective of this study is to illustrate, demonstrate and exemplify the potential advantages of high infrared spectral resolution observations data analysis with partially scanned interferograms, through the exploitation of IASI data. Towards this objective, we have used a forward/inverse methodology, which we refer to as φ-IASI, whose mathematical aspects and validation has been largely documented (see e.g. Amato et al. (2002); Carissimo et al. (2005); Grieco et al. (2007); Masiello et al. (2002; 2003; 2004); Masiello & Serio (2004); Masiello et al. (2009; 2011); Serio et al. (2000)).

The remote sensing of atmospheric minor and trace gases from nadir looking instrumentation on board polar satellites is not a new subject. Among many others we quote here the experience with the Japanese IMG (Interferometric Monitoring of Greenhouse Gases) (Lubrano et al., 2004), the American AIRS (Atmospheric Infrared Radiometer Sounder) (Chahine et al, 2005; 2008; McMillan et al., 2005) and the European IASI (Boynard et al., 2009; Clarisse et al., 2008; 2009; Clerbaux et al., 2009; Crevoisier et al., 2009; Grieco et al., 2011; Ricaud et al., 2009). However, this work is focused mostly on the novel methodology of FTS*PSI rather than particular applications, which nevertheless are here considered to exemplify the use of the procedure for the remote sensing of minor and trace gases.

The study is organized as follows. Section 2 is mainly devoted to the methodological aspects: in section 2.1 we present the IASI data, whereas the description of the fundamentals of the partially scanned interferogram approach is presented in sections 2.2, 2.2.1 and 2.3. Application to IASI for the retrieval of CO_2, CO, CH_4 and N_2O is discussed in sections

3, 4, 5 and 6, respectively. Section 3, which covers the case of CO_2 is mainly intended to exemplify the overall methodology. Results for CO_2 concerning the quality and validation of the methodology have been recently presented in Grieco et al. (2011). These results will not be shown here again, but only summarized for the benefit of the reader in section 3.3. However, the material shown in the two methodological sections, 3.1, and 3.2 is largely complementary to that presented in Grieco et al. (2011) and addresses important aspects, which generalize the application of the methodology to the full earth disk. The results for CO, CH_4 and N_2O, shown in sections 4, 5 and 6, are presented here for the first time and have not been covered in previous publications by the authors. Conclusions are drawn in section 7.

2. Background

2.1 IASI data

For the purpose of simulations we use the Chevalier data set (Chevalier, 2001). This set is mostly used to define pairs of IASI spectra and atmospheric state vectors.

Real IASI observations and related atmospheric state vectors come from the 2007 Joint Airborne IASI Validation Experiment (JAIVEx) campaign (JAIVEx, 2007) over the Gulf of Mexico. We have a series of 6 spectra for 29 April 2007, 16 spectra for 30 April 2007, and finally 3 spectra for 04 May 2007. The spectra were recorded for clear sky fields of view, selected based on high resolution satellite imagery from AVHRR (Advanced Very High Resolution Radiometer) on MetOp (Meteorological Operational Satellite) and in-flight observations. Furthermore, the data for 29 April 2007 correspond to a nadir IASI field view, whereas those for the other two days to a field of view of 22.50 degrees.

Dropsonde and ECMWF (European Centre for Medium range Weather Forecasts) model data, which have been used to have a best estimate of the atmospheric state during each MetOp overpass, have been prepared and made available to us by the JAIVEx team. The JAIVEx dataset contains dropsonde profiles of temperature, water vapor, ozone, and carbon monoxide which extend up to 400 hPa. Above 400 hPa only ECMWF model data are available. The atmospheric state vectors used in this study use JAIVex dropsonde data up to 400 hPa supplemented by collocated ECMWF forecasts of temperature, water vapor and ozone from 400 hPa to 0.1 hPa (corresponding to about 65 km).

For the purpose of atmospheric gas retrieval we have also used IASI data for a monthly acquisition on July 2010 over the Mediterranean area. These data set has been used to produce monthly maps for the four gases at hand. The spectra were checked for clear sky using the IASI stand alone cloud detection scheme developed by Grieco et al. (2007); Masiello et al. (2002; 2003; 2004); Serio et al. (2000).

2.2 Basic aspects of FTS⋆PSI

The principles of FTS⋆PSI have been recently discussed in Grieco et al. (2011), here we limit to show the basic aspects. FTS⋆PSI is a particular application of Fourier spectroscopy, which today counts widespread applications in many fields. Fourier spectroscopy fundamentals can be found in appropriate textbooks (see e.g. Bell (1972)). A summary of the basic definitions, which are relevant to our methodology is now presented.

In Fourier spectroscopy the spectrum, $r(\sigma)$ (with σ the wavenumber) and the interferogram, $I(x)$ (with x the optical path difference) constitute a Fourier pair defined by the couple of equations (see e.g. Bell (1972)),

$$r(\sigma) = \int_{-\infty}^{+\infty} I(x)\exp(-2\pi i\sigma x)dx \tag{1}$$

$$I(x) = \int_{-\infty}^{+\infty} r(\sigma)\exp(2\pi i\sigma x)d\sigma \tag{2}$$

with i the imaginary unit. The spectrum and the interferogram are in practice band-limited functions, therefore taking into account that the interferogram is sampled up to a given maximum optical path difference, x_{max}, we modify Eq. (1) by introducing the data-sampling window, $W(x)$

$$r(\sigma) = \int_{-\infty}^{+\infty} W(x)I(x)\exp(-2\pi i\sigma x)dx \tag{3}$$

with

$$W(x) = \begin{cases} 1 \text{ for } \mid x \mid \leq x_{max} \\ 0 \quad \text{otherwise} \end{cases} \tag{4}$$

where $|\cdot|$ means absolute value, x_{max} is the maximum optical path difference.

The maximum optical path difference, x_{max} also determines the sampling rate, $\Delta\sigma$ within the spectral domain. According to the Nyquist rule, the relation is

$$\Delta\sigma = \frac{1}{2x_{max}}, \quad \Delta x = \frac{1}{2(\sigma_2 - \sigma_1)} \tag{5}$$

where $\sigma_2 - \sigma_1$ is the spectral band-width. For IASI we have $\sigma_1 = 645$ cm^{-1}, $\sigma_2 = 2760$ cm^{-1}, hence $x_{max} = 2$ cm and $\Delta\sigma = 0.25$ cm^{-1}.

The fact that IASI is apodized (see e.g. Amato et al. (1998)) is of no concern here. For IASI we just consider the interferogram obtained from the calibrated, apodized spectrum. One could also consider to obtain the interferogram corresponding to each IASI band one at a time. This is appropriate, e.g., when dealing with a given gas whose absorption bands are confined within a single IASI band.

According to the Shannon-Whittaker sampling theorem (e.g. see (Robinson & Silvia, 1981)), in case we want to re-sample the spectrum at a sampling rate lower than the original, we just have to introduce in Eq. (3) a data-sampling window with its new cutting point at $x_\tau < x_{max}$. The new sampling rate, $\Delta\sigma_\tau$ appropriate to x_τ involves again the Nyquist rule, we have $\Delta\sigma_\tau = (2x_\tau)^{-1}$.

2.2.1 The couple partial interferogram, difference spectrum

For a nadir viewing instrument such as IASI, one of the most prominent feature within the observations is the surface emission. This emission allows us to retrieve skin temperature and has information about surface emissivity. However, in case we are interested in retrieving atmospheric parameters, such as minor gases, we would like to have information only about atmospheric emission, since the surface emission interferes with the signal we want to exploit for the retrieval process.

As opposite to gas emission, surface emission varies smoothly with the wave number, σ. Thus, if we consider two spectral samplings, $\Delta\sigma_1$, $\Delta\sigma_2$, which are close each other but small enough so that the surface emission has been resolved at either samplings, then the difference spectrum

$$d(\sigma) = r_1(\sigma) - r_2(\sigma) \tag{6}$$

contains mostly the atmospheric emission alone. In Eq. 6, r_i is the spectrum at sampling $\Delta\sigma_i$, with $i = 1, 2$.

In reality, the above differentiation cannot get completely rid of the surface radiation, because the corresponding emission, which reaches the Top of the Atmosphere (TOA) is modulated by the total transmittance function, which is itself a highly oscillatory function. However, in window regions where the transmittance approximates the unity, and where gases have only weak absorption bands, the difference $d(\sigma)$ yields a signal, which is mostly the result of atmospheric emission alone.

It is important here to stress that to compute the difference spectrum, $d(\sigma)$ we do not need to measure the spectrum twice. The operation can be done by simply Fourier transforming one single *partial* interferogram. In fact, with reference to Eq. 6 and provided that, $\Delta\sigma_i > \Delta\sigma$ $(i = 1, 2)$, and with $\Delta\sigma$ defined by Eq. 5, that is the sampling rate corresponding to the maximum optical path difference, we have that $d(\sigma)$ corresponds to the interferogram, $I(x)$ computed over the range $[x_1, x_2]$, with

$$x_1 = (2\Delta\sigma_2)^{-1}; \quad x_2 = (2\Delta\sigma_1)^{-1} \tag{7}$$

where, to fix the idea we have assumed $\Delta\sigma_1 < \Delta\sigma_2$. Therefore, $d(\sigma)$ can be computed by Fourier transforming the *partial* interferogram, $I_p(x)$ defined by

$$\begin{cases} I_p(x) = 0 & \text{for} \quad x < x_1 \\ I_p(x) = I(x) & \text{for} \quad x_1 \le x \le x_2 \\ I_p(x) = 0 & \text{for} \quad x_2 < x \le x_{max} \end{cases} \tag{8}$$

The left and right zero-padding ensures that the difference spectrum is defined on the same σ-grid as that corresponding to the sampling rate $\Delta\sigma$.

For the purpose of Fourier Transform computations, the above concept of partial interferogram is better formalized by considering an appropriate data-sampling window with a lower and upper truncation point (Grieco et al., 2011; Kyle, 1977),

$$\tilde{W}(x) = \begin{cases} 1 \text{ for } x_1 \le \mid x \mid \le x_2 \\ 0 \quad \text{otherwise} \end{cases} \tag{9}$$

This data-sampling window removes from the spectrum all those broad features, which are represented by interferometric radiances below the left truncation point, x_1. In fact we have

$$\tilde{W}(x) = W_2(x) - W_1(x) \tag{10}$$

where W is the box-car window as defined in Eq. (4) and where the under scripts 1 and 2 identify the window function with cutting points, x_1 and x_2, respectively. Because the Fourier transform is linear, within the spectral domain, the double-truncation data-sampling window is equivalent to take the difference of Eq. 6.

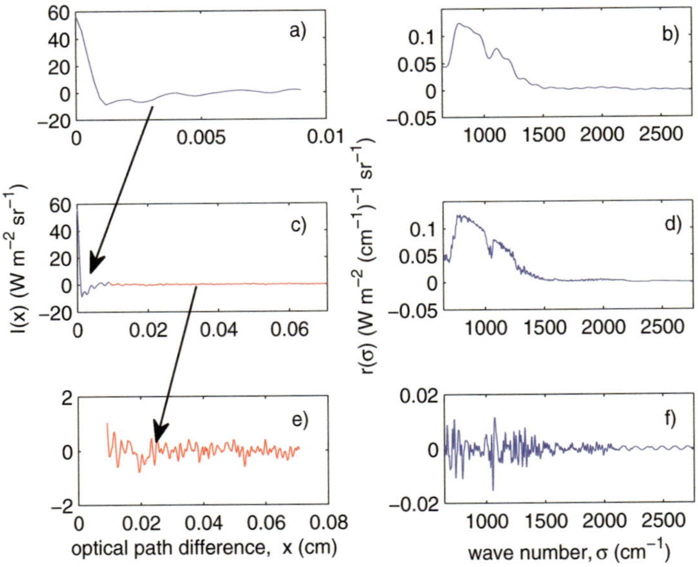

Fig. 1. Examples of couples (interferogram, spectrum) for different optical path differences. The couple a) b) refers to $x = 0.1$ cm, the couple c) d), to $x = 0.2$ cm. Couple e) f) exemplifies that the partial interferogram in the range $[0.1, 0.2]$ cm corresponds to the difference of the spectra d)-b).

An example of partial interferogram is provided in Fig. 1, where it is also exemplified that the difference spectrum enhances the atmospheric emission, while the broad surface emission is almost zeroed.

Another important characteristic of FTS⋆PSI is that the difference spectrum can be observed with enhanced signal-to-noise ratio with respect to the whole spectrum, $r(\sigma)$, that is to the spectrum corresponding to the maximum optical path difference. According to a well-known results of Fourier spectroscopy (Pichett & Strauss, 1972), the spectral noise variance is proportional to the interferogram bandwidth $\Delta x = x_2 - x_1$. In case we consider the full interferogram, Δx is exactly the maximum optical path difference, which for IASI is 2 cm. As a consequence, the variance of $d(\sigma)$ is simply computed in case we know the variance of $r(\sigma)$, that is the radiometric noise affecting the spectral radiances.

The radiometric noise for $r(\sigma)$ we have used in this paper is the IASI L1C radiometric noise, which is shown in Fig. 2. Let $\varepsilon(\sigma)$ be the radiometric noise affecting the spectrum, $r(\sigma)$. For the difference spectrum $d(\sigma)$, the noise is

$$\varepsilon_\Delta(\sigma) = \varepsilon(\sigma)\sqrt{\frac{x_2 - x_1}{x_{max}}} \tag{11}$$

where the underscript Δ refers to the difference spectrum.

Strictly speaking, Eq. 11 applies to the case of uncorrelated noise, that is in case the noise affecting the original spectrum, $r(\sigma)$ is truly random. However, this is not the case for IASI,

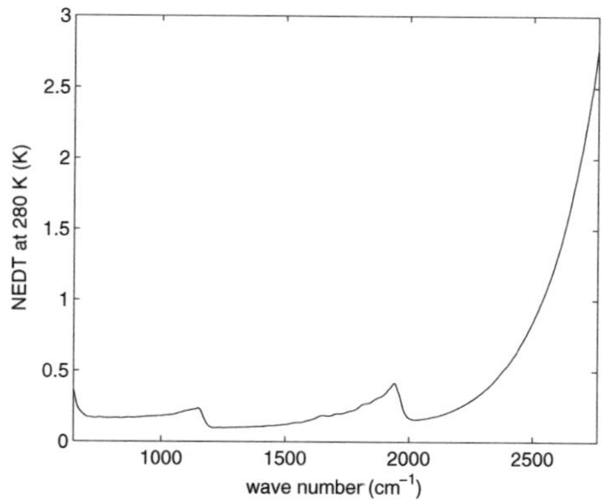

Fig. 2. IASI level 1C radiometric noise in terms of Noise Equivalent Difference Temperature (NEDT) at as scene temperature of 280 K.

because IASI data are Gaussian apodized (Amato et al., 1998). However, in section 3 we will see that, at least for IASI, Eq. 11 provides a good reference to have the order of magnitude of the noise reduction in the transformed difference-spectrum space.

However, whatever the kind of noise (correlated or uncorrelated) in the original spectrum space may be, to obtain $d(\sigma)$ we use zero padding in order not to modify the IASI sampling of 0.25 cm^{-1}. This means that the noise in the d-spectrum space is correlated.

2.3 Sensitivity of $d(\sigma)$-channels to the atmospheric state vector

When properly defined, the difference spectrum, $d(\sigma)$ is expected to be less sensitive to the atmospheric state, but the given parameter whose signal we want to amplify, than the original spectrum, $r(\sigma)$.

To quantitatively analyze this aspect we introduce an obvious measure of the sensitivity of $d(\sigma)$ to a given parameter. This measure makes use of the Jacobian derivative.

For each given channel or wavenumber, σ, the noise-normalized sensitivity to a generic parameter, $\mathbf{X}$, is defined and computed according to

$$\mathbf{S}_{X,\Delta}(\sigma) = \left(\frac{1}{\varepsilon_\Delta(\sigma)} \frac{\partial d(\sigma)}{\partial \mathbf{X}} \right) \tag{12}$$

where $\varepsilon_\Delta(\sigma)$ is the radiometric noise affecting the difference spectrum, $d(\sigma)$. The above sensitivity expression says how large is the noise-normalized variation of the signal at σ to a unitary perturbation of $\mathbf{X}$. Note that in general, $\mathbf{X}$ can a be a scalar (e.g. surface temperature and emissivity, in which case the sensitivity is a scalar itself, function of the wave number, σ), or a vector (e.g., temperature profile, water vapour profile), in which case the sensitivity is a vector itself, function of the wave number.

For purpose of comparison, the above sensitivity can be also defined for the case of the original IASI spectrum, $r(\sigma)$

$$\mathbf{S}_X(\sigma) = \left(\frac{1}{\varepsilon(\sigma)} \frac{\partial r(\sigma)}{\partial \mathbf{X}} \right) \tag{13}$$

3. Application to CO_2

As said in section 1, for the case of CO_2 we have that a small range around the optical path difference at 0.66 cm is mostly dominated by CO_2 emission. The resulting CO_2 signature is clearly visible in Fig. 3, which shows a partial interferogram in the range 0.65 to 0.68 cm. Transforming back to the spectral domain the partial interferogram, we obtain the difference

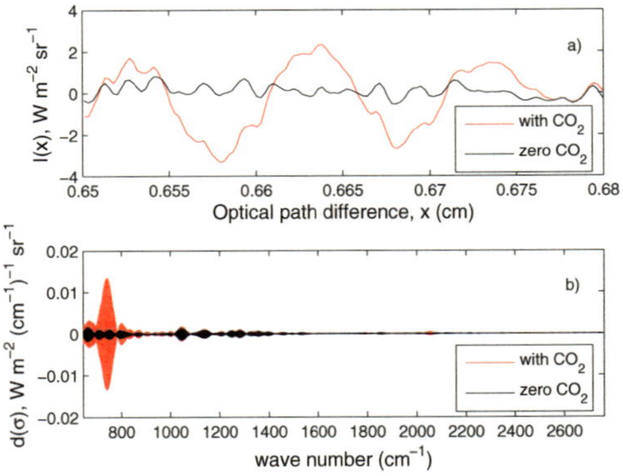

Fig. 3. a) Example of partial interferogram in the range $[x = 0.65, x = 0.68]$ cm for a case of CO_2 load equal to 0 and 385 ppmv, respectively; b) the difference spectrum $d(\sigma)$. The full IASI spectral coverage 645 to 2760 cm^{-1} has been considered.

spectrum shown in 3(b), where the CO_2-lines periodic pattern is clearly amplified.

Figure 3(b) suggests that $d(\sigma)$ is mostly a function of the CO_2 amount alone. This can be checked in simulation through generation and analysis of difference-spectra corresponding to diverse amount of CO_2.

3.1 The case of noise free radiances

Radiative transfer calculations were performed based on the atmospheric states summarized in Fig. 4. With reference to the tropical model of atmosphere shown in Fig. 4 (panels (a), (b) and (c)), difference spectra were generated for sixteen different columnar amounts, $q_{CO_2}(i)$, $(i = 1 \ldots 16)$ of CO_2 ranging from 300 to 450 ppmv with a step of 10 ppmv. To perform the radiative transfer calculations, we had to assume a model of dependence on altitude for the CO_2 mixing ratio. As shown in Fig. 4, this was assumed to be constant with altitude. A search of the channels that are best linearly correlated with the CO_2 columnar amount shows that there are many and a few of this are largely insensitive to surface emissivity and temperature. Four of these good channels have been identified and they correspond to the wave numbers $\sigma_1 = 783.75$ cm^{-1}, $\sigma_2 = 809.25$ cm^{-1}, $\sigma_3 = 976.75$ cm^{-1}, $\sigma_4 = 2105$ cm^{-1}, respectively.

Let $d_j(i)$ be the difference spectrum ordinate corresponding to the wave number σ_j (with $j = 1 \ldots 4$) and the CO_2 columnar amount, $q_{CO_2}(i)$, (with $i = 1 \ldots 16$), respectively. For each wave number, σ_j, we consider the standardized quantity,

$$D_j(i) = \frac{d_j(i) - \mu_j}{s_j} \quad (14)$$

with μ_j and s_j the mean and standard deviation of the sample, $\{d_j(i)\}_{i=1\ldots16}$, respectively.

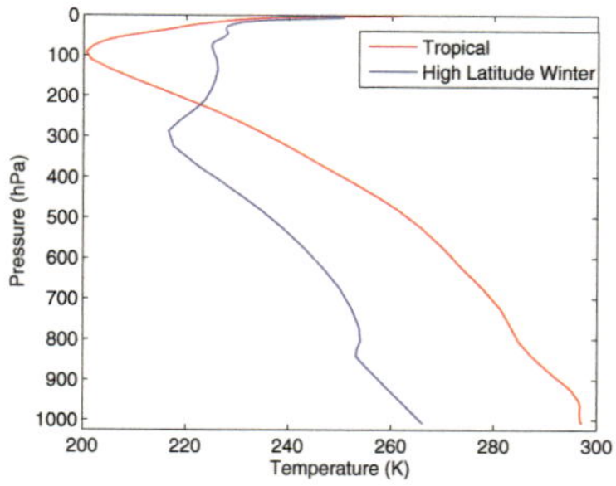

(a) Temperature profile

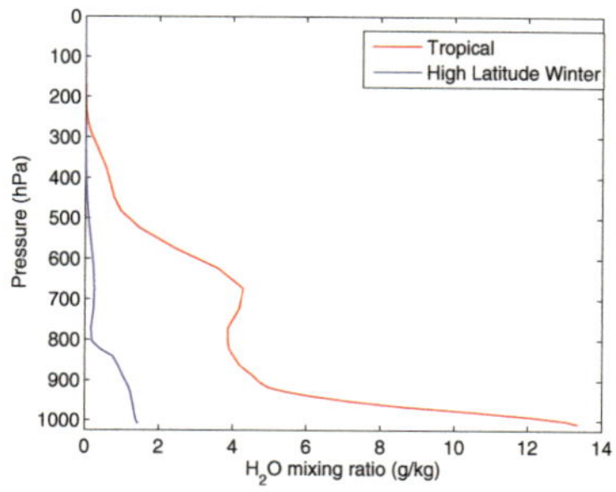

(b) H_2O profile

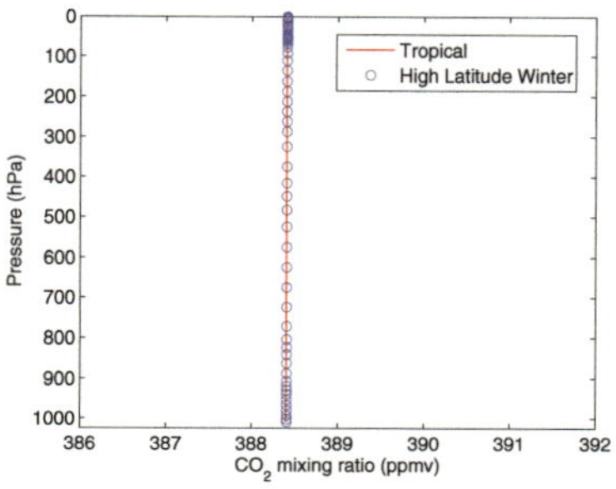

(c) CO_2 uniform profile

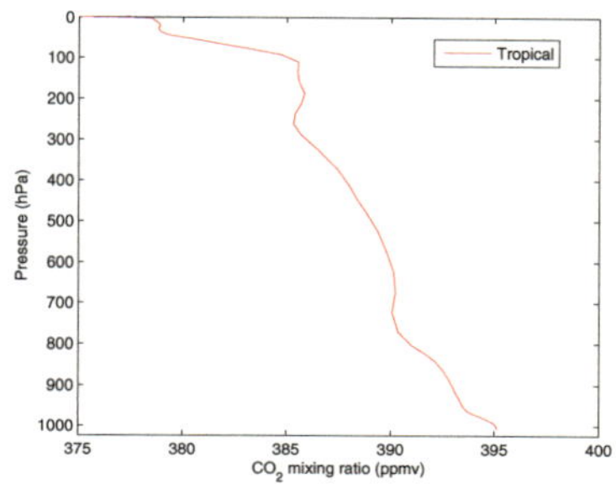

(d) CO_2 non-uniform profile

Fig. 4. The two reference atmosphere models used to simulate IASI synthetic radiances. The models belong to two opposite weather conditions: tropical and High Latitude Winter. As a rule, for the reference state, the CO_2 profile is assumed constant with altitude (as shown in (c)). For checking a possible dependence of the methodology on the shape of the CO_2 profile, the analysis has also considered a case of CO_2 profile non uniform with altitude (shown in (d)); this case was used in combination with the tropical model alone.

Figure 5(a) exemplifies, for the case of the channel at $\sigma_1 = 783.75$, that the relation between D_j and q_{CO_2} is perfectly linear, that is

$$q_{CO_2} = a_j D_j + b_j \tag{15}$$

The solid lines shown in Fig. 5 are the linear best fit to the data points. It is important to stress that the regression coefficients a_j and b_j depend on the Field of View angle. Unless otherwise stated, the results shown in this section apply to the nadir angle. In addition, it is also important to stress that even for noisy-free radiances the estimation of the column-integrated CO_2 from Eq. 15 has a residual uncertainty. This uncertainty is of the order od 1-1.5 ppmv and can be thought of as a sort of error inherited from the linearization of the dependence of q_{CO_2} on d_j. The exact amount of this uncertainty can depend on the wave number. Actually, the search of the *linear* channels have been done by computing the linear regression standard error at each single channel and choosing only those with standard error less than 1.5 ppmv. The typical accuracy or estimation error of q_{CO_2} estimated by Eq. 15 for the four channels at hand is shown in Tab. 1.

Table 1 (see the accuracy for the noiseless case) also shows that a linear relation holds for the remaining three channels, $\sigma_1 = 809.25$ cm^{-1}, $\sigma_3 = 976.75$ cm^{-1}, $\sigma_4 = 2105$ cm^{-1}.

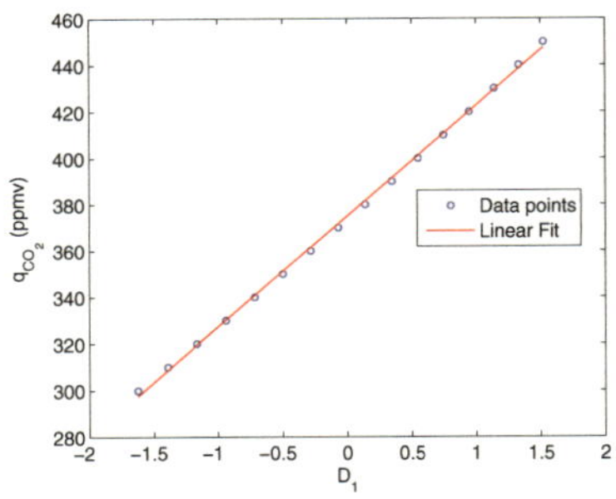

(a) Tropical, CO_2 unifrom

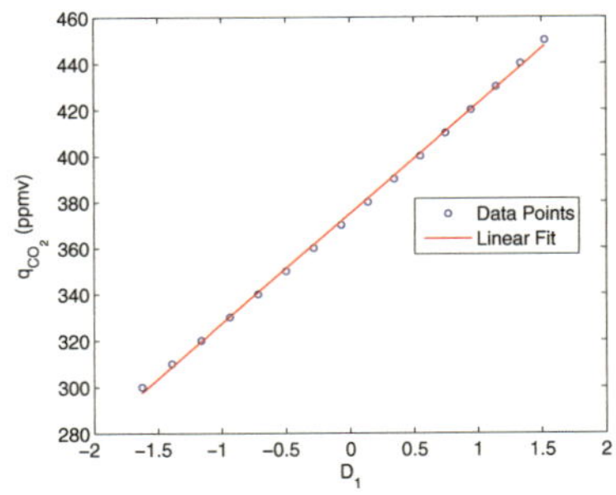

(b) Tropical, CO_2 non unifrom

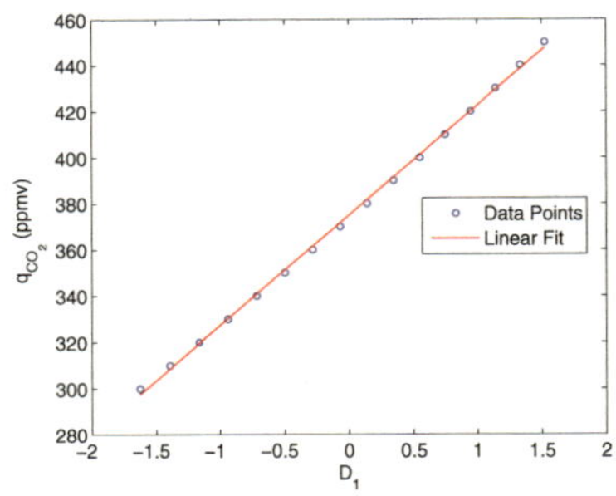

(c) HLW, CO_2 unifrom

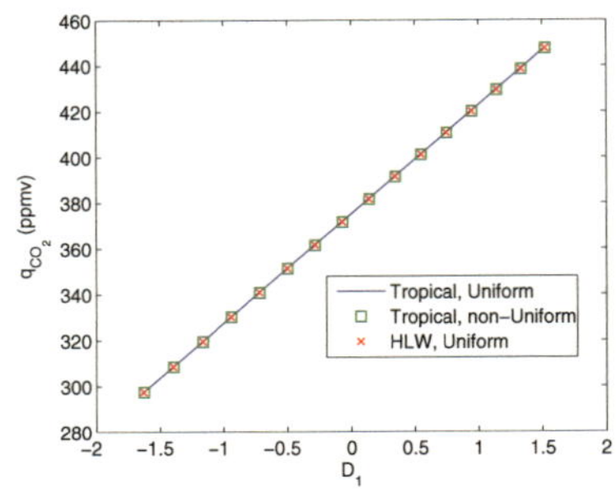

(d) The three linear best fits in (a), (b) and (c) are intercompared.

Fig. 5. Exemplifying the dependence of q_{CO_2} on D_j for the channel at $\sigma_1 = 783.75$ cm^{-1} as a function of the atmosphere model (tropical and High Latitude Winter, HLW) and shape of the CO_2 profile.

Channel (cm^{-1})	Accuracy (Noise-free case) (ppmv)	Accuracy (Noisy case) (ppmv)
783.75	1.4	17.5
809.25	1.5	10.8
976.75	1.2	28.7
2105	1.1	26.2

Table 1. Accuracy of the column-integrated CO_2 amount estimated through the linear fit of Eq. 15 for the case of noise-free and noisy radiances.

More important here is the fact that the functional relation between q_{CO_2} and D_j does not depend on the assumed shape for the CO_2 profile. To check this dependence we have re-done the calculations, but now with a CO_2 mixing ratio profile, which is not constant with altitude. This altitude-varying profile (shown in Fig. 4(d)) represents a realistic situation. In fact, it is the result of the ECMWF analysis Engelen et al. (2009) corresponding to the date (29 April 2007) and location of the JAIVEx experiment. Figure 5(b) shows that the functional relation remains perfectly linear.

Moreover, the shape of the functional relation is completely independent of the state vector, as well. In fact, if we redo the calculations, but now with the High Latitude Winter model of atmosphere (see Fig. 4 from (a) to (c)), we obtain the result shown in Fig. 5(c), which says that the dependence of q_{CO_2} on D_j is linear whatever the atmospheric state vector may be.

In addition, not only the linear shape does not change with the state vector and the assumed shape of the CO_2 profile, it is exactly the same (as it is shown in Fig. 5(d)). That is the regression coefficients remain constant under a varying state vector, even in the CO_2 profile shape, which means that Eq. 15 has a general or universal validity.

Once again, we stress that although Fig. 5 focuses on the channel at $\sigma_1 = 783.75$ cm^{-1}, the same results hold for all the four channels, σ_j, $j = 1, \ldots, 4$. The linearity for these channels is likely to be a results of the fact that they correspond to wing regions of strong absorption bands, or to moderate absorption bands. In general, channels inside strong absorptions bands show a behavior, which is largely non linear.

It is important to stress that while the linear dependence of q_{CO_2} on D_j is universal, this is not true for the case in which we consider the difference spectrum, d_j, since the standardization parameters, μ_j and s_j do depend on the state vector.

Since in practice we observe d_j, in case we have one single IASI spectrum, we are faced with the problem of computing μ_j and s_j to perform the standardization, which yield, D_j (see Eq. 14). The only way to go in this case is to compute a series of synthetic IASI spectra with different load of CO_2. The synthetic spectra have to be computed on the basis of the possibly best estimate of the atmospheric state vector corresponding to the given IASI observation. Thus, for *weather* applications in which we need a space-time resolved CO_2 estimation, we do need information on the atmospheric state vector.

For *climate* analysis, we need to average over many weather states. Thus, the fact that q_{CO_2} has an universal linear dependence on D_j means that, by considering averages, we get rid of the dependence on the state vector. In practice, if we compare d_j-values properly averaged on climate time scales, these values are directly proportional to q_{CO_2}-values averaged over the same climate time scales. As an example if we take all the IASI clear sky sea-surface tropical spectra for the year, say Y_1 and compute the mean value of d_j for whatever j (call this average $< d_{j,Y_1} >$), and redo this operation for a second year Y_2, the difference $< d_{j,Y_2} > - < d_{j,Y_1} >$ is directly proportional to the variation of the CO_2 over the time span $Y_2 - Y_1$. To accurately estimate CO_2 changes on a climate time span is an easy task from satellite observations, provided we use FTS*PSI.

3.2 Noisy radiances

With radiances affected by noise we can again just use Eq. 15 to estimates the CO_2 columnar amount. However, now the variability of the estimate is expected to increase according to the noise affecting IASI radiances.

In case of noisy radiances, taking into account Eq. 11, the accuracy can be computed through the usual rule of variance propagation. With reference to the basic Eq. 15, we have

$$\mathrm{var}\left(q_{CO_2}\right) = (a_j^2/s_j^2)\mathrm{var}\left(d_j\right) \tag{16}$$

where var(·) stand for variance.

Figure 6 shows the standard deviation, $\varepsilon_{\Delta}(\sigma)$ of the noise affecting the difference spectrum, $d(\sigma)$ under the assumptions of IASI correlated noise (Fig. 6(a)) and IASI uncorrelated noise (Fig. 6(b)).

These computations have been done in simulation. First, we generated samples of random numbers with zero mean and standard deviation according to the IASI radiometric noise figures in Fig. 2. Second, these samples were passed through the same series of calculations as those that transform $r(\sigma)$ to $d(\sigma)$. In case we consider the effect of correlation, random numbers were generated according to the full IASI level 1C covariance matrix. The noise shown in Fig. 2 corresponds to the square root of the diagonal of the IASI covariance matrix.

Figure 6 also provides a comparison with the standard deviation obtained from a direct use of Eq. 11, with the appropriate band factor equal to $\sqrt{2/(0.68 - 0.65)} = 8.165$. It is seen that the comparison between simulation and Eq. 11 is perfect for the IASI uncorrelated noise. For the real case of IASI correlated noise, Eq. 11 slightly underestimates the standard deviation of the difference spectrum, $d(\sigma)$. Here and in the following the noise affecting $d(\sigma)$ has been computed by considering correlated IASI noise at level 1C. Using the standard deviation

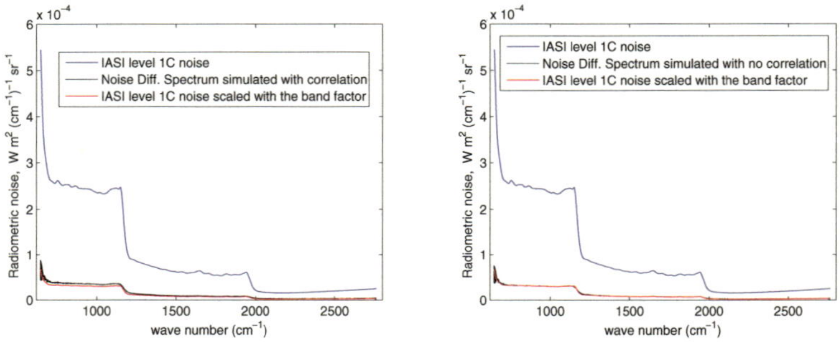

(a) with correlation (b) truly random

Fig. 6. Standard deviation of the noise affecting the difference spectrum in case of a partial interferogram extending from 0.65 to 0.68 cm. The computations have been made with IASI correlated noise (a) and IASI uncorrelated noise (b). A comparison is also provided with the standard deviation calculated with Eq. 11.

expected for the difference spectrum (Fig. 6(a)) together with Eq. 16, we can easily compute the accuracy (that is the square root of var (q_{CO_2})) of the linear regression at each of the four channels at hand. The accuracy is shown in Tab. 1, which allows us to compare the results with the noise-free case.

Form Tab. 1 it is seen that the regression error increases of a factor 5 to 25, depending on the channel. However, we can optimally average the four different estimates, in order to improve the accuracy. To do so, it is important to realize that the d_j-channels may be correlated, so that we have to consider their proper covariance matrix in order to compute the optimal average.

Let $\mathbf{C}_d$ be the covariance matrix of the four channels. The size of $\mathbf{C}_d$ is 4×4 and it can be computed in simulation or directly by transforming the IASI covariance matrix from the radiance space to the difference-radiance space. Whatever we do, this matrix can be computed in advance and stored for later applications, therefore it can be thought of as being a known parameter. Based on simulations, which made use of the IASI level 1C covariance matrix, we

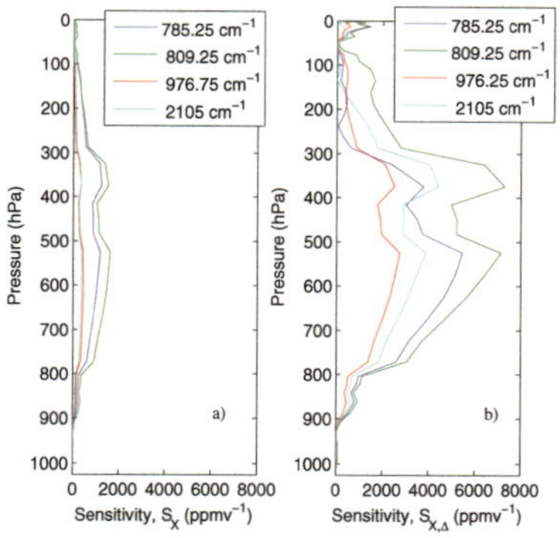

Fig. 7. For the four channels listed in the legend, panel a) shows the sensitivity of $r(\sigma)$ and panel b) that of $d(\sigma)$ to the CO_2 mixing ratio profile.

have that the couple (d_1, d_2) is slightly correlated with a correlation of 0.34, whereas the other combinations are truly un-correlated.

For a given IASI observation, let us suppose that we have computed through Eq. 15, the CO_2 amount, say q_j, corresponding to the four channels at σ_j, $j = 1, \ldots, 4$. Let us define $\mathbf{q} = (q_1, \ldots, q_4)^t$, where the super-script t denotes transpose operation. Let $\mathbf{A}_d$ be the 4×4 diagonal matrix, whose diagonal elements are the regression coefficients $(a_j/s_j)^2$. Then the covariance matrix, $\mathbf{C}_q$ of the vector $\mathbf{q}$ is the product $\mathbf{C}_q = \mathbf{A}_d \mathbf{C}_d \mathbf{A}_d^t$ and the optimal Least Squares estimation of, $\bar{q}_{CO_2}$ from the four estimates is given by

$$\bar{q}_{CO_2} = \frac{\mathbf{1}^t \mathbf{C}_q^{-1}}{\mathbf{1}^t \mathbf{C}_q^{-1} \mathbf{1}} \mathbf{q} \tag{17}$$

with the variance, $\mathrm{var}(\bar{q}_{CO_2})$ given by

$$\mathrm{var}(\bar{q}_{CO_2}) = \left(\mathbf{1}^t \mathbf{C}_q^{-1} \mathbf{1} \right)^{-1} \tag{18}$$

where $\mathbf{1} = (1, 1, 1, 1)^t$. Combining this way the four estimates, we have that the accuracy improves to the value of $\approx \pm 7$ ppmv.

One could say, why not to use more channels to get the accuracy close to the inherent limit of 1.5 ppmv? Well, the problem is that the more channels we use, the more the correlation among channels themselves increase. With high correlated channels, the optimal estimation does not improve so much in comparison to the channel with the best accuracy. Furthermore, another important point, which brings us to the next section, is the sensitivity of the channels to the state vector. Since we have only access to a (possibly) best estimate of the atmospheric state corresponding to the given IASI observation, we need to make sure to select channels, which are loosely sensitive to the state vector, but CO_2.

In this respect, the four channels, we are considering so long, show a very high sensitivity to CO_2. This is seen from Fig. 7, which compares $\mathbf{S}(\sigma)$ to $\mathbf{S}_\Delta(\sigma)$ for the case of a tropical model of the atmosphere. We see that the sensitivity to CO_2 improves of a factor 4-5 when passing from $r(\sigma)$ to $d(\sigma)$. Conversely, the sensitivity to the temperature profile (not shown for the sake of brevity) largely decreases when transforming from $r(\sigma)$ to $d(\sigma)$. The sensitivity analysis shows that the d_j-channels are sensitive to a large part of the troposphere, from to 900 to 100 hPa, which for the case of a tropical model of atmosphere encompasses all the troposphere above the Planetary Boundary Layer. Conversely, the d_j channels are almost insensitive to what happen in the boundary layer either for CO_2 or e.g. temperature, which is good because the lower troposphere is where we expect to have a larger uncertainty to the atmospheric state. In other words, also in case we implement the technique with a state vector which largely differs from the *truth* in the lower troposphere, we can still have valuable estimates for CO_2 provided the state vector is sufficiently accurate for the rest of the troposphere.

3.3 Application to IASI data for CO_2 estimation

The mechanics of the procedure to estimate the columnar amount of CO_2 can be summarized as follow,

- for a given IASI spectrum, choose the possibly best estimate of the state vector corresponding to the IASI overpass.
- With this state vector perform the radiative transfer calculations to yield synthetic IASI spectra (normally 4 are enough) corresponding to different CO_2 columnar amounts in the range 300 to 450 ppmv.
- Transform the IASI spectra to difference spectra in order to estimate the regression coefficients of Eq. 15.
- Compute the observed d_j values from the IASI spectrum and input them to Eq. 15 to obtain the estimate at each channel.
- Finally, combine these estimates according to Eq. 17 and get the accuracy from Eq. 18.

The best estimate of the state vector can be obtained in many modes e.g. colocated ECMWF analysis, radiosonde, inversion of IASI spectral radiances. For the work here shown, the state vector is normally obtained from IASI spectral radiances themselves. As said in section 1 we use a retrieval methodology, which we refer to as φ-IASI, whose mathematical aspects and validation have been largely documented (see e.g. Amato et al. (2002); Carissimo et al. (2005); Grieco et al. (2007); Masiello & Serio (2004); Masiello et al. (2009; 2011)).

To exemplify the procedure Fig. 8 shows the CO_2 columnar amount estimated for the 25 IASI soundings of the JAIVEx experiment. Figure 8(a) also provides a comparison with the ECMWF analysis (Engelen et al., 2009). Although the two analysis are largely in agreement, we see that ECMWF tends to overestimate the CO_2 amount in comparison to the findings provided in this work. A comparison with aircraft in situ profiles (Grieco et al., 2011) shows that the analysis provided by our methodology is that correct. Finally, for illustrative purposes Fig. 8(b) shows a monthly map of CO_2 computed for the Mediterranean area for the month of July 2010. This has been obtained by processing IASI spectra for July 2010 for sea surface. The spectra were checked for clear sky using the IASI stand alone cloud detection developed by Grieco et al. (2007); Masiello et al. (2002; 2003; 2004); Serio et al. (2000).

Figure 8(b) clearly shows that the CO_2 amount is not a constant in the atmosphere. The North-to-South-East trend evidenced in the figure is consistent with the general circulation of the Mediterranean area, and is likely a consequence of the summer African anticyclone, which in July begins to extend its influence in the Mediterranean area. This possible effect is currently under investigation with IASI data.

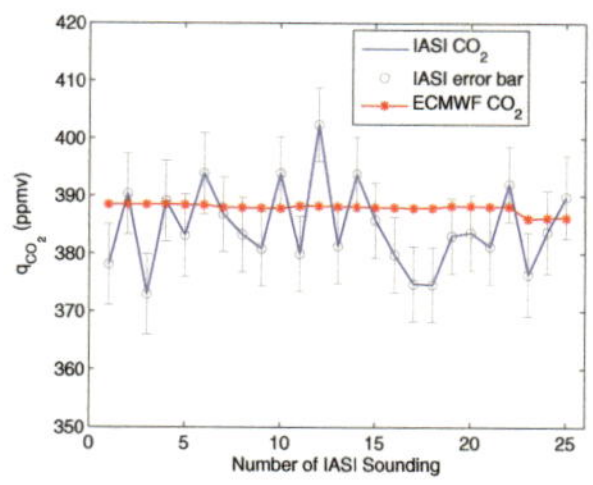

(a) JAIVEx case study

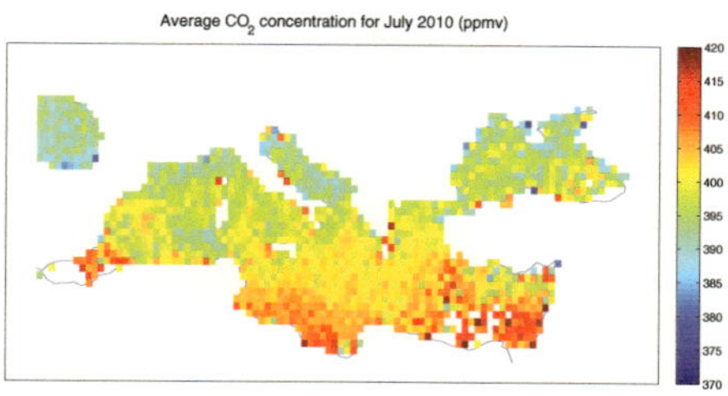

(b) Mediterranean case study

Fig. 8. (a)- CO_2 amount estimated from IASI (this work) and ECMWF analysis. (b)- IASI CO_2 for July 2010 over the Mediterranean area.

However, it is also important to stress that with a polar satellite such as METOP/1, the time-space data coverage is not uniform over the Mediterranean area, so that the spatial gradient has to be considered with some care and its assessment needs a suitable transport model. However, our finding is in agreement with similar maps derived by the Atmospheric Infrared Radiometer Sounder (AIRS) for the month of July (Chahine et al, 2008).

4. Application to CO

As for the case of CO_2, CO has well defined and known rotation transitions, which yield an absorption band, centered in between the atmospheric window at 4.67 μm (2142 cm^{-1}). Because CO is a linear molecule, its strongest absorption features are regularly spaced and yield a periodic pattern, whose period is ≈ 4 cm^{-1} (see Fig. 9). For this reason, the

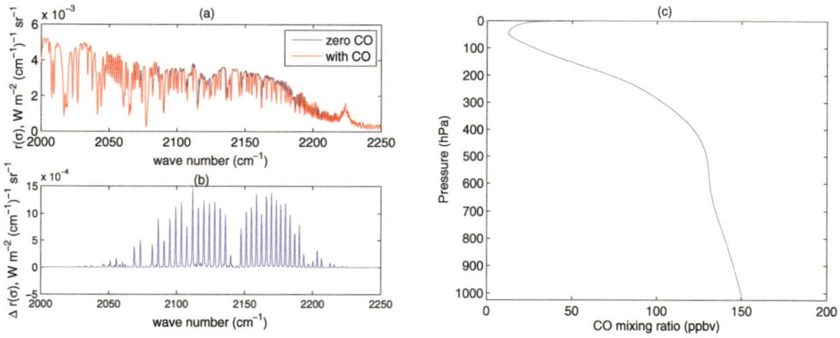

Fig. 9. The figure shows (a) two synthetic IASI spectra in the spectral region of CO absorption; one of the spectra has been calculated with zero load of CO. (b) The difference between the two evidences the sinusoidal appearance of the CO absorption features. Panel (c) shows the CO reference profile used for radiative transfer calculations.

interferogram has to show a characteristic CO-beating at $x = 1/4\text{cm} = 0.25$ cm. We consider a partial interferogram extending from 0.21 to 0.31 cm for a width $\Delta x = 0.1$ cm. With the choice $\Delta x = 0.1$, the factor of noise reduction of Eq. 11 is approximately 4.45. In addition, since the CO band at 4.67 μm is completely covered by IASI band 3 (this extends from 2000 to 2760 cm^{-1}), the interferogram has been built up for band 3 alone. Finally, the reference CO profile is assumed from climatology and is shown in Fig. 9(c). The columnar amount is 109.12 ppbv (parts per billion per volume).

Figure 10 shows a detail of the interferogram in the optical path difference range 0.21 to 0.31 cm. The interferogram has been computed for a tropical model of atmosphere with the CO reference profile, shown in 9(c). A comparison with a case of zero CO load is also provided in the same figure. The corresponding difference spectra are shown in Fig. 10(b).

Unlike the case for CO_2, we have that in the range 0.21 to 0.31 cm, the interferogram gets much signal from the other emission atmospheric sources, which means that the CO signal may be masked from other emitting gases. Moreover, IASI band 3 is that of the three bands with lower signal-to-noise ratio.

However if we look at the difference spectrum, we see that around 2150 cm^{-1} there is a spectral region where the signal is almost determined by the CO signal alone. Once again this behaviour stresses that the difference spectrum can isolate emission feature of a given gas. Needless to say a search for *d*-channels which are mostly sensitive to CO shows that they

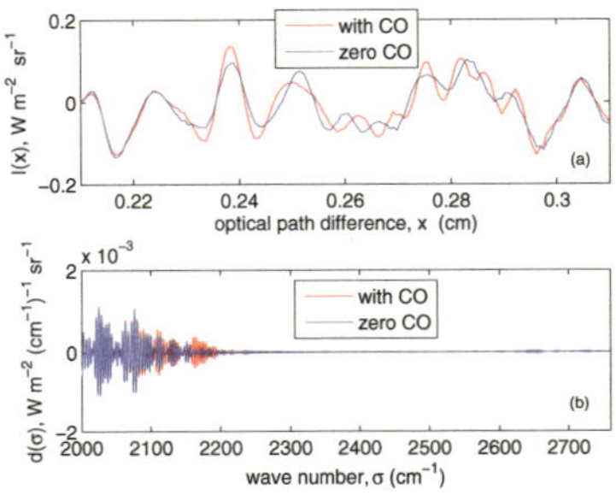

Fig. 10. (a) Inteferogram in the range 0.21 to 0.32 cm for a case of a tropical model of atmosphere; (b) the corresponding *d*-spectrum for the IASI band 3.

are in the range 2150 to 2200 cm^{-1}. Three channels in this range, namely at 2191.25, 2193, 2195 cm^{-1}, provide estimates of the CO columnar amount, q_{CO} whose accuracy is in between 16 to 19 ppbv. Considering that the present average valus of atmospheric CO is around 109 ppbv, they allow an estimate with an accuracy within 15 to 20%. However, the channels are strongly correlated, therefore a methodology as that shown for CO_2 is not possible in this case. For this reason, we have focused on one single channel, that at 2195 cm^{-1}, which achieves the better retrieval performance for CO.

In a case of noise-free radiances the dependence of q_{CO} on $d(\sigma_1) \equiv d_1$, with $\sigma_1 = 2195$ cm^{-1} is parabolic on the full range, which spans from $q_{CO} = 0$ to $q_{CO} = 328$ ppbv. In terms of standardized quantities, the quadratic relation is

$$q_{CO} = a_1 D_1^2 + b_1 D_1 + c_1 \tag{19}$$

where the regression coefficients can be computed through simulation of synthetic radiances belonging to different values of CO columnar amount, q_{CO}. Doing so, we have found the result shown in Fig. 11, where we summarize the quadratic best fit for the two reference models of atmosphere: tropical and High Latitude Winter. As in the case of CO_2, we see that despite the large difference in the state vector, the quadratic fit is accurate for both models. In case we use the standardized difference-radiance, D_1 the regression coefficients do not depend on the atmospheric model. However, as in the case of CO_2 the standardization parameters, μ_1 and s_1 do depend on the state vector. In case of noise-free radiances the quadratic fit of Eq. 19

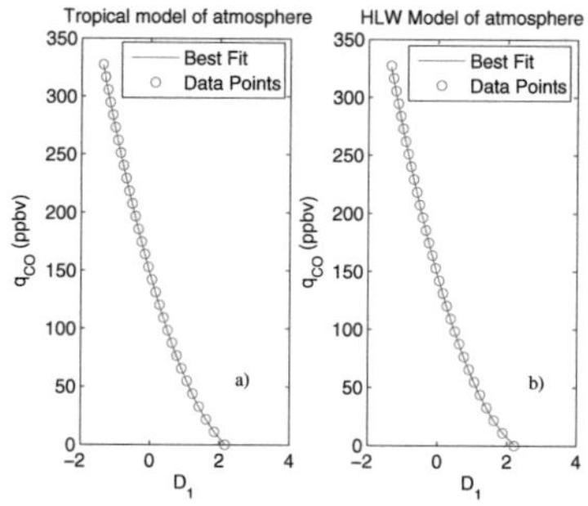

Fig. 11. Quadratic best fit of q_{CO} vs D_1 for (a) a tropical model of atmopshere and (b) a High Latitude Winter (HLW) model of atmopshere.

provides estimates for q_{CO} within ± 0.50 ppbv. In case of noisy radiances, the accuracy can be computed by the usual rule of variance propagation directly from Eq. 19, we have

$$\mathrm{var}(q_{CO}) = \left(2\frac{a_1}{s_1^2}d_1 + \frac{b_1}{s_1}\right)^2 \mathrm{var}(d_1) \tag{20}$$

For a tropical model of atmosphere the typical standard deviation of q_{CO} estimated by Eq. 19 is $\approx$ 16 ppbv (around 15% of its present climatological value). This figure increases to about 25 ppbv in a case of a High Latitude Winter air mass. These figures hold for one single IASI observation. The noise can be halved by considering an average over the $2\times$ 2 pixel mask of the IASI FOV geometry.

As done for CO_2, the sensitivity of the d_1-channel to CO variations has been computed with the help of Eq. 12 and it is compared to that of the equivalent IASI spectrum-channel, $r(\sigma_1) \equiv r_1$ in Fig. 12. It is seen that the difference spectrum improves the sensitivity of a factor two and more. The figure also allows us to get insight into understanding the atmospheric pressure-range at which the retrieval approach is sensitive. As for the case of CO_2, it is seen that the sensitivity extends to a broad range in between 900 to 100 hPa, therefore extending form the Planetary Boundary Layer to the upper troposphere. In contrast to the results of Fig. 12, the sensitivity analysis shows that the d_1 channel at hand is poorly sensitive to other atmospheric parameters. This analysis is not shown here for the sake of brevity.

4.1 Application to IASI data

The mechanism of the procedure to estimate CO columnar amount from IASI observations is the same as that illustrated for CO_2 in section 3.3. During the JAIVEx experiment the CO profile was recorded by airborne in situ profiles (JAIVEx, 2007) recorded with the commercial gas instrument AL 5002 VUV Fast Fluorescence CO Analyser (produced by Aerolaser GmbH). The analyser employs the measurement of the fluorescence of CO when exposed to UV light at a wavelength of 150 nm, which is proportional to the concentration of CO. The measurements covered the lower-middle troposphere (1000 to 400 hPa) and extended to the upper part of the atmosphere based on climatology.

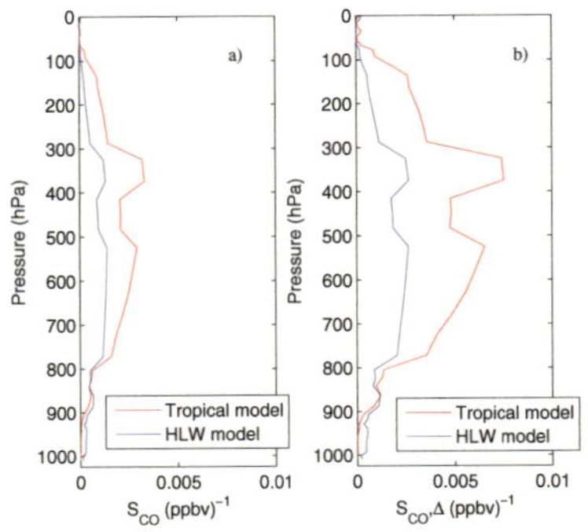

Fig. 12. Panel on the left: (a) Sensitivity of the IASI spectrum channel at 2195 cm^{-1} to CO variations for the case of two atmospheric models; (b) as in (a), but now the sensitivity is computed for the corresponding channel of the difference-spectrum.

These CO profiles are shown in Fig. 13 and compared to the CO reference profile we have used to perform all the radiative transfer calculations needed to estimate the regression coefficients. It is seen from Fig.13 that the reference profile largely differs form those observed in the lower part of the troposphere. Below 400 hPa, the agreement is excellent, just because we use the same climatology as that used by the JAIVEx team. From Fig. 13 it is clearly

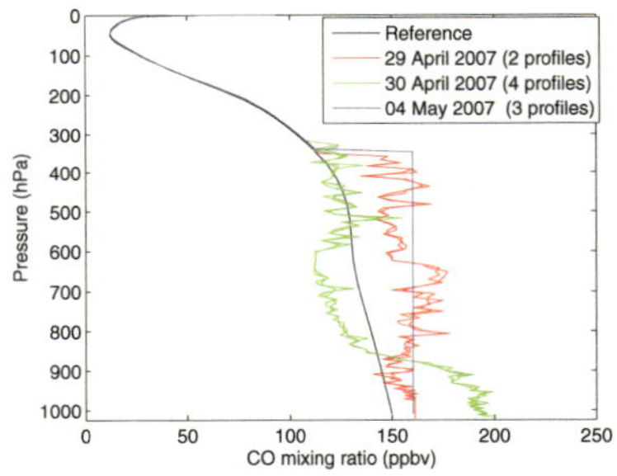

Fig. 13. CO profiles for two days of the JAIVEx experiemnt and comparison with the CO reference profile within the retrieval methodology to estimate the CO columnar amount.

seen that the CO profiles for 04 May 2007 are just a crude interpolation of that on 29 April 2007. Nevertheless, they have been included in the comparison with columnar CO retrieved from IASI for completeness and because these profiles constitute for that day the best in situ estimate of the CO profile.

Having said that, we see that the JAIVEx experiment provides a case study in which the CO reference profile is different from the supposedly correct CO profile corresponding to the JAIVEx campaign (see Fig.13). Thus, we have a case study in which the shape of the profile, and not only the CO integrated amount, differs form that of the reference profile. This situation allows us to check the sensitivity of the methodology to the shape of the CO profile.

The results of our methodology applied to the 25 IASI soundings during the JAIVEx experiment are shown in Fig. 14(a) along with the estimation of the columnar amount from in situ measurements. This last estimate has been obtained by integrating the CO mixing ratio profile (see Fig.13) for each day of the JAIVEx experiment. For each day, the corresponding in situ estimates (2 for 29 April 2007, 4 for 30 April 2007, and 3 for 04 May 2007) have been averaged and these three average values are shown as flat lines in Fig. 14(a). The comparison

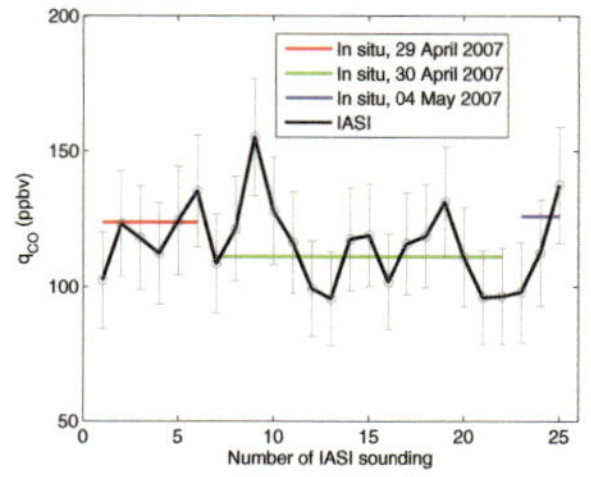

(a) JAIVEx case study

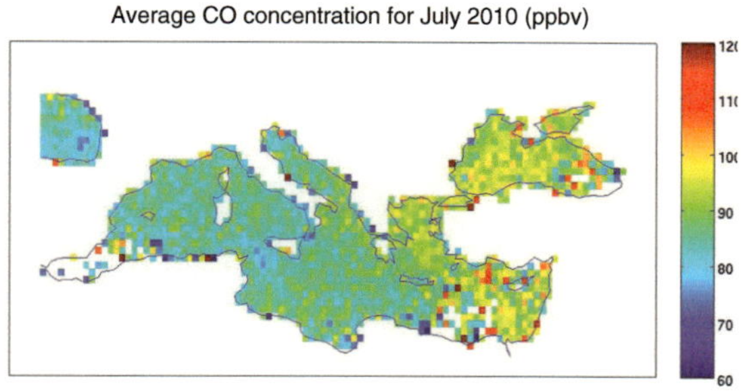

(b) Mediterranean case study

Fig. 14. (a)- CO integrated amount estimated from IASI (this work) and in situ observations. (b)- IASI CO for July 2010 over the Mediterranean area.

provided in Fig. 14 shows that IASI agrees with in situ observations in evidencing a slight decrease of the CO load for the target area on 30 April 2007 in comparison to those over passed by IASI on 29 April and 04 May, respectively. This is also evidenced if we compare the average values for the day 29 April 2007 and 30 April 2007. For 29 April 2007 we have a mean value of (119.6 $\pm$ 7.0) ppbv for IASI against a value of 123.7 ppbv estimated from in situ observations. For the day 30 April 2007, both IASI and in situ observations show a lower value for the CO load, IASI (114.2 $\pm$ 4.2) ppbv, in situ 111.0 ppbv.

Finally, as done for CO_2, for illustrative purposes, Fig. 14(b) shows a monthly map of CO computed over the Mediterranean area for the month of July 2010.

5. Application to CH_4

Unlike CO and CO_2, methane is not a linear molecule, therefore we have no particular hint from its structure about which interval of the interferogram is most sensitive to the variation of this gas. However, methane together with water vapour, is the main absorber within IASI band 2, which means that if we consider the interferogram of IASI band 2 alone, we should be able to isolate a suitable portion of the interferogram signal, which is mostly dominated by CH_4. By trial and error this interval has been identified in the segment 1.34-1.352 cm, for a bandwidth of 0.0120 cm. With this reduced bandwidth, according to Eq. 11, we have a noise reduction within the difference spectrum of 12.90. Actually, because of the effect of IASI noise correlation, the reduction factor is even higher. If we consider that for IASI band 2 we have the better signal-to-noise ratio, we have that methane is the gas, which we can retrieve with the highest stability and accuracy. In particular the channels in the spectral segment 1210 to 1220 cm^{-1} exhibit the poorest sensitivity to the state vector, but methane.

A good channel is that at $\sigma = 1210.75\ cm^{-1}$. The regression relation between the channel and the methane columnar amount is a polynomial of third order. The regression error is 0.01 ppmv in case of noise-free radiances and $\approx$ 0.1 ppmv in case of noisy radiances. The regression relation is invariant with the state vector as it is shown in Fig. 15. The CH_4 reference profile we use for the radiative transfer calculation is that shown in Fig. 15(d), which gives a columnar amount for methane of 1.65 ppmv. The sensitivity, $S_{CH_4,\Delta}(\sigma)$ of the d-spectrum (channel at $\sigma = 1210.75\ \text{cm}^{-1}$) to methane is shown in 15(c) for the tropical and High Latitude Winter models of atmosphere, whose main atmospheric parameters have been shown in Fig. 4

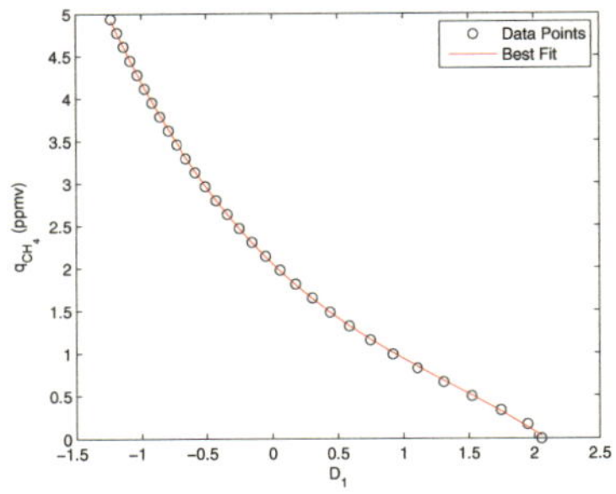

(a) tropical model

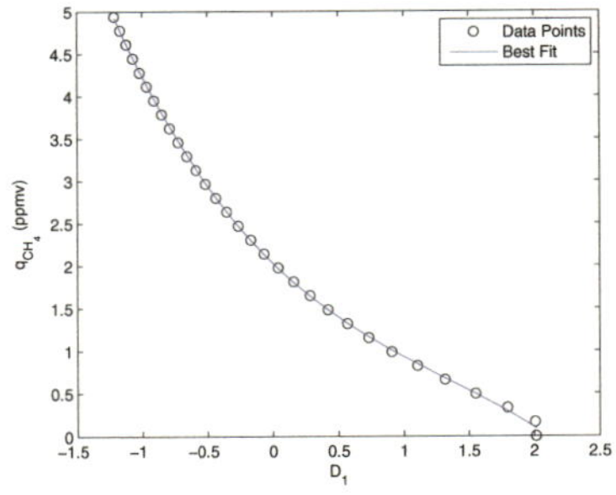

(b) High Latitude Winter model

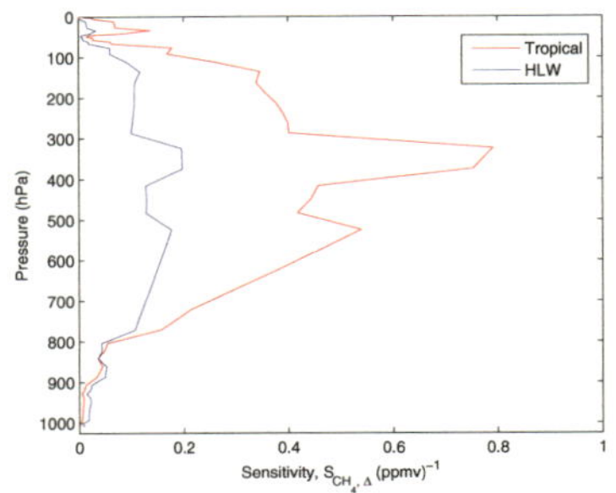

(c) Sensitivity to CH_4

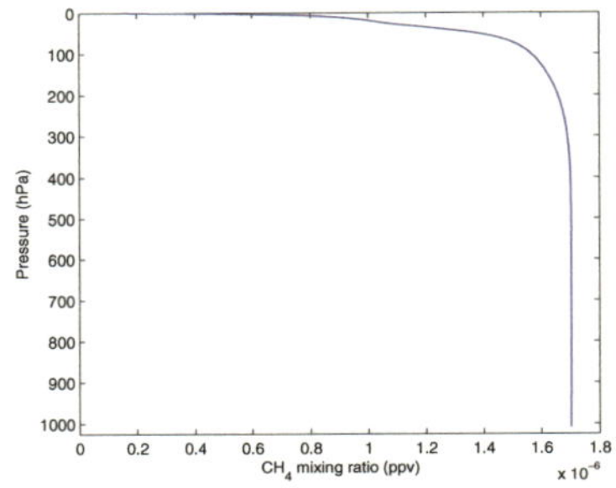

(d) CH_4 reference profile

Fig. 15. The polynomial fit at $1210.75\ \text{cm}^{-1}$ exemplified for two models of atmosphere, (a) and (b); the sensitivity to methane (c); the reference methane mixing ratio profile (d).

5.1 Application to IASI data

The mechanism of the procedure to estimate CH_4 columnar amount from IASI observations is the same as that illustrated for CO_2 in section 3.3. The procedure for methane has been applied to the 25 IASI spectra and the results are shown in Fig. 16(a). We see that the columnar amount is very stable and varies in between 1.60-1.90 ppmv with an average of (1.70 ± 0.02) ppmv. We remember that this data were acquired in 2007. Today the average global value of methane is credited of a value equal to 1.74 ppmv (Blasing, 2011). During the JAIVEx experiment there were no in situ observations of methane. However, we can perform a consistency check about the observed and computed variability. According to our procedure, the accuracy of the estimates is $\approx$ 0.094 ppmv. Because the JAIVEx case study consider a limited target area, we have to expect a very low variability as far as the columnar amount of CH_4 is considered. This means that the variability we see in Fig. 16(a) has to be largely due to random fluctuations, therefore the standard deviation of the 25 IASI estimates has to be consistent with the computed accuracy of 0.094 ppmv. In fact, this standard deviation gives the value 0.0945 ppmv.

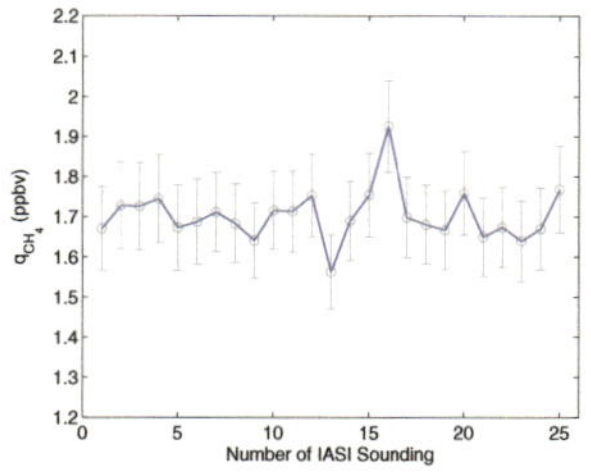

(a) JAIVEx case study

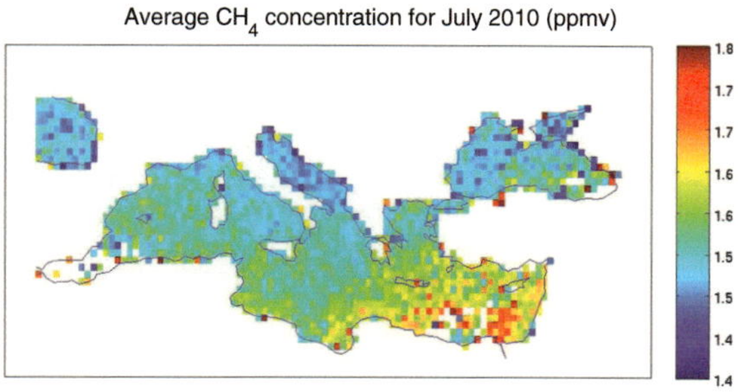

(b) Mediterranean case study

Fig. 16. (a)- CH_4 integrated amount estimated from IASI. (b)- IASI CH_4 for July 2010 over the Mediterranean area.

Finally, as done for CO_2 and CO, for illustrative purposes Fig. 16(b) shows a monthly map of CH_4 computed over the Mediterranean area for the month of July 2010. Also for CH_4, the map clearly shows a gradient in the North-to-South direction, which is consistent with the general circulation of the Mediterranean area.

6. Application to N_2O

As for the methane, N_2O is not a linear molecule, therefore partial interferograms, which are capable of enhancing the variations of this gas with respect to those of other dominant atmospheric parameters, have to be be judiciously found by careful inspections of synthetic interferogram signals generated as a function of N_2O columnar amount, q_{N_2O}. Using this strategy we have found that the partial interferogram in the range 1.06-1.08 cm for a width of 0.02 cm is largely sensitive to N_2O. With this reduced bandwidth, according to Eq. 11, we have a noise reduction within the difference spectrum of 10. Actually, because of the effect of IASI noise correlation, the reduction factor is even higher. However, we have to consider that N_2O absorption insists within the IASI band 3, which is that with the worse signal-to-noise ratio. With this in mind we have that accuracy with which we can estimate q_{N_2O} is of the order of 10% at the level of single channels. In addition, *good* channels tend to be strongly correlated, therefore there is no advantage in trying to combine them to improve the final accuracy.

The regression relation, which fits to the data with an error of less than 3 ppbv is a polynomial of fourth order. The polynomial is independent of the atmospheric sate vector as it is shown in Fig. 17 which exemplifies the polynomial regression for the case of the *d*-channel at 2165.50 cm^{-1}.

The N_2O reference profile we use for the radiative transfer calculation is that shown in Fig. 17(d), which gives a columnar amount for N_2O of 306.5 ppbv. The sensitivity, $S_{N_2O,\Delta}(\sigma)$ of the *d*-spectrum (channel at $\sigma = 2204.5\ cm^{-1}$) to N_2O is shown in 17(c) for the tropical and High Latitude Winter models of atmosphere, whose main atmospheric parameters have been shown in Fig. 4

6.1 Application to IASI data

The philosophy and mechanism of the procedure to estimate N_2O columnar amount from IASI observations is the same as that illustrated for CO_2 in section 3.3. For N_2O, the procedure

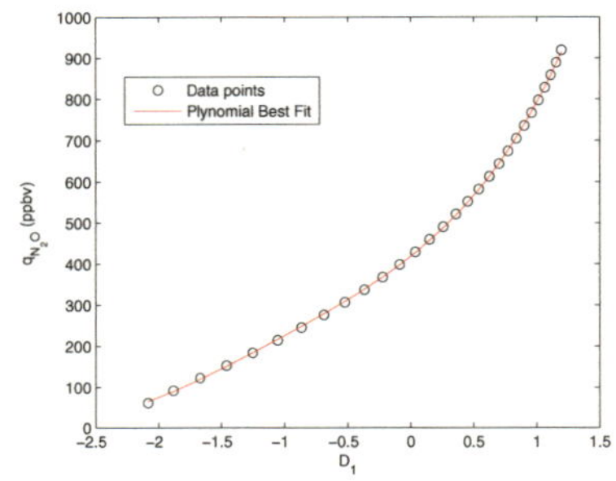

(a) tropical model

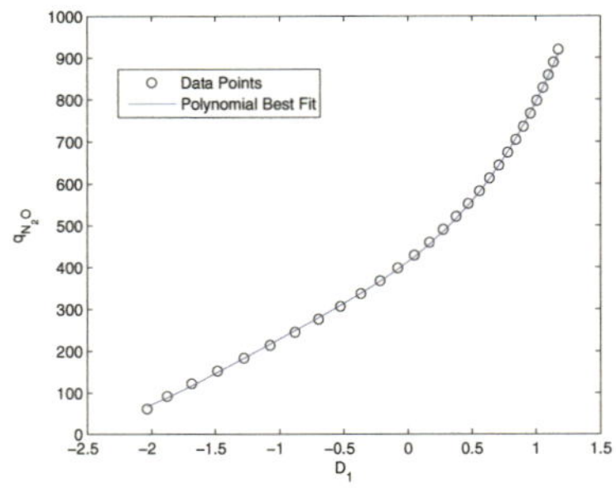

(b) High Latitude Winter model

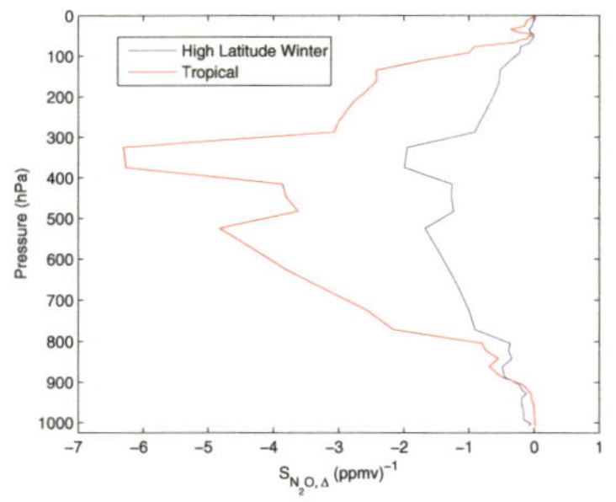

(c) Sensitivity to N_2O

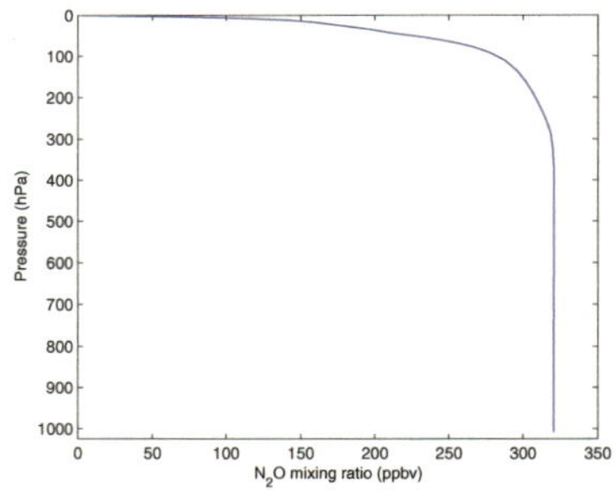

(d) N_2O reference profile

Fig. 17. The polynomial fit at 2165.50 cm^{-1} exemplified for two model of atmosphere, (a) and (b); the sensitivity to N_2O (c); the reference N_2O mixing ratio profile (d).

has been applied to the 25 IASI spectra and the results are shown in Fig. 18(a). We see that the columnar amount varies in between 250-450 ppbv with a peak-to-peak variability of about 200 ppbv. The average value over the 25 soundings is 303 ppbv. This is a bit lower than the today average global mean value ($\approx$ 323 ppbv (Blasing, 2011)) estimated based on in situ observations at ground level stations. However, N_2O may be characterized by large anomaly from its mean global value even on monthly time scales (Blasing, 2011; Lubrano et al., 2004; Ricaud et al., 2009). However, if we consider the monthly map of N_2O computed

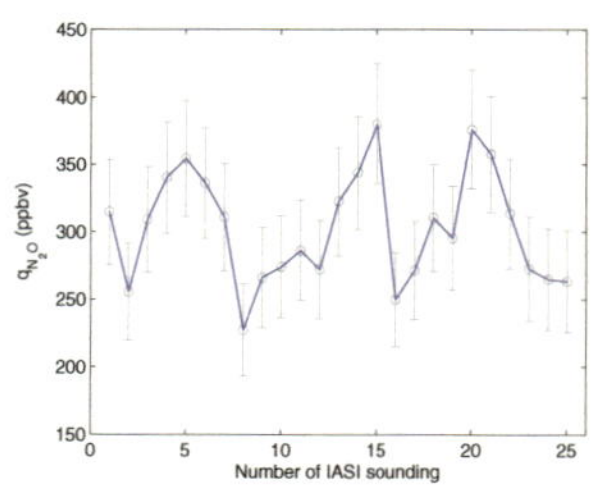

(a) JAIVEx case study

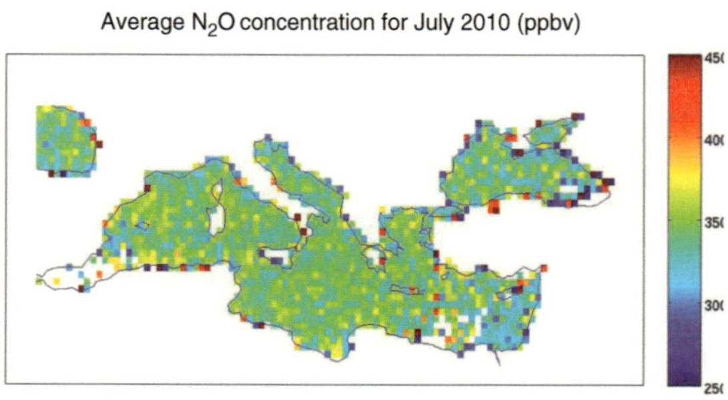

(b) Mediterranean case study

Fig. 18. (a)- N_2O integrated amount estimated from IASI. (b)- IASI N_2O for July 2010 over the Mediterranean area.

over the Mediterranean area for the month of July 2010 (see Fig. 18(b)), we see that the average concentration comes closer to the monthly values normally observed with in situ observation at ground level stations.

7. Conclusion

We have described and presented the basic aspects of Fourier Spectroscopy with Partially Scanned Interferogram and exemplified its application to the retrieval of minor and trace gases in the atmosphere. Observations from the IASI instrument have been considered and the technique has been applied to estimate the columnar amount of CO_2, CO, CH_4 and N_2O.

The retrieval algorithms we have implemented rely on simple polynomial regression relations, whose coefficients, once properly standardized, are independent of the atmospheric state. However, the technique needs the standardization parameters, which may depend on the state vector. Thus, the procedure still relies on the availability of a suitable a-priori best estimate of the atmospheric state vector. We have shown that this best estimate can be confidently obtained by previously inverting IASI radiances for skin temperature, and temperature, water vapour and ozone profiles. The accuracy of this *best estimate* can be by far lower than that expected from IASI radiances themselves. We have shown that uncertainty about the temperature profile of the order of $\pm$2K along the profile are easily tolerated. For water vapour we can easily tolerate uncertainties of more than $\pm$20%, along the profile. Furthermore, the technique is largely insensitive to the surface emission.

FTS*PSI is a truly novel methodology as far as its applications to high spectral resolution infrared observations is concerned. The tools we have presented in this work have not been particularly optimized. Nevertheless, their applications to atmospheric gases yielded impressive results for accuracy and quality, which are unprecedented if compared to those normally obtained with the usual machinery of inverting spectral radiances. We think that the capability of the methodology has been only scratched at the surface, and we hope this study can soon attract attention and stimulate new research studies, which can hopefully exploit the many facets of the tool.

8. Acknowledgements

IASI has been developed and built under the responsibility of the Centre National d'Etudes Spatiales (CNES, France). It is flown onboard the Metop satellites as part of the EUMETSAT Polar System. The IASI L1 data are received through the EUMETCast near real time data distribution service. We thank Dr Stuart Newman (Met Office) for providing the JAIVEx data. The JAIVEx project has been partially funded under EUMETSAT contract Eum/CO/06/1596/PS. The FAAM BAe 146 is jointly funded by the Met Office and the Natural Environment Research Council. The US JAIVEx team was sponsored by the National Polar-orbiting Operational Environmental Satellite System (NPOESS) Integrated Program Office (IPO) and NASA.

9. References

Amato, U.; De Canditiis, D.; Serio, C. (1998). Effect of apodization on the retrieval of geophysical parameters from Fourier-Transform Spectrometers. *Appl. Opt.*, 37, 6537-6543.

Amato, U.; Masiello, G.; Serio, C.; Viggiano, M. (2002). The σ-IASI code for the calculation of infrared atmospheric radiance and its derivatives. *Environ. Model. Software*, 17, 651-667.

Bell, R.J. (1972). *Introductoiy Fuorier Transform Spectroscopy*, Acad.Press, New York.

Blasing, T.J. (2011), Recent Greenhouse Gas Concentrations, DOI: 10.3334/CDIAC/atg.032

Boynard, A.; Clerbaux, C.; Coheur, P.-F.; Hurtmans, D.; Turquety, S.; George, M.; Hadji-Lazaro, J.; Keim, C.; Mayer-Arnek, J. (2009). Measurements of total and tropospheric ozone from the IASI instrument: comparison with satellite and ozonesonde observations, *Atmos. Chem. Phys.*, 9, 6255-6271.

Carissimo, A.; De Feis, I.; Serio, C. (2005). The physical retrieval methodology for IASI: the δ-IASI code. *Environ. Model. Software*, 20, 1111–1126.

Chevalier, F. (2001) *Sampled Database of 60 Levels Atmospheric Profiles from the ECMWF Analysis*; Technical Report; ECMWF EUMETSAT SAF programme Research Report 4; ECMWF: Shinfield Park, Reading, UK.

Chahine M.T.; Barnet, C.; Olsen, E.T.; Chen, L.; Maddy, E. (2005). On the determination of atmospheric minor gases by the method of vanishing partial derivatives with application to CO_2, *Geophys. Res. Lett.*, 32, L22803, doi:10.1029/2005GL024165.

Chahine, M. T.; Chen, L.; Dimotakis, P.; Jiang, X.; Li, Q.; Olsen, E. T.; Pagano, T.; Randerson, J.; Yung Y. L. (2008). Satellite remote sounding of mid-tropospheric CO_2, *Geophys. Res. Lett.*, 35, L17807, doi:10.1029/2008GL035022.

Clarisse, L.; Coheur, P. F.; Prata, A. J.; Hurtmans, D.; Razavi, A.; Phulpin, T.; Hadji- Lazaro, J.; Clerbaux, C. (2008). Tracking and quantifying volcanic SO2 with IASI, the September 2007 eruption at Jebel at Tair, *Atm. Chem. Phys.*, 8, 7723-7734.

Clarisse L., Clerbaux, C.; Dentener, F.; Hurtmans, D.; Coheur, P.-F. (2009). Global ammonia distribution derived from infrared satellite observations, *Nature Geosci.*, 2, 479- 483, doi:10.1038/ngeo551

Clerbaux C.; A. Boynard, A.; Clarisse, L.; M. George, M.; Hadji-Lazaro, J.; Herbin, H.; Hurtmans, D.; Pommier, M.; Razavi, A.; Turquety, S.; Wespes, C.; Coheur, P.-F. (2009). Monitoring of atmospheric composition using the thermal infrared IASI/Metop sounder, *Atmos. Chem. Phys.*, 9, 6041-6054.

Crevoisier, C.; Chédin, A.; Matsueda, H.; Machida, T.; Armante, R.; Scott, N. A. (2009). First year of upper tropospheric integrated content of CO_2 from IASI hyperspectral infrared observations. *Atmos. Chem. Phys.*, 9, 4797-4810, doi:10.5194/acp-9-4797-2009.

Engelen, R.J.; Serrar, S.; Chevallier, F. (2009). Four dimensional data assimilation of Atmospheric CO_2 using AIRS observations. *J. Geophys. Res.*, 114, D03303, doi:10.1029/2008JD010739.

Gardiner, T.; Mead, M. I.; Garcelon, S.; Robinson, R.; Swann, N.; Hansford, G. M.; Woods, P. T.; Jones, R. L. (2010). A lightweight near-infrared spectrometer for the detection of trace atmospheric species. *Rev. Sci. Instrum.* 81, 083102; doi:10.1063/1.3455827 (11 pages).

Goldstein, H. W.; Grenda, R. N.; Bortner, M. H.; Dick, R. (1978). CIMATS: a correlation interferometer for the measurements of atmospheric trace species. *Proceedings of the 4th Joint Conference on Sensing of Environmental Pollutants*. American Chemical Society, 586-589.

Grieco, G.; Masiello, G.; Matricardi, M.; Serio, C.; Summa, D.; Cuomo, V. (2007). Demonstration and validation of the φ-IASI inversion scheme with NAST-I data, *Q. J. R. Meteorol. Soc.*, 133, 217-232.

Grieco, G.; Masiello, G.; Serio, C. (2010). Interferometric vs Spectral IASI Radiances: Effective Data-Reduction Approaches for the Satellite Sounding of Atmospheric Thermodynamical Parameters. *Remote Sens.* 2, 2323-2346.

Grieco, G.; Masiello, G.; Serio, C.; Jones, R.L.; Mead, M.I. (2011). Infrared Atmospheric Sounding Interferometer correlation interferometry for the retrieval of atmospheric gases: the case of H_2O and CO_2. *Appl. Opt.* 50, 4516-4528. doi:10.1364/AO.50.004516

FAAM, "Joint Airborne IASI Validation Experiment (JAIVEX)" http://badc.nerc.ac.uk/data/jaivex/

Kyle, T.G. (1977). Temperature soundings with partially scanned interferograms. *Appl. Opt.*, 16/2, 326-332.

Lubrano, A.M.; Masiello, G.; Matricradi, M.; Serio, C.; Cuomo, V. (2004). Retrieving N_2O from nadir-viewing infrared spectrometers. *Tellus B*, 56(3), 249 - 261.

Masiello, G.; Matricardi, M.; Rizzi, R.; Serio, C. (2002). Homomorphism between Cloudy and Clear Spectral Radiance in the 800-900-cm^{-1} Atmospheric Window Region. *Appl. Opt.* 41, 965-973. doi:10.1364/AO.41.000965

Masiello, G.; Serio, C; Shimoda, H. (2003). Qualifying IMG tropical spectra for clear sky. *Journal of Quantitative Spectroscopy and Radiative Transfer.* 77/2, 131-148. doi:10.1016/S0022-4073(02)00083-3

Masiello, G.; Serio, C.; Cuomo, V. (2004). Exploiting Quartz Spectral Signature for the Detection of Cloud-Affected Satellite Infrared Observations over African Desert Areas. *Appl. Opt.* 43, 2305-2315. doi:10.1364/AO.43.002305

G. Masiello, G.;& Serio, C. (2004). Dimensionality-reduction approach to the thermal radiative transfer equation inverse problem. *Geophys. Res. Lett.*, 31, L11105, doi:10.1029/2004GL019845.

Masiello, G.; Serio, C.; Carissimo, A.; Grieco, G. (2009). Application of ϕ-IASI to IASI: retrieval products evaluation and radiative transfer consistency. *Atmos. Chem. Phys.*, 9, 8771-8783.

Masiello, G.; Matricardi, M.; Serio, C. (2011). The use of IASI data to identify systematic errors in the ECMWF forecasts of temperature in the upper stratosphere. *Atmos. Chem. Phys.*, 11, 1009-1021, doi:10.5194/acp-11-1009-2011.

K. Masuda, K.; Takashima, T.; Takayama, Y. (1988). Emissivity of pure and sea waters for the model sea surface in the infrared window regions, *Remote Sens. Environ.*, 24, 313-329.

McMillan W. W., C. Barnet, L. Strow, M. T. Chahine, M. L. McCourt, J. X. Warner, P. C. Novelli, S. Korontzi, E. S. Maddy, S. Datta (2005), "Daily global maps of carbon monoxide from NASA's Atmospheric Infrared Sounder", Geophys. Res. Lett., 32, L11801, doi:10.1029/2004GL021821.

Pichett, H.M.; & Strauss, H.L. (1972). Signal-to-Noise ratio in Fourier spectrometry, *Analytical Chemistry*, 44(2), 265-270.

Ricaud, P.; Attié, J.-L.; Teyssédre, H.; El Amraoui, L.; Peuch, V.-H.; Matricardi, M.; Schluessel, P. (2009). Equatorial total column of nitrous oxide as measured by IASI on MetOp-A: implications for transport processes, *Atmos. Chem. Phys.*, 9, 3947-3956, doi:10.5194/acp-9-3947-2009.

Robinson, E.A.;& Silvia, M.T. (1981). *Digital Foundation of Time Series Analysis: Wave-Equation Space-Time Processing*, Vol. 2 Holden-Day, San Francisco.

Serio, C.; Lubrano, A.M.; Romano, F.; Shimoda, H. (2000). Cloud Detection Over Sea Surface by use of Autocorrelation Functions of Upwelling Infrared Spectra in the 800-900-cm^{-1} Window Region. *Appl. Opt.* 39, 3565-3572, doi:10.1364/AO.39.003565

Smith, W.L.; Howell, H.B.; Woolf, H.M. (1979). The use of interferometric radiance measurements for the sounding the atmosphere. *Appl. Opt.*, 36, 566-575.

12

Identification of Intraseasonal Modes of Variability Using Rotated Principal Components

Michael Richman and Andrew Mercer
The University of Oklahoma and Mississippi State University
USA

1. Introduction

Documenting mid-tropospheric global-scale circulation is important to climate modelers. Models are applied to capture the most basic statistics of the flow (e.g., annual and season means and variability). As model physics improve, the goal is to extend the accuracy of the models to shorter, more societally relevant time scales (e.g., monthly average flow). Within a month or season, recurrent modes are present. Comparing such observed flows to modeled counterparts can provide an arbiter of success. There is a long history is isolating such patterns in meteorology using teleconnections (correlation patterns). Some of the initial work by Namias (1980) was used to create atlases of teleconnections at every gridpoint. This approach has the merit of completeness; however, the redundancy in patterns from adjacent gridpoints is inefficient, given present day computational power. An extension of the Namias approach to selected base points provided extensive documentation of Northern Hemisphere winter teleconnection patterns of sea level pressure and 500 mb height (Wallace & Gutzler, 1981).

A second approach to identifying these variability modes uses eigenvectors to filter the correlation structures into recurrent patterns that are localized. To document the characteristic flow patterns and their time dependent statistics, an objective methodology is applied to portray the localization of the centers of action in each mode. Often, rotated principal component analysis (RPCA) is employed to decompose the aforementioned correlation structure of the flow to obtain information on the morphology of the flow patterns, their associated time behavior and the variability of the total flow associated with each pattern. The eigenvectors are scaled to create principal components that are linearly transformed or rotated to exploit the local structure within the domain, identifying physically meaningful circulation patterns.

Barnston & Livezey (1987; hereafter referred to as BL87) present an extensive catalogue of mid-tropospheric patterns using this approach. Their work is somewhat limited by use of a specific rotation (Varimax) that enforces an orthogonal transformation matrix from the principal components to their rotated counterparts. Additionally, they did not test rigorously the number of eigenmodes, selecting ten modes for each analysis. Selecting too few eigenmodes can result in multiple patterns being merged on each eigenvector retained. Moreover, if too many patterns are selected, the circulation modes can be fragmented.

In this work, we extend the eigenvector-based approach by relaxing the orthogonality constraint and test each analysis for the optimal number of eigenvectors to retain and rotate. While BL87 used cutting-edge analyses for the 1980's, the limited availability of data (35 years) and computational power (limiting the number of gridpoints to 358) are suboptimal by current standards. Innovation in the analysis procedure in recent years combined with newer data sets, the availability of 63 years of data and much denser grids to document the climate system are strong motivators to re-examine Northern Hemisphere geopotential modes of variability.

2. Data and methodology

2.1 Description of the dataset

To conduct a study on hemispheric teleconnectivity, it is critical to have high-resolution, global gridded data with a long period of record. The NCEP/NCAR reanalysis (NNRP – Kalnay et al., 1996) dataset, which is formulated on a 2.5° x 2.5° latitude-longitude grid, includes over 200 meteorological variables at up to 17 vertical levels at 6 hour intervals from 1948 - present. Variables in the NNRP were rated by Kalnay et al. (1996) based on the influence of observations, model derived data, and climatology in their formulation. As this study is concerned with mid-tropospheric intraseasonal modes of variability, 500 hPa Northern Hemispheric geopotential heights from the NNRP (rated most reliable by Kalnay et al., 1996) were used in the analysis.

To maintain consistency with the methodology of BL87, monthly averages of 500 hPa heights were formulated from 63 years of the NNRP (1948-2010). The domain of interest used by BL87 did not include height data south of 20°N latitude, primarily owing to the lack of variability in the tropics in mid-level height data. To maintain consistency, the same 20°N threshold for the southern-most latitude line was used in the present study. Fig. 1a shows the study domain and the grid spacing associated with the NNRP.

The NNRP are provided on a latitude-longitude grid; hence, gridpoints will converge as the domain extends poleward. This convergence causes artificial inflation of correlation values at higher latitudes. To mitigate this issue, the data were analysed objectively using a Barnes

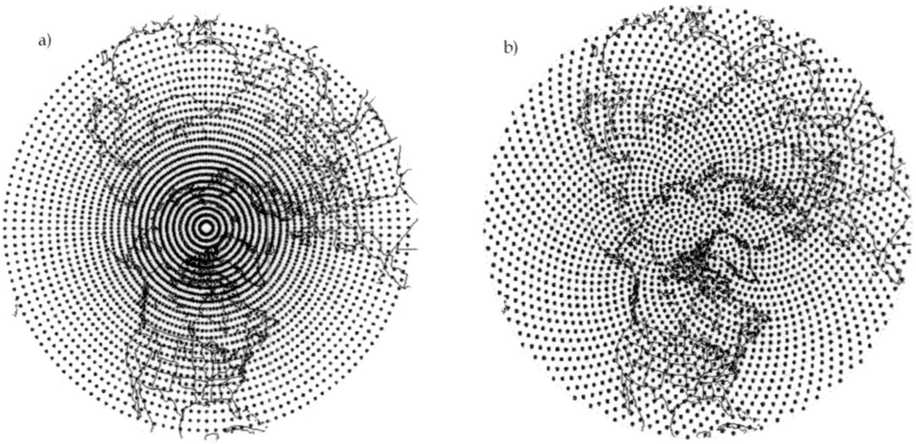

Fig. 1. The NNRP grid (panel a) and the interpolated Fibonacci grid (panel b).

analysis (Barnes, 1964) to place the data on a Fibonacci grid (Swinbank & Purser, 2006) that, by definition, has equal grid spacing. Such a grid will not result in artificial inflation of the correlations between the 2303 gridpoints (Fig. 1b). To ensure no data extrapolation at lower latitudes, the spacing associated with the Fibonacci grid was kept identical to that at the equator. Interpolation error of the Barnes analysis was calculated at less than 1%. The Fibonacci grid defines the domain and is used in the computation of the correlation matrix.

2.2 RPCA methodology

Variation of geopotential height is a function of latitude. To avoid biasing the analyses and to permit smaller, but equally important, variation in the southern regions of the domain to be represented equally, a correlation matrix (**R**), rather than variance-covariance matrix, was computed. Using the correlation matrix standardizes the data to a mean of zero and standard deviation of one. Hence, all the subsequent analyses are standardized anomalies (**Z**) from the mean. The correlation matrix was formed among the grid points by summing over the 63 year sample for January or July, providing a representative month in the cold or warm season, respectively. The correlation matrix for January (July) was decomposed into a square matrix of eigenvectors (**V**) and associated diagonal matrix of eigenvalues (**D**), given by the equation

$$\mathbf{R} = \mathbf{V}\mathbf{D}\mathbf{V}^{T} \tag{1}$$

The rank of the eigenvector matrix is equal to the smaller of the number of gridpoints or number of observations minus 1. Because there were 63 observations, only 62 eigenvalues were nonzero and 62 eigenvectors were extracted. The goal of this stage of the analysis is to create a set of basis vectors that compress the original variability in **R** into a new reference frame. It is possible to plot the elements of each vector (**V**) on spatial maps; however, the patterns in **V** do not result in any localization of the spatial variance, nor do they represent well the variability in **R** (Richman, 1986). The eigenvectors were scaled by the square root of the corresponding eigenvalue to create principal component loadings (**A**). Doing so permits the data to be expressed as

$$\mathbf{Z} = \mathbf{F}\mathbf{A}^{T} \tag{2}$$

where the vectors in **F** represent the new set of basis functions, known as principal component scores and **A** is the matrix of weights that relates the original standardized data (**Z**) to **F**. The vectors in **A** contain elements that are regression coefficients between **Z** and **F**.

Many of the 62 dimensions represent small-scale signals (sub-planetary scale) that have variance properties indistinguishable from noise, associated with very small eigenvalues. We truncate the number of principal components to represent only that variance associated with planetary scale wavetrains. To accomplish this goal, a two-step process is applied. First, the magnitudes of the eigenvalues are examined and those with relatively large eigenvalues are retained to yield a subset of l principal component loading vectors. The value l is selected by implementing the scree test (Wilks, 2011) to provide a visual estimate of the approximate number of non-degenerate eigenvectors to retain. At this stage, a number of roots, l, is selected to be liberal, intentionally representing more than the ideal number of roots, k, associated with the large-scale signal. To assess the coherent signal, the vectors of **A** are linearly transformed to a new set of vectors, **B**, known as rotated principal component loadings, which simplify or localize the hemispheric wavetrains to agree with the correlation structure of the data. This process can be summarized by the equation

$$\mathbf{B} = \mathbf{A}\mathbf{T}^{-1} \quad (3)$$

where **T** is an invertible transformation matrix that represents a rotation of the reference frame into a position that results in the maximal simplification that the data permit. The rotation algorithm used in this analysis is Promax (Richman, 1986). Promax PC scores allow for correlations between the vectors in **F**. The goal of the rotation is to identify, in the vectors of **B**, height anomaly patterns that recur often during the month of January (July). The spatial properties of each vector in **B** are a function of the number of rotated PC vectors retained (l). Therefore, it is critical to select the optimal number of vectors, k. To accomplish that goal, the matrix **A** is transformed to **B** for a variable number of PC vectors retains (i.e., 2 to l). Each solution yields a different set of patterns. We desire a set $k < l$ that capture as much coherent large scale signal as possible that matches the patterns embedded within **R**. The one set of k PC loadings that relates best to the correlation matrix generating them is determined and the number of PCs retained is set to k. The process, outlined in Richman & Lamb (1985), and refined in Richman (1986), selects each vector of **B** and identifies the location or gridpoint with the highest absolute PC loading and matches the rotated PC loading vector to the corresponding correlation vector using the congruence coefficent (Richman, 1986). The corresponding correlation vector in **R** is a teleconnection pattern. Hence, this method incorporates the logic of the traditional teleconnection approach, providing an objective procedure to selecting k. For January, the Promax solution with the optimum match occurred as $k = 8$ and for July, $k = 4$. The average absolute congruence match was found to be 0.904 Promax in January and 0.895 for Promax in July.

Owing to the linear decomposition of **R**, signs of the coefficients, in each vector of **B** can be multiplied by -1 with no loss of interpretation. Because positive values exceeding ~ +0.25 correspond to ridges (Richman & Gong, 1999), negative values of ~ −0.25 correspond to troughs, multiplication by -1 reverses the interpretation of the troughs and ridges. Furthermore, the sign of the PC score is multiplied by the sign of the PC loading to obtain a physical interpretation of any monthly map (Compagnucci & Richman, 2008). For example, a negative anomaly in a PC loading map multiplied by a negative PC score gives the interpretation of a positive height anomaly.

3. Results

3.1 January

For the January analysis, 8 principal components are retained and rotated. The patterns represent a decomposition of the flow into modes of variability. The total variability extracted is 70.0 percent.

3.1.1 January mode descriptions

Pattern 1 - The West Pacific / North Pacific Oscillation (WP/NPO; Fig. 2) is characterized by a center near the Kamchatka Peninsula at 60°N, 160°E and an oppositely signed east-west Pacific anomaly band at 30°N, centered on the dateline. This oscillation accounts for 10.3% of the total January 500 hPa variance. The congruence coefficient (Richman, 1986) between the RPC loading vector and the point teleconnection pattern that has a basepoint location coinciding with the largest magnitude RPC loading (hereafter referred to as the pattern/teleconnection congruence) is -0.91, suggesting a close match between the RPC and correlation structure. Wallace & Gutzler (1981) identified this pattern in the 500 hPa geopotential heights during boreal winter. They found the thickness pattern consistent with a cold-core equivalent barotropic structure. BL87 found WP/NPO in a 700 hPa analysis of 1950–84 monthly winter height anomalies. Hsu & Wallace (1985) investigated the temporal structure of WP variability at subseasonal time scales, relating the anomalies to Rossby wave dispersion. The NPO has been associated with Alaskan blocking events (Renwick & Wallace, 1996) and modulation of the Pacific storm track (Lau 1988; Rogers 1990). Linkin & Nigam (2008) claim the WP pattern is a basic analog to the North Atlantic Oscillation and has impact on the weather in the Pacific Northwest, especially in coastal regions, in the south-central Great Plains, and on marginal sea ice zones in the Arctic. Our analysis of the pattern time series (Fig. 3a) suggests no significant trend and no year to year persistence in the autocorrelation function (ACF; Fig. 3b) However, the year to year January variability in the WP/NPO (Fig 3a) shows decadal nonstationarity as there exist several periods where the mean is significantly different from zero (Table 1) and the variance structure undergoes dramatic rapid year to year variability in the first three decades of the analysis with lower frequency variability in the past three decades (Table 2).

Pattern 2 - Subtropical zonal winter pattern (SZW; Fig. 4) has a unique morphology with one major anomaly centered over western China (34°N, 90°E). Nearly all regions south of 35°N have negative RPC loadings, suggesting a zonal pattern in the subtropics. There are a number of secondary centers of the same sign extending to the west across northern Africa and in the southwestern US. The pattern accounts for 9.4 percent of the total 500 hPa variance in January. The pattern/teleconnection congruence is -0.92, a close match between the RPC and correlation structure. The time series of PC scores for SZW is nonstationary with a strong inverse linear trend that implies the sign of the anomalies has reversed over the 63 year period (Fig. 5a). There is a sharp discontinuity around 1980. The statistical significance of the linear trend line (Fig. 5a) is $p=6x10^{-6}$, which supports the idea of strong height rises in the subtropical regions. The ACF (Fig. 5b) has a unique pattern with multi-year persistence, with 5 of the first 8 lags significant, owing to the trend in subtropical heights. The decadal analysis of the mean RPC scores strengthens the idea of a reversal in anomaly pattern as the scores are positive in the first three decades and negative in the latter three (Table 1). The decadal variance is below the overall mean of 1 in every decade, suggesting that this pattern persists substantially from year to year (Table 2).

Pattern 3 - The Northern Asian pattern (NA; Fig. 6) has its main center of action close to the North Pole at 80°N and 40°E. Secondary centers of opposite sign are situated over Mongolia at 50°N 100°-120°E and just west of the United Kingdom. Esbensen (1984) identified a similar pattern, although the main center was shifted south of the position shown herein. This mode explains 8.3% of the January height variance. The pattern/teleconnection congruence is 0.95, representing a very close match to the traditional teleconnection pattern. The time series (Fig. 7a) has no discernible trends but has periods of high variance in the 1960's and the 1990's -2000's (Table 2). The ACF (Fig. 7b) shows no year to year persistence.

Pattern 4 - The eastern Atlantic teleconnection (AE; Fig. 8), accounting for 6.5% of the variance, has a dipole at ~ 40°W with one center at longitude 30°W and a second elongated

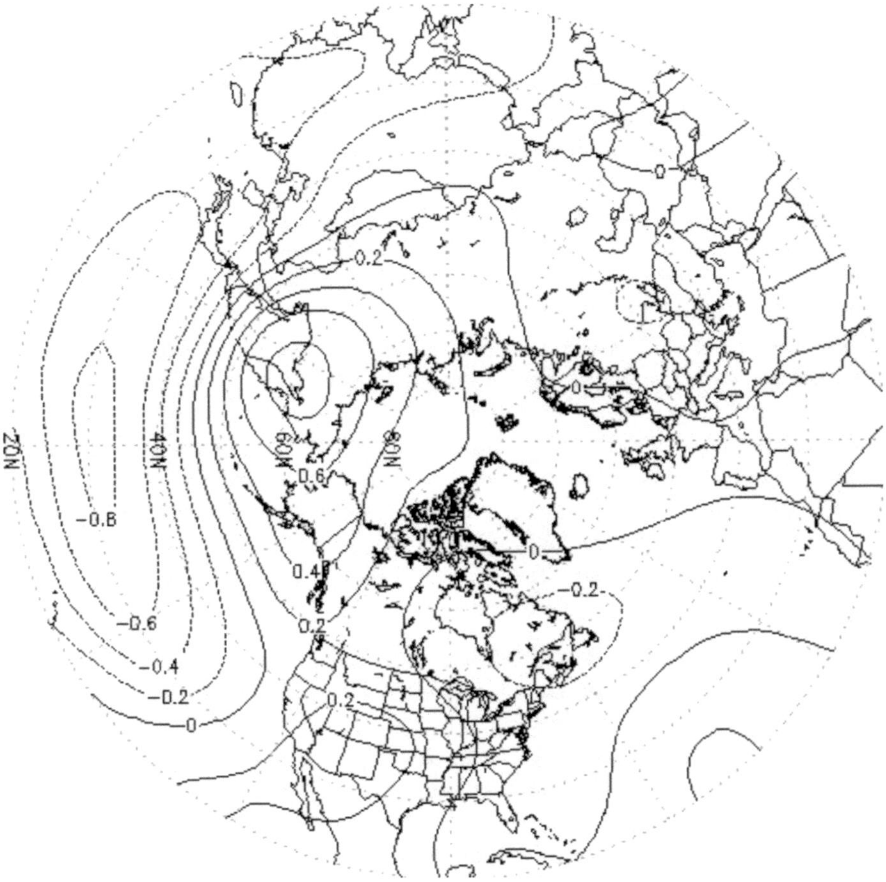

Fig. 2. The WP/NPO January pattern.

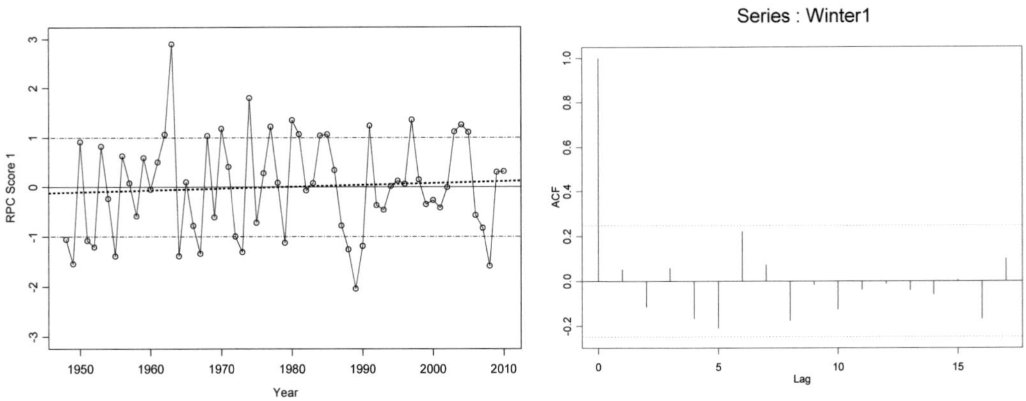

(a) (b)

Fig. 3. WP/NPO PC score time series (panel a) and autocorrelation function (ACF – panel b). Panel a has horizontal lines at ± 1, to assist in identification of extreme scores, and a linear regression line, shown as dotted line. Panel b indicates the correlation versus lag. The white noise bandwidth is identified as dotted lines on panel b.

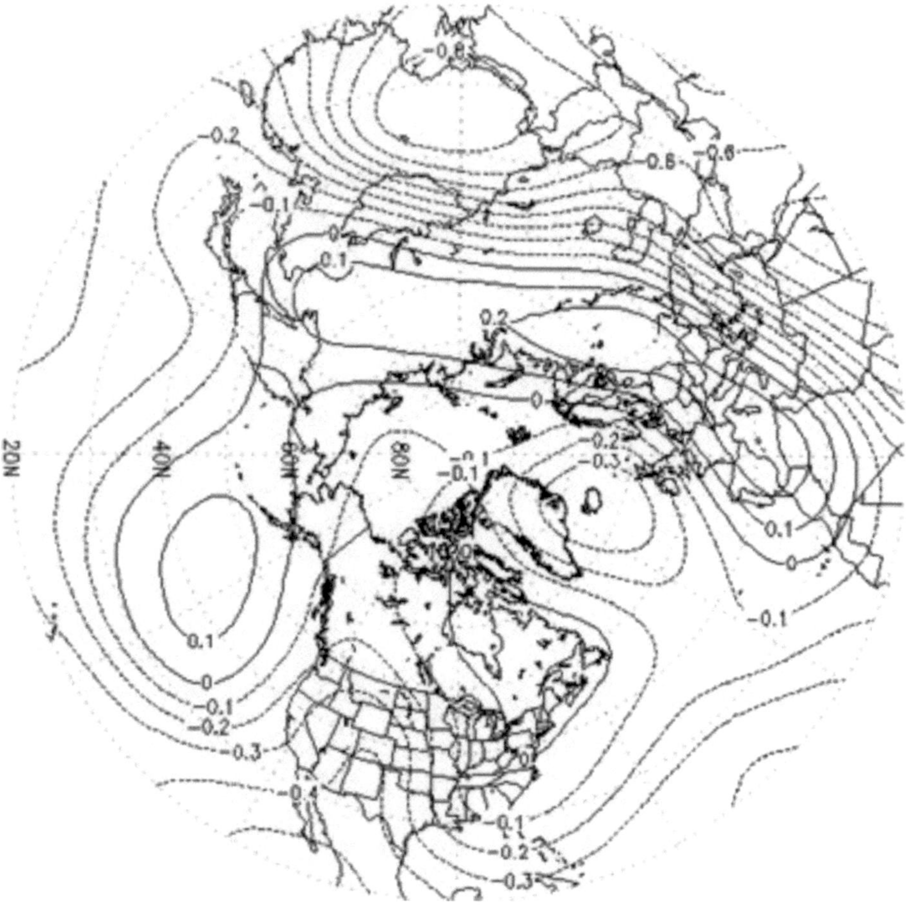

Fig. 4. The SZW January pattern.

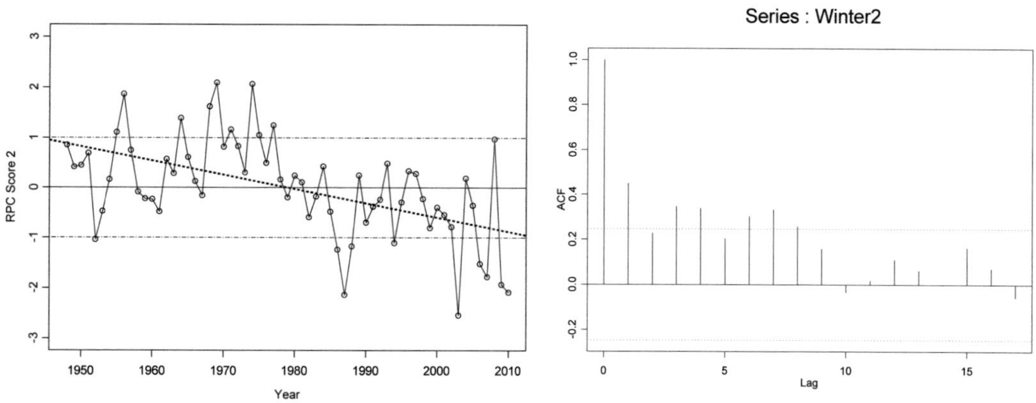

Fig. 5. SZW PC score time series (panel a) and ACF (panel b).

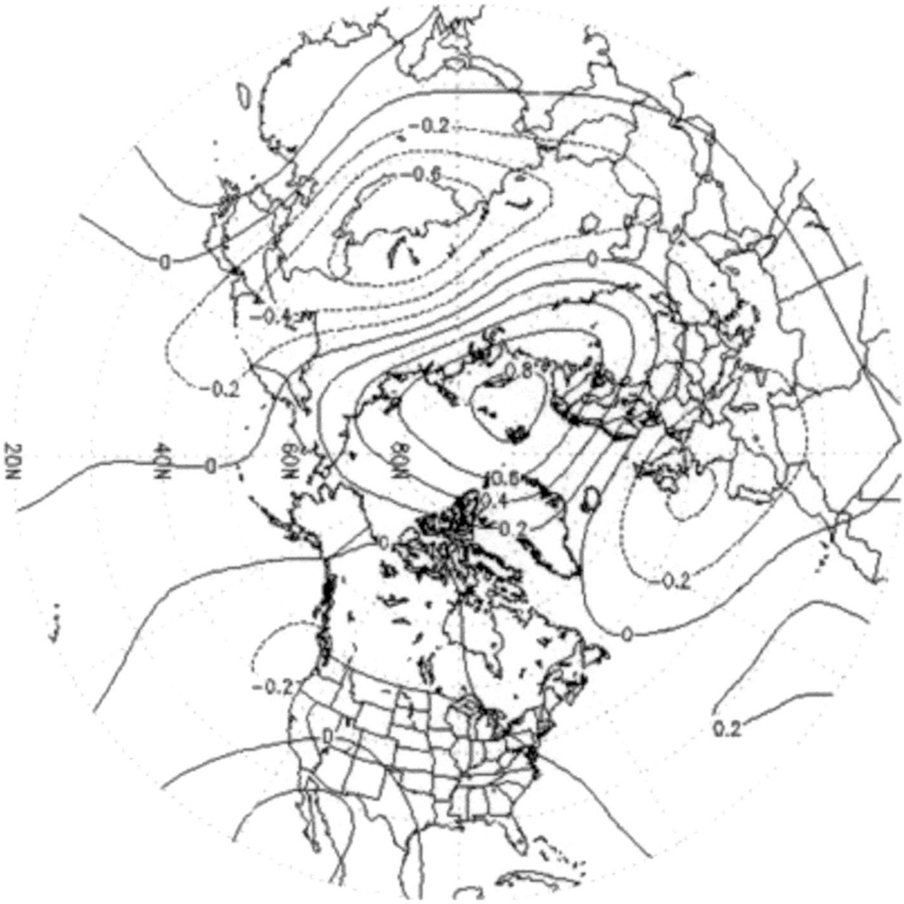

Fig. 6. The NA January pattern.

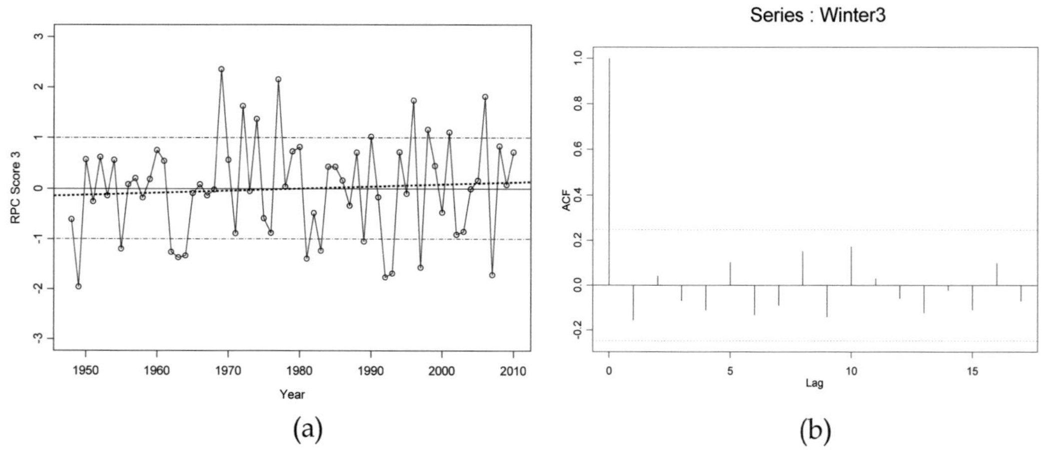

Fig. 7. NA PC score time series (panel a) and ACF (panel b).

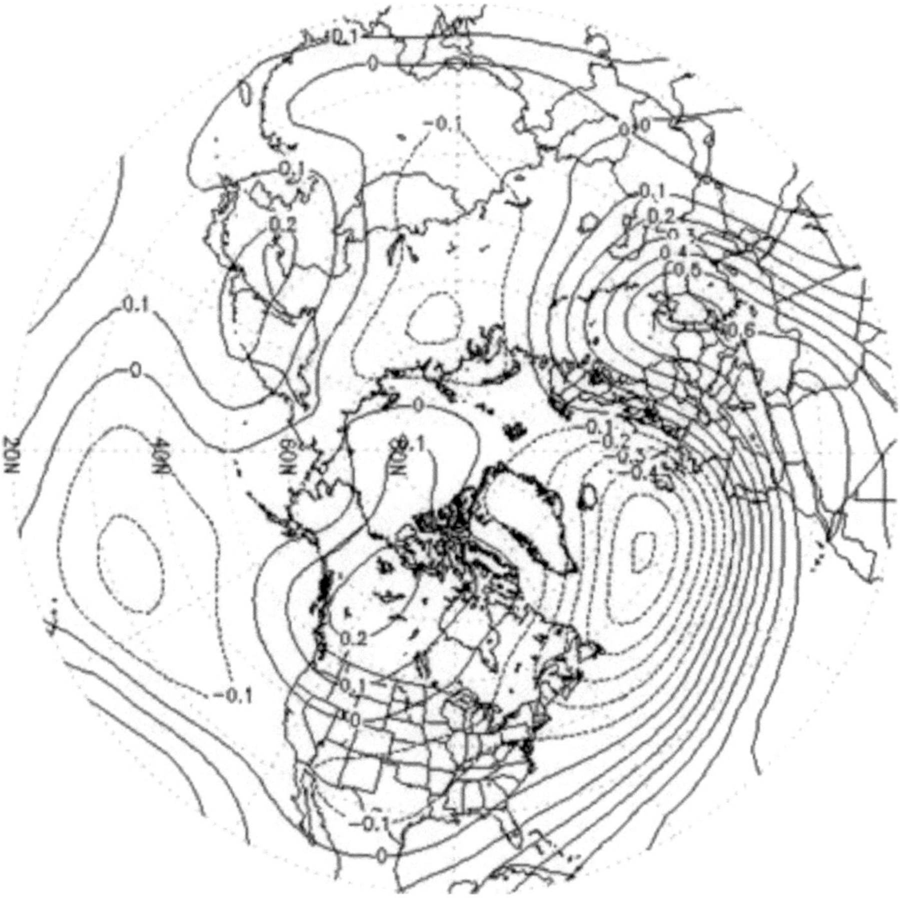

Fig. 8. The AE January pattern.

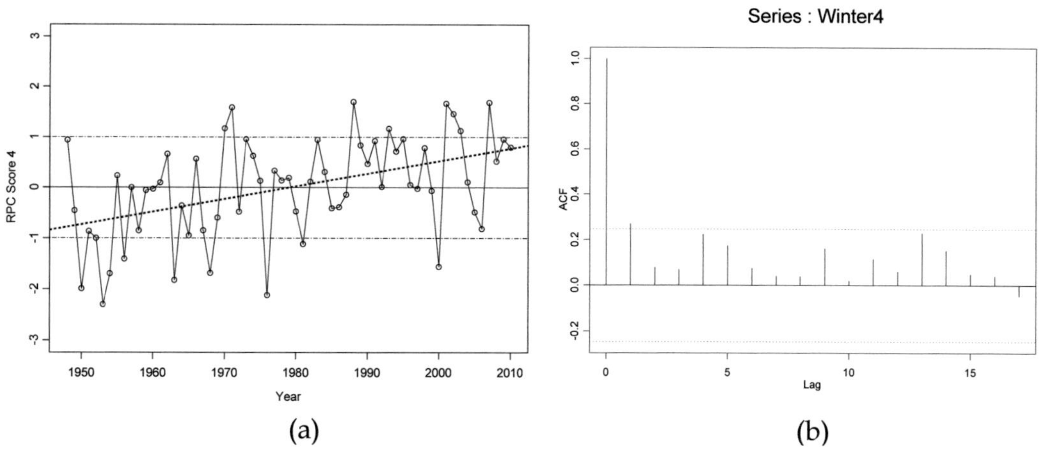

Fig. 9. AE PC score time series (panel a) and ACF (panel b).

center of opposite sign with maxima at 30°E, with a strong gradient in PC loadings. The pattern in Europe is elongated and extends southeast over North Africa and the subtropical western Atlantic Ocean with a secondary center at 30°N, 0°. Lau (1988) discusses how this pattern modulates Atlantic storm tracks in Europe through modifying the strength of the westerlies. BL87 identify a version of the AE pattern with more of a north-south dipole. The pattern/teleconnection congruence of AE is 0.81, representing a fairly close match to the traditional teleconnection pattern. The time series (Fig. 9a) has a strong positive trend (p=0.00002). The decadal mean analysis (Table 1) indicates nonstationarity with mostly negative scores during the 1950's–1960's, then near-zero means for the next twenty years and positive values after. The variance of the time series (Table 2) shows periodic fluctuations with low variability in the 1990's. Analysis of the persistence of the time series for the AE (Fig. 9b) indicates possible slight lag 1 autocorrelation although, by chance, one year in twenty is expected to exceed the white noise level.

RPC Mean	1948-1957	1958-1967	1968-1977	1978-1987	1988-1997	1998-2007
RPC1	-0.40	0.10	0.23	0.31	-0.25	0.12
RPC2	0.48	0.18	1.17	-0.39	-0.25	-0.88
RPC3	-0.21	-0.28	0.56	-0.09	-0.22	0.07
RPC4	-0.85	-0.36	-0.01	-0.08	0.68	0.39
RPC5	-0.06	0.64	0.49	-0.06	-0.39	-0.58
RPC6	0.76	-0.28	0.37	-0.67	0.01	-0.15
RPC7	0.03	0.50	-0.10	0.57	-0.67	-0.31
RPC8	0.04	0.08	0.40	-0.25	0.15	-0.35

Table 1. RPC mean scores by decade for January

RPC Variance	1948-1957	1958-1967	1968-1977	1978-1987	1988-1997	1998-2007	Row Mean
RPC1	0.93	1.63	1.17	0.69	1.14	0.59	1.03
RPC2	0.65	0.31	0.37	0.61	0.36	0.69	0.50
RPC3	0.70	0.60	1.53	0.59	1.57	1.20	1.03
RPC4	1.08	0.59	1.47	0.31	0.31	1.27	0.84
RPC5	0.35	0.60	1.09	2.15	0.79	0.33	0.89
RPC6	1.49	0.54	1.38	1.14	0.42	0.46	0.91
RPC7	1.19	0.98	1.16	0.52	1.12	0.59	0.93
RPC8	1.02	1.07	1.09	1.21	0.84	0.40	0.94
Column Mean	0.93	0.79	1.16	0.90	0.82	0.69	

Table 2. RPC variance by decade for January. Row means are the average RPC variance and column means are the decadal variance.

Pattern 5 - The North Atlantic Oscillation (NAO ; Fig. 10) is a well-documented pattern exhibiting a dipole centered over Iceland and the Azores, explaining 9.3% of the January variance. The center in the subtropics extends well into the United States with a secondary maxima over the southern Plains. A weaker center appears in central Russia near latitude 60°N, 80°E. This pattern is known to control the strength of the westerlies (BL87). The pattern/teleconnection congruence is -0.91, representing a close match to the traditional teleconnection pattern. The time series (Fig. 11a) has a significant negative trend (p=0.02) largely arising from the positive values in the 1960s–1970's and negative values in the 1990s–2000's. The decadal analysis (Table 1) confirms nonstationary behavior with 20 year means (1958-1977) of +0.57 followed by -0.485 for the period 1988-2007. The variance of the time series (Table 2) undergoes extreme fluctuations in the 1970's and 1980's, with strikingly less variance in the adjacent periods. No year to year persistence is noted in the ACF time series analysis (Fig. 11b).

Pattern 6 - The Pacific North American pattern (PNA; Fig. 12), accounting for 8.0% of the variance, is another well-established teleconnection (Esbensen, 1984; Hsu & Wallace, 1985; BL87). The structure is defined by four centers of action. In the RPC loadings, the tropical center is located at 20°N, 165°W and another of opposite sign north at 50°N 175°E. This center has the largest magnitude PC loadings in excess of 0.8. The North American centers are at 50°N, 120°W and an elongated area anchored at 30°N 90°W. This Rossby wavetrain is controlled by the strength of the westerlies (BL87). The pattern/teleconnection congruence is -0.90, representing a close match to the teleconnection pattern. The time series (Fig. 13a) is nonstationary with a significant negative trend (p=0.05), arising largely from the positive values in the first decade of the analysis and near-zero or negative means in the last three decades (Table 1). The variance of the time series (Table 2) indicates aperiodic fluctuations in the variance, most noteworthy during 1948-1957. The time series of the PNA sugest no persistence (Fig. 13b).

Pattern 7 - The Eurasian pattern (EA; Fig. 14) is a high latitude patterns with four centers of action. In the RPC loadings, the primary centers are over the North Sea (58°N, 4°E) and northern Kazakhstan (50°N, 60°E). A secondary center with slightly lower magnitude RPC loadings exists over northeast China (44°N, 125°E) with another lower magnitude center over northern Canada's Northwestern Passages (75°N, 120°W). EA explains 11% of the winter height variance. The pattern is similar to the BL87 Eurasian pattern type 2. The pattern/teleconnection congruence is -0.95, representing a very close match to the traditional teleconnection pattern. The time series (Fig. 15a) has a marginally significant negative trend (p=0.10) arising from the near-zero to positive values through around 1990, followed by more negative values (Table 1). The variance of the time series (Table 2) indicates nonstationarity with alternating bursts of variability on a decadal time scale. The ACF of the time series for the EA (Fig. 15b) indicates very minor persistence in lag 1 and another negative value outside the white noise spectrum in year 12. It is likely that the latter is a statistical artifact.

Pattern 8 - The tropical Northern Hemisphere mode (TNH; Fig. 16) spans the North American sector from the eastern Pacific Ocean (40°N, 135°W) linked to a higher latitude center in Ontario, Canada just west of Hudson Bay (52°N, 85°W) and an elongated subtropical center in the western Atlantic, well off the southeast coast of the US (28°N, 70°W). The mode accounts for 7.2% of the variance in January 500 hPa height fields. The

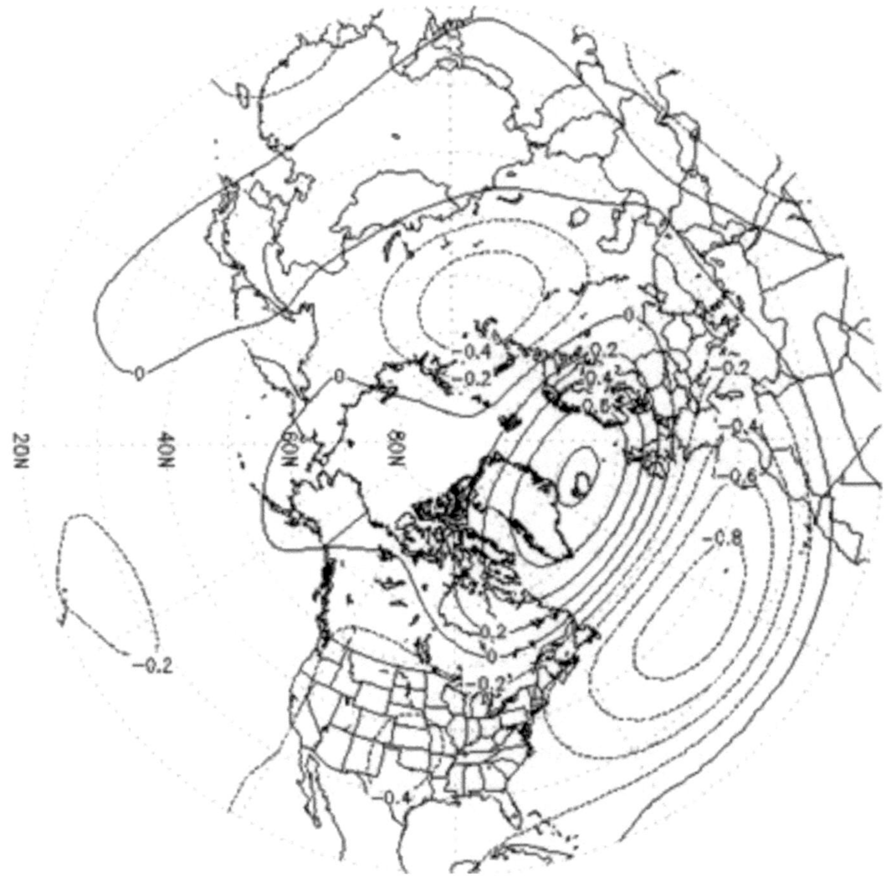

Fig. 10. The NAO January pattern.

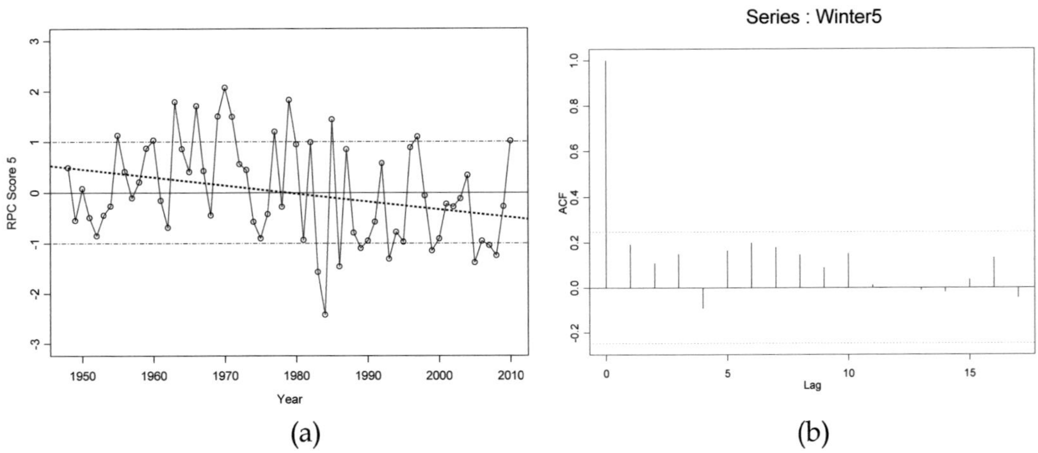

Fig. 11. NAO PC score time series (panel a) and ACF (panel b).

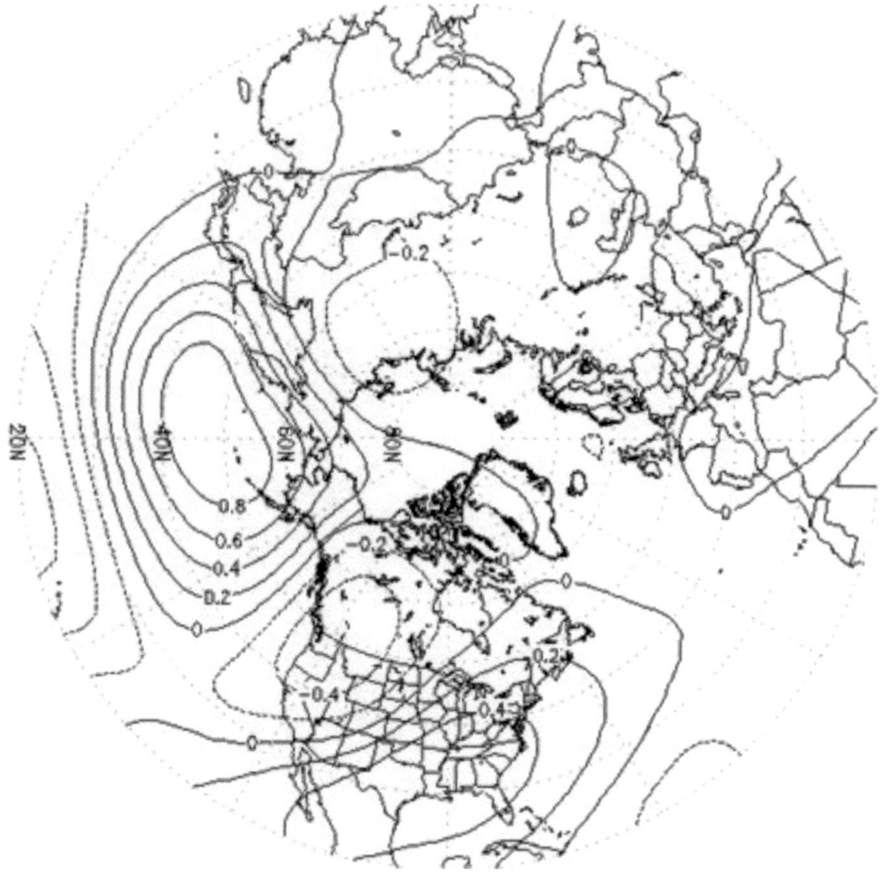

Fig. 12. The PNA January pattern.

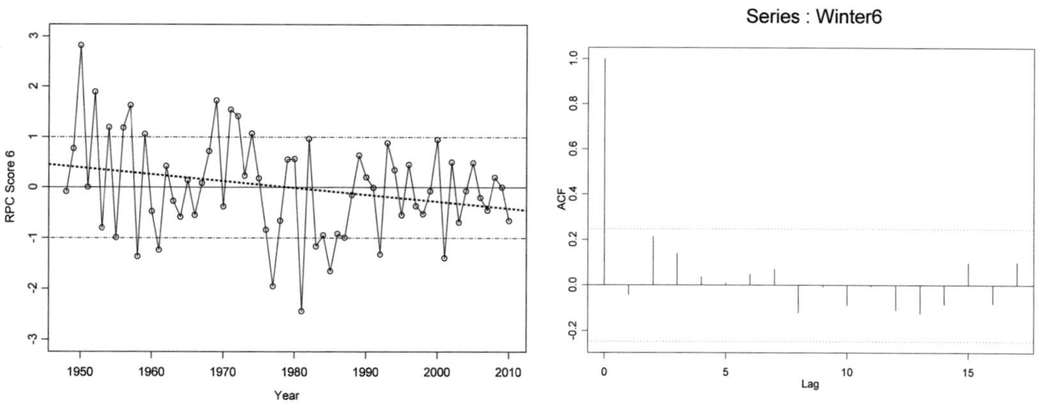

(a) (b)

Fig. 13. PNA PC score time series (panel a) and ACF (panel b).

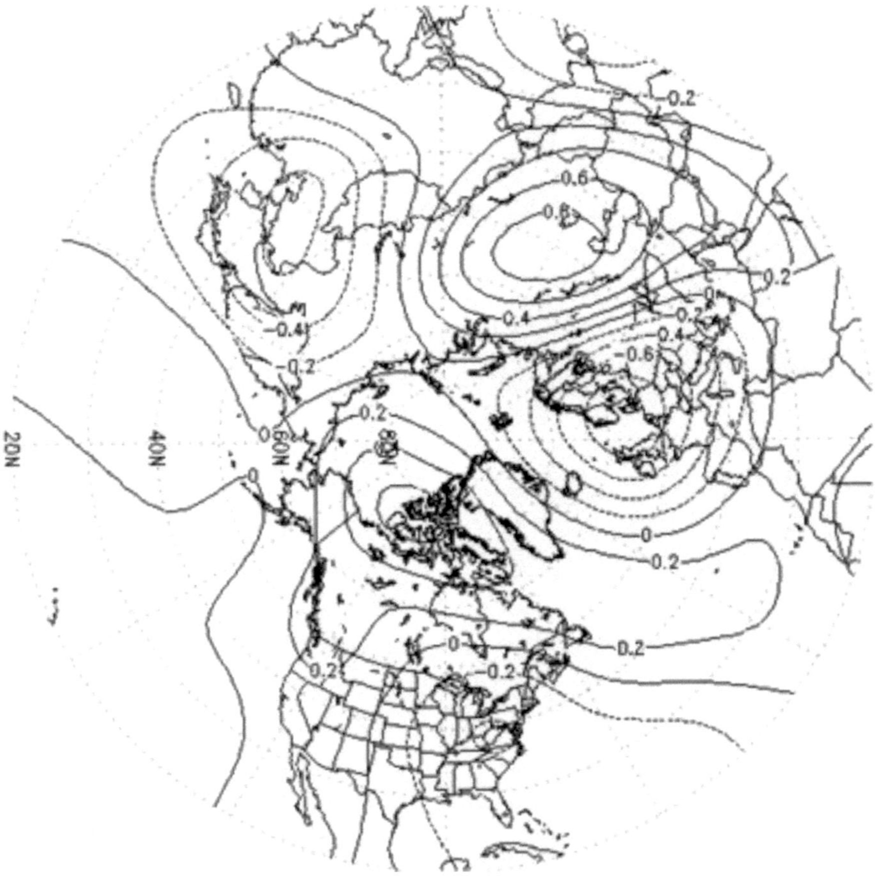

Fig. 14. The EA January pattern.

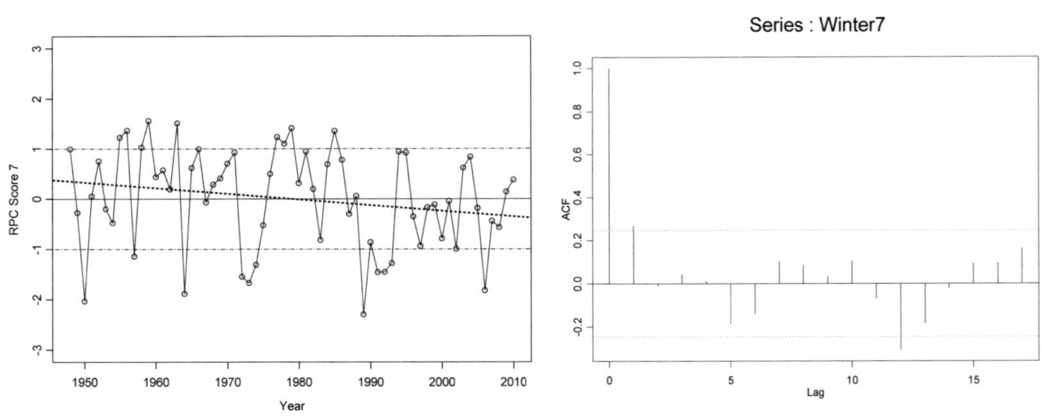

(a) (b)

Fig. 15. Eurasian PC score time series (panel a) and ACF (panel b).

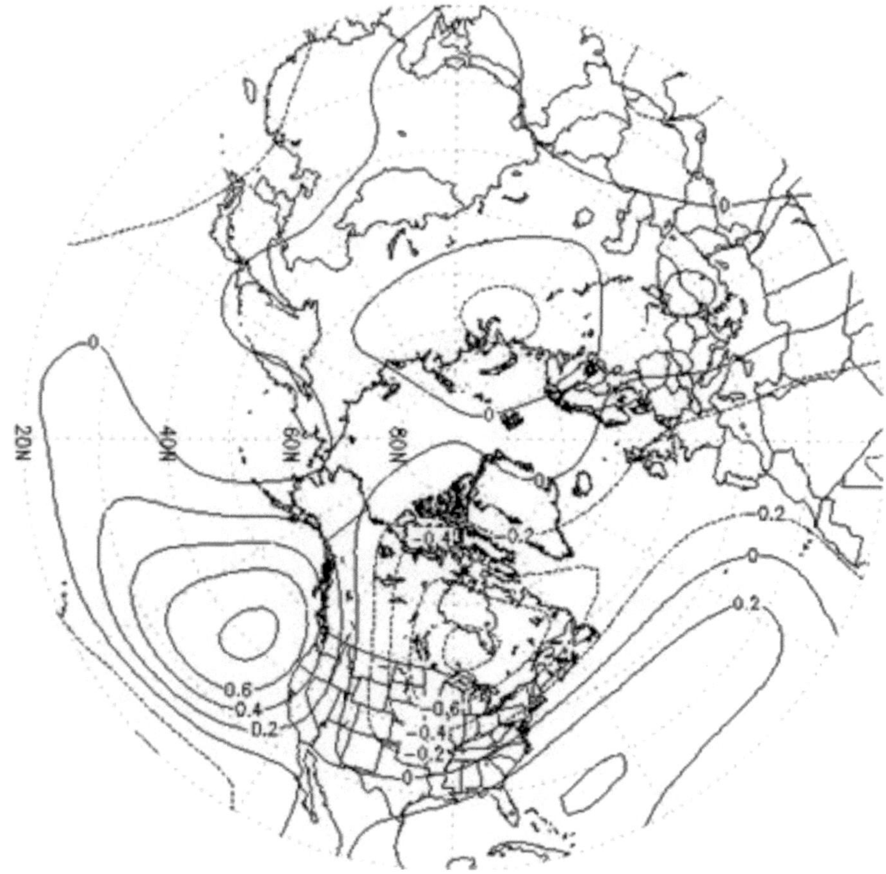

Fig. 16. The TNH January pattern.

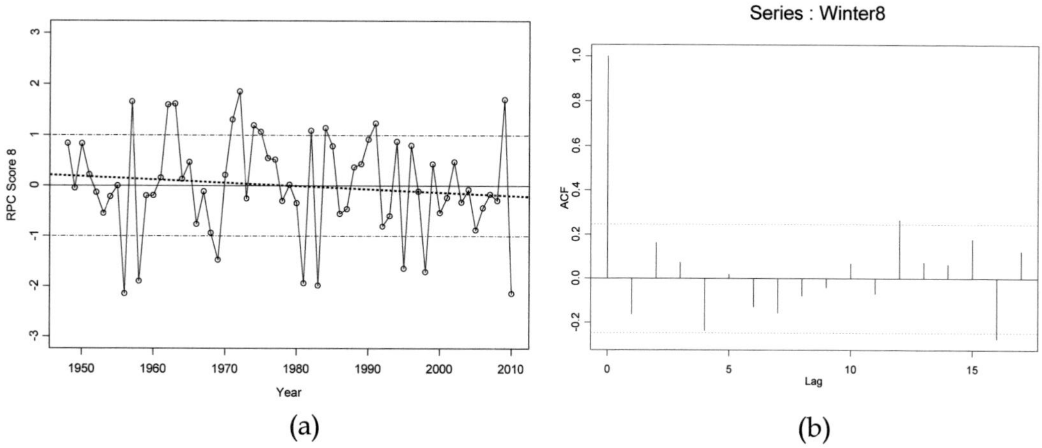

Fig. 17. PNA PC score time series (panel a) and ACF (panel b).

mode shown herein is nearly identical to the TNH pattern identified in BL87. Our pattern to teleconnection congruence is -0.88, representing a close match to the traditional teleconnection pattern. The time series (Fig. 17a) has a slight negative trend that is not statistically significant. Analysis of decadal means (Table 1) indicates one time period (1968-1977) that had a noteworthy positive value. The variance of the time series (Table 2) is nonstationarity with higher variability prior to 1988 and decreasing variability in the following two decades. The ACF of the time series (Fig. 17b) contains no significant autocorrelations at short lags and two values that barely exceed the white noise bandwidth beyond lag 10, likely statistical artifacts.

3.1.2 Intercorrelation of the time series for the January modes

One advantage of applying an oblique rotational algorithm to the height data is that it provides additional information on the degree of correlation among the modes. This information is shown in Table 3. Although most off-diagonal values are near-zero, SZW's time series was slightly positively correlated with the time series for the PNA. Similarly, the NAO time series was positively correlated with the Eurasian pattern. This was the largest magnitude correlation found (+.32). The time series for the PNA was weakly negatively related to the Eurasian pattern and weakly positively related to that of the TNH.

RPC	W1	W2	W3	W4	W5	W6	W7	W8
W1	1.00	0.06	0.04	-0.18	0.13	-0.12	0.29	0.09
W2	0.06	1.00	0.05	-0.23	0.21	0.28	-0.01	0.10
W3	0.04	0.05	1.00	-0.04	0.12	0.17	0.08	0.04
W4	-0.18	-0.23	-0.04	1.00	-0.05	-0.15	-0.07	0.12
W5	0.13	0.21	0.12	-0.05	1.00	0.01	0.32	0.06
W6	-0.12	0.28	0.17	-0.15	0.01	1.00	-0.29	0.29
W7	0.29	-0.01	0.08	-0.07	0.32	-0.29	1.00	-0.16
W8	0.09	0.10	0.04	0.12	0.06	0.29	-0.16	1.00

Table 3. Correlations of the January RPC time series.

3.2 July

The July analysis has 4 eigenmodes retained. The RPCs represent a decomposition of the flow into modes of variability extracting 46.8 percent of the monthly height variance.

3.2.1 July mode descriptions

Pattern 1 - The Subtropical Zonal Summer pattern (SZS; Fig. 18) portrays a zonal ring of one sign extending from the south edge of the domain north to approximately 40°N around the Northern Hemisphere. The pattern/teleconnection congruence is 0.98, indicating the PC is nearly identical to the teleconnection mode. Furthermore, the SZS extracts a very large amount of variance explained (24.9%) compared to any other mode (winter or summer) in our analyses. The time series (Fig. 19a) reveals the main signal is a strong trend of increasing heights that is highly statistically significant (p=8.2x10^{-12}). The ACF (Fig. 19b) illustrates

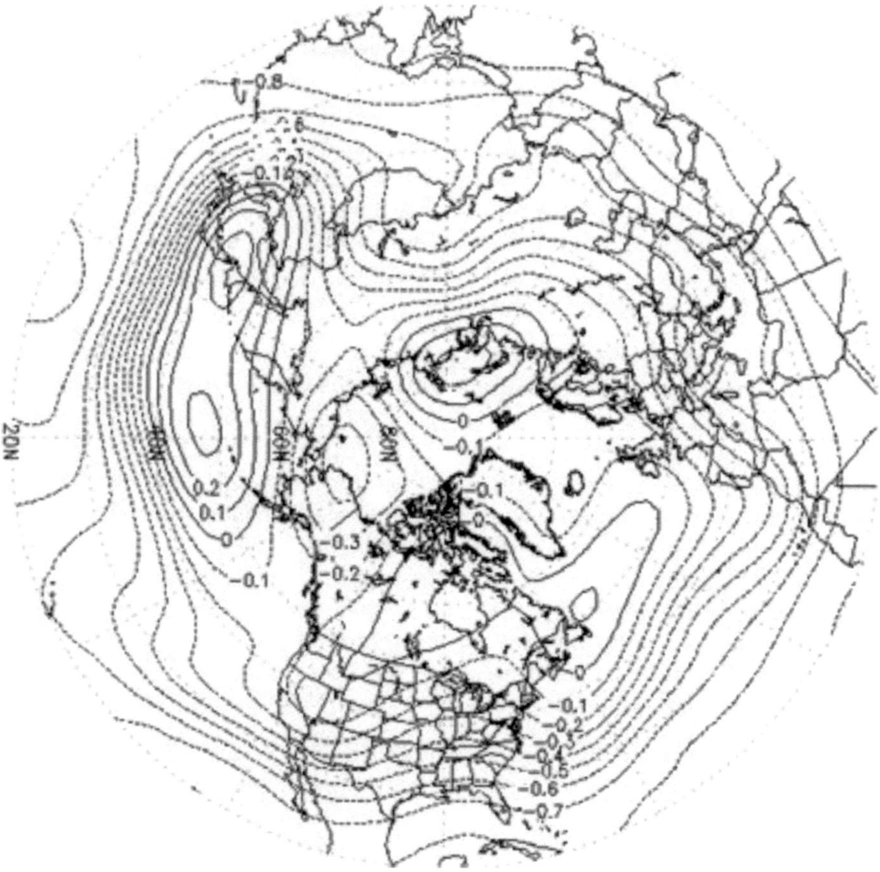

Fig. 18. The SZS July pattern.

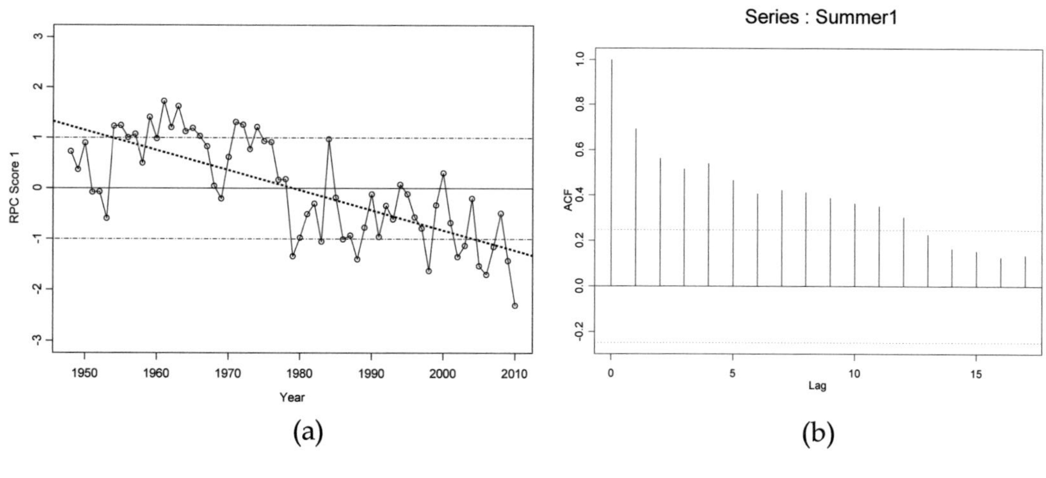

Fig. 19. SZS PC score time series (panel a) and ACF (panel b).

the relentless increase in subtropical summer heights with significant lag correlations to twelve years. The time series for the SZS mode (Fig. 19a) and the winter subtropical zonal mode (Fig. 5a) correlate at r=+0.60 (not shown), suggesting seasonal persistence. The SZW and SZS may be manifestations of the same process when the seasonal migration of the ITCZ is considered. Analysis of the decadal means (Table 4) show nonstationarity with thirty years of highly positive values followed by 30 years of highly negative values. As the RPC loadings in the tropics are negative, the negative means correspond to above normal geopotential heights. The variance statistics (Table 5) indicate extremely low variability in every decade, strengthening the argument that the trend, which accounts for 53% of the time series variance, is the major feature of this pattern.

Pattern 2 - The North Pacific mode (NP; Fig. 20) is a complicated wave number 4 pattern, with both north-south and east-west dipoles, that accounts for 7.3% of the variance. The north Pacific dipole has a double center at 59°N, 150°E and 39°N, 158°W. North of that pair of centers is a second set with opposite sign (82°N, 170°E and 51°N, 130°W). There are many other localized positive and negative maxima, most notably over Ontario, Canada, off the mid-Atlantic coast, over the central Atlantic, over the Mediterranean Sea, Greenland, Russia and northern India. The pattern/teleconnection congruence is 0.86, a large value considering the complexity of the pattern. The time series (Fig. 21a) indicates the main signal is a very slight upward trend (not statistically significant). The NP ACF (Fig. 21b) is unremarkable. Analysis of the decadal means (Table 4) shows values near zero except during the 1998-2007 decade where there was a large positive mean. The variance statistics (Table 5) indicate low variability, coupled with the positive mean, in the 1998-2007 decade.

Pattern 3 - The North American-European mode (NAE; Fig. 22) is a Great Circle pattern with centers over Ontario, Canada (45°N, 80°W), north Greenland (82°N, 45°W), the United Kingdom (52°N, 0°) and the Mediterranean Sea (35°N, 20°E) and describes 6.8% of the variance. The pattern to teleconnection congruence is 0.83, suggesting a fair match between the RPC and the teleconnection pattern. The time series (Fig. 23a) has virtually no trend with low variance in the middle of the data period. The ACF (Fig. 23b) shows no year to year persistence. Decadal means of the NAE (Table 4) were close to zero for all decades. In contrast, the variance was low in the late 1960's to the late 1980's, flanked by periods of much higher variance in a 20 year high, 20 year low, 20 year high sequence.

Pattern 4 - The East Asian-North American pattern (EANA; Fig. 24) is wave number 4 pattern extracting 7.8% of the July variance. There is an elongated center extending from eastern China into the central Pacific Ocean, near 40°N latitude. The pattern becomes more chaotic over North America with two centers at high latitude (~ 75°N) across Canada to south of Iceland. Additionally, there is a center over northern California with one of opposite sign over Illinois, and a third with another sign change in the northwest Atlantic, off the coast of New England. The pattern/teleconnection congruence is -0.91, conveying a very good match with the parent correlations. The time series (Fig. 25a) indicates a significant downward trend over the full period (p=0.03); however, close examination reveals an upward trend from 1948-1966 (piecewise regression trend in that period had p=0.002), little trend in the years 1967-1990 followed by a strong downward trend from 1991-2010 (piecewise regression tend p=.01). The ACF (Fig. 25b) shows no year to year persistence. Analysis of the decadal means (Table 4) indicates nonstationarity with a highly negative mean from 1998-2007. Additionally, during that period, the variance statistics (Table 5) had very low values, with consistently negative RPC scores.

Fig. 20. The NP July pattern.

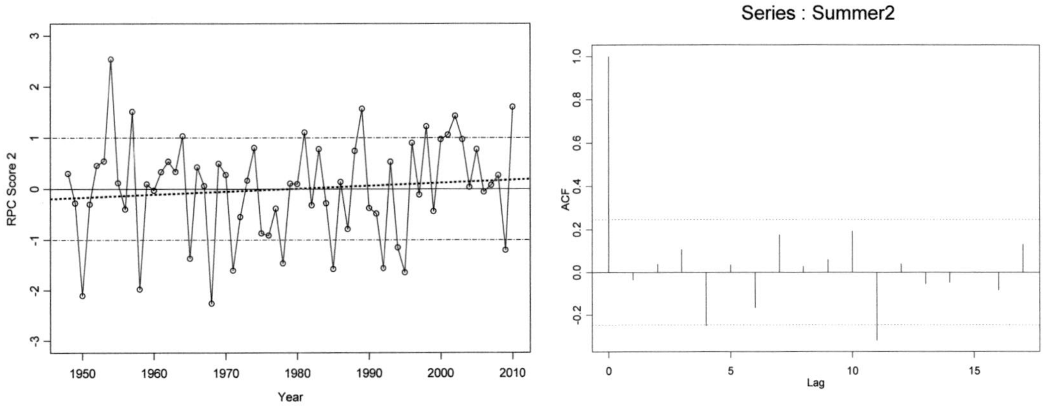

Fig. 21. NP PC score time series (panel a) and ACF (panel b).

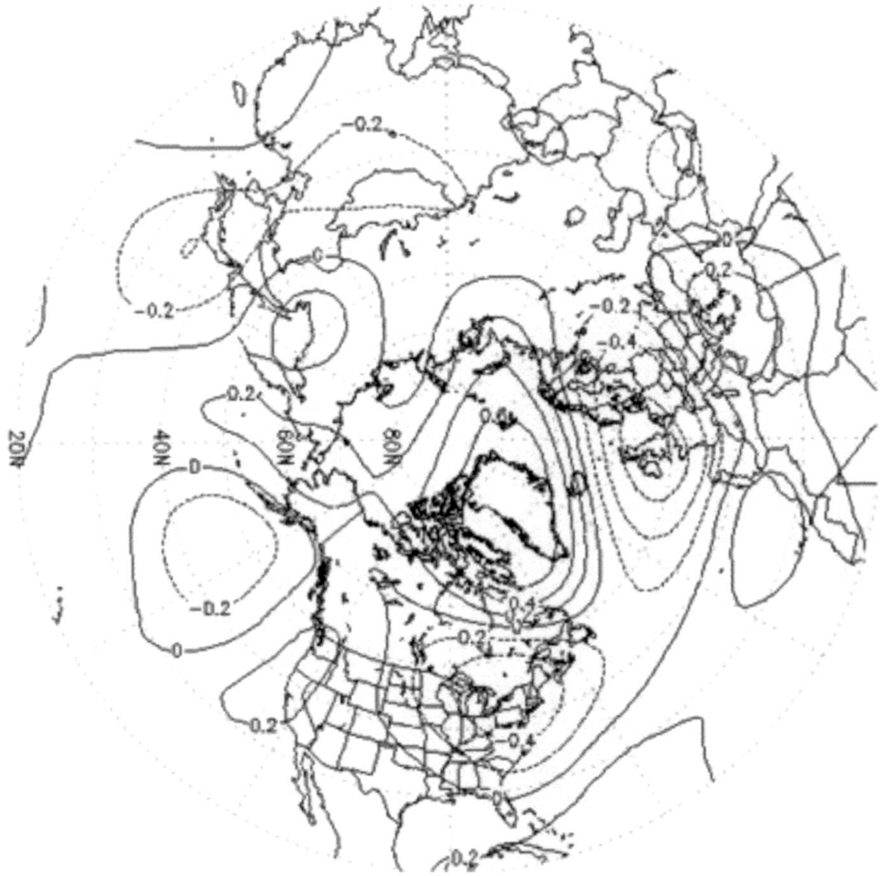

Fig. 22. The NAE July pattern.

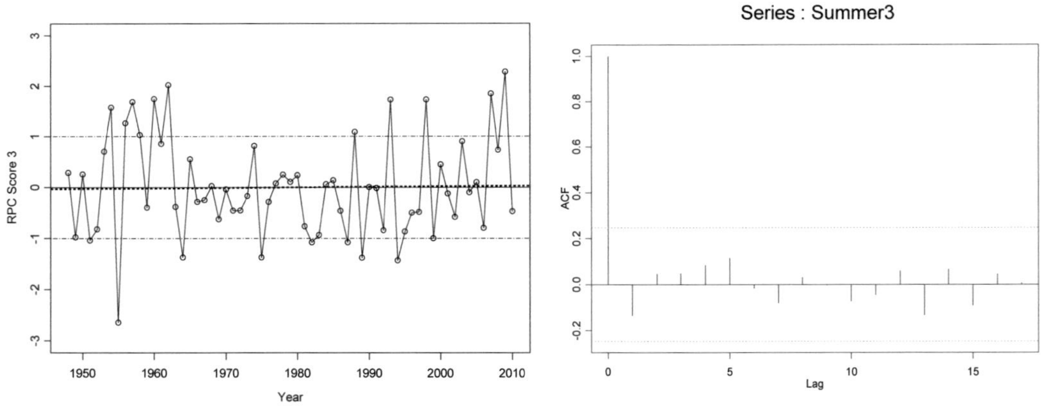

(a) (b)

Fig. 23. NAE PC score time series (panel a) and ACF (panel b).

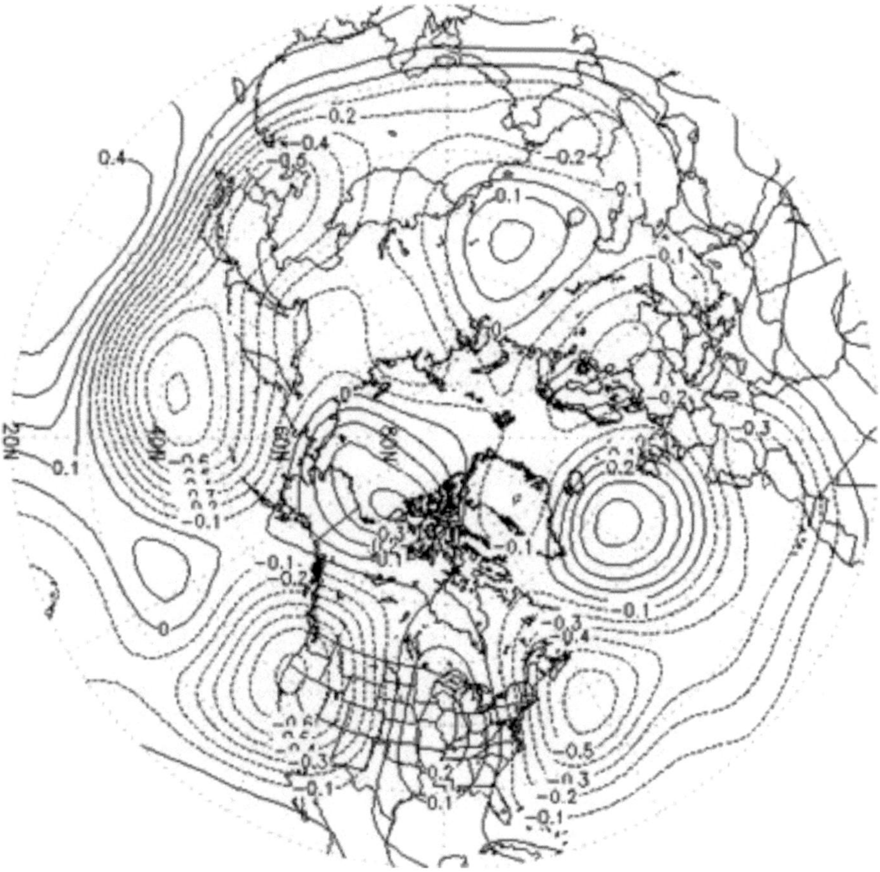

Fig. 24. The EANA July pattern.

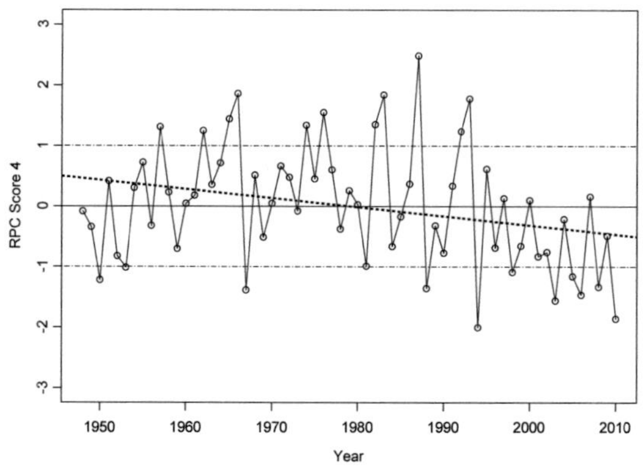

(a)

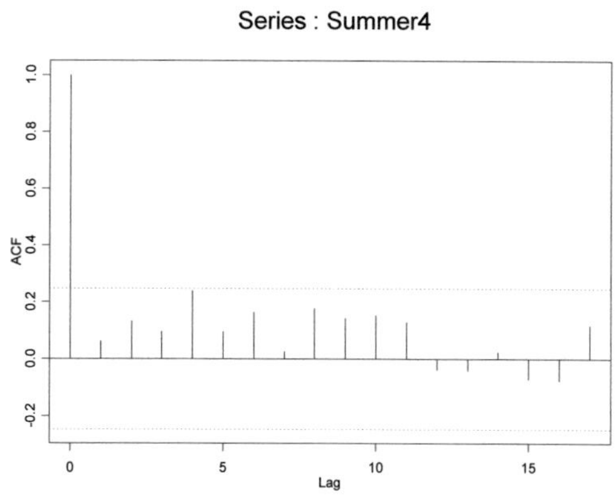

(b)

Fig. 25. EANA PC score time series (panel a) and ACF (panel b).

RPC Mean	1948- 1957	1958-1967	1968-1977	1978-1987	1988-1997	1998-2007
RPC1	0.58	1.16	0.70	-0.52	-0.56	-0.94
RPC2	0.24	-0.05	-0.48	-0.22	-0.16	0.61
RPC3	0.03	0.35	-0.25	-0.35	-0.27	0.24
RPC4	-0.10	0.40	0.51	0.41	-0.10	-0.75

Table 4. Same as Table 1, but for July

RPC Variance	1948-1957	1958-1967	1968-1977	1978-1987	1988-1997	1998-2007	Row Mean
RPC1	0.41	0.13	0.29	0.49	0.20	0.46	0.33
RPC2	1.50	0.84	0.93	0.76	1.18	0.41	0.94
RPC3	1.92	1.15	0.32	0.32	1.04	0.98	0.96
RPC4	0.65	0.96	0.38	1.28	1.35	0.37	0.83
Column Mean	1.12	0.77	0.48	0.71	0.94	0.56	

Table 5. Same as Table 2, but for July.

3.2.2 Intercorrelation of the time series for the July modes

After computing oblique PC score correlations (Table 6), only 1 mode pair (RPCs 1 and 4) were significantly correlated (+0.35 for SZS and EANA). Inspection of the associated spatial maps (Figs. 18 and 24) shows Pacific and Atlantic Ocean areas had similar loading patterns.

RPC	S1	S2	S3	S4
S1	1.00	-0.19	0.00	0.35
S2	-0.19	1.00	0.13	-0.16
S3	0.00	0.13	1.00	-0.02
S4	0.35	-0.16	-0.02	1.00

Table 6. Same as Table 3, but for July.

4. Conclusion

RPCA was used to filter the monthly 500 hPa geopotential heights in January and July for 1948 – 2010 to establish the dominant modes of variability over that period. For January, 8 modes, accounting for 70% of the total variance, were extracted. The majority of the patterns that emerged in these analyses had been identified in other studies, published mainly in the 1980's. However, several new patterns emerged in both months and the variance extracted by each pattern showed major shifts when the recent height data were included. RPCs that highlighted subtropical height anomalies were found to be more important when post-1980's data were included and statistically significant linear trends emerged on 4 of the 8 winter patterns with one additional marginally significant pattern. Moreover, the variability of all patterns had marked decadal variation. The July analyses were even more striking as two of the four modes has significant trends in their RPC scores. The leading mode for summer, SZS, extracted a massive 24.9% of the total variance. That pattern was zonally symmetric with the largest negative RPC loadings in the subtropics. The time series had an exceedingly large downward trend, with a p=8.2X10^{-12}, indicating large height rises in the subtropics. As this mode had a pattern/teleconnection congruence of -0.98, it is thought to be very robust and not an artifact of the multivariate analysis. Three other patterns were found for the July analysis and, together, the four account for 46.8% of the 500 hPa variability. Taken collectively, these findings suggest that the dominant 500 hPa modes are robust but nonstationary.

It is hoped that climate modelers will decompose their model outputs for 1948-2010 atmospheres into RPC modes and compare those model modes to the ones documented herein. Differences in the model modes, time series and persistence statistics from the observed modes document the details of how model climatologies differ. Inspection of the specific disparities may help to identify correctable problems in such models.

5. Acknowledgment

A portion of Michael Richman's time was supported by the Cooperative Institute for Mesoscale Meteorological Studies. Andrew Mercer's time was supported in part by the Northern Gulf Institute.

6. References

Barnes, S. (1964). A technique for maximizing details in numerical weather-map analysis. *Journal of Applied Meteorology* , Vol.3, No. 4, (August 1964), pp. 396–409, ISSN 1558-8424

Barnston, A. & Livezey, R. (1987). Classification, seasonality and persistence of low-frequency atmospheric circulation patterns. *Monthly Weather Review*, Vol.115, No.6, (June 1987), pp. 1083-1126, ISSN 0027-0644

Compagnucci, R. & Richman, M. (2008). Can principal component analysis provide atmospheric circulation or teleconnection patterns?, *International Journal of Climatology*, Vol.28, No.6, (May 2008), pp. 703-726, ISSN 0899-8418

Esbensen, S. (1984). A comparison of intermonthly and interannual teleconnections in the 700 mb geopotential height field during the northern hemisphere winter. *Monthly Weather Review*, Vol.112, No.10, (October 1984), pp. 2016-2032, ISSN 0027-0644

Hsu, H. & Wallace, J. (1985). Vertical structure of wintertime teleconnection patterns. *Journal of the Atmospheric Sciences*, Vol.42, No.16, (August 1985), pp. 1693-1710, ISSN 0022-4928

Kalnay, E., Kanamitsu, M., Kistler, R., Collins, W., Deaven, D., Gandin, L., Iredell, M., Saha, S., White, G., Woollen, J., Zhu, Y., Leetmaa, A., Reynolds, R., Chelliah, M., Ebisuzaki, W., Higgins, W., Janowiak, J., Mo, K., Ropelewski, C., Wang, J., Jenne, R. & Joseph, D. (1996). The NCEP/NCAR 40-year reanalysis project. *Bulletin of the American Meteorological Society*, Vol.77, No.3, (March 1996), pp. 437-471, ISSN 1520-0477

Lau, N. (1988). Variability of the observed midlatitude storm tracks in relation to low frequency changes in the circulation pattern. *Journal of the Atmospheric Sciences*, Vol.45, No.19, (October 1988), pp. 2718-2743, ISSN 0022-4928

Linkin , M. & Nigam, S. (2008). The North Pacific Oscillation–West Pacific teleconnection pattern: mature-phase structure and winter impacts. *Journal of Climate*, Vol.21, No.9, (May 2008), pp. 1979-1997, ISSN 0894-8755

Namias, J. (1980). Causesof some extreme Northern Hemisphere climate anomalies from summer 1978 through the subsequent winter. *Monthly Weather Review*, Vol.108, No.9, (September 1980), pp. 1333-1346, ISSN 0027-0644

Renwick, J. & Wallace, J. (1996). Relationships between North Pacific blocking, El Niño, and the PNA pattern. *Monthly Weather Review*, Vol.124, No.9, (September 1996), pp. 2071–2076, ISSN 0027-0644

Richman, M. (1986), Review article: rotation of principal components. *International Journal of Climatology*, Vol.6, No.3, (May 1986), pp. 293-335, ISSN 0899-8418

Richman, M. & Gong, X. (1999). Relationships between the definition of hyperplane width to the fidelity of principal component loading patterns. *Journal of Climate*, Vol.12, No.6, (June 1999), pp. 1557-1576, ISSN 0894-8755

Richman, M. & Lamb, P. (1985). Climatic pattern analysis of three- and seven-day summer rainfall in the central United States: some methodological considerations and a regionalization. *Journal of Climate and Applied Meteorology*, Vol.24, No.12, (December 1985), pp. 1325-1343, ISSN 1558-8424

Rogers, J. (1990). Patterns of low-frequency monthly sea level pressure variability (1899–1986) and associated wave cyclone frequencies. *Journal of Climate, Vol.3, No.12,* (December 1990), pp. 1364–1379, ISSN 0894-8755

Swinbank, R. & Purser, J. (2006). Fibonacci grids: a novel approach to global modelling. *Quarterly Journal of the Royal Meteorological Society*, Vol.132, No.619, (July 2006), pp. 1769-1793, ISSN 1477-870X

Wallace, J. & Gutzler, D. (1981). Teleconnections in the geopotential height field during the Northern Hemisphere winter. *Monthly Weather Review*, Vol.109, No.4, (April 1981), pp. 784-812, ISSN 0027-0644

Wilks, D. (2011). *Statistical Methods in the Atmospheric Sciences, Volume 100,* (Third Edition), Academic Press, ISBN 0123-85022-3, Burlington, MA, United States